VOLUME ONE HUNDRED AND NINE

ADVANCES IN PROTEIN CHEMISTRY AND STRUCTURAL BIOLOGY

Structural and Mechanistic Enzymology

VOLUME ONE HUNDRED AND NINE

Advances in PROTEIN CHEMISTRY AND STRUCTURAL BIOLOGY

Structural and Mechanistic Enzymology

Edited by

TATYANA KARABENCHEVA-CHRISTOVA

Department of Applied Sciences,
University of Northumbria at Newcastle,
Newcastle upon Tyne, United Kingdom

ELSEVIER

ACADEMIC PRESS

An imprint of Elsevier

Academic Press is an imprint of Elsevier
50 Hampshire Street, 5th Floor, Cambridge, MA 02139, United States
525 B Street, Suite 1800, San Diego, CA 92101-4495, United States
The Boulevard, Langford Lane, Kidlington, Oxford OX5 1GB, United Kingdom
125 London Wall, London, EC2Y 5AS, United Kingdom

First edition 2017

ISBN: 978-0-12-811876-4
ISSN: 1876-1623

For information on all Academic Press publications
visit our website at https://www.elsevier.com/books-and-journals

Publisher: Zoe Kruze
Acquisition Editor: Helene Kabes
Senior Editorial Project Manager: Helene Kabes
Production Project Manager: Surya Narayanan Jayachandran
Senior Cover Designer: Alan Studholme

Typeset by SPi Global, India

CONTENTS

Contributors *ix*
Preface *xi*

1. Collagenolytic Matrix Metalloproteinase Structure–Function Relationships: Insights From Molecular Dynamics Studies **1**
Tatyana G. Karabencheva-Christova, Christo Z. Christov, and Gregg B. Fields

1. Matrix Metalloproteinases 2
2. MD Simulation Approaches 2
3. MD Simulation of the Collagen Triple Helix 5
4. MD Simulation of the MMP Cleavage Site in Collagen 6
5. MD Consideration of Water Surrounding Collagen 9
6. MD of MMP-1 or MMP-8 Interaction With Collagen 10
7. MD of MMP-2 Interaction With Collagen 17
8. Mechanism of Collagenolysis 19

Acknowledgments 20
References 21

2. Computational Glycobiology: Mechanistic Studies of Carbohydrate-Active Enzymes and Implication for Inhibitor Design **25**
Andrew P. Montgomery, Kela Xiao, Xingyong Wang, Danielle Skropeta, and Haibo Yu

1. Introduction 26
2. Computational Methods 31
3. Puckering of Carbohydrate Structures 36
4. Identification of Catalytically Competent Protonation States of Key Residues 40
5. Clarification of Catalytic Mechanisms 48
6. Transition-State Analogues as Inhibitors of CAZymes 54
7. Conclusions and Future Directions 65

Acknowledgments 66
References 66

3. Computational Biochemistry—Enzyme Mechanisms Explored **77**
Martin Culka, Florian J. Gisdon, and G. Matthias Ullmann

1. Introduction 78
2. Structural Models 81

3. Calculating Enzymatic Mechanisms 91
4. Going Beyond the Exploration of the Reaction Paths 104
References 105

4. A Paradigm for C—H Bond Cleavage: Structural and Functional Aspects of Transition State Stabilization by Mandelate Racemase 113

Stephen L. Bearne and Martin St. Maurice

1. Introduction 115
2. Structure and Function Within the MR Subgroup 117
3. MR Catalysis: An Uphill Battle 125
4. Meeting the Challenges: Solutions Afforded by MR 126
5. Mandelate Racemase Structure: Ligand Binding Within a Catalytic Cage 133
6. Evolutionary Relationships: A Mechanistic Link to Aldolases? 148
7. Concluding Remarks 151
Acknowledgments 152
References 152

5. Striking Diversity in Holoenzyme Architecture and Extensive Conformational Variability in Biotin-Dependent Carboxylases 161

Liang Tong

1. Introduction 162
2. Structures of Conserved Components in Biotin-Dependent Carboxylases 166
3. Extensive Diversity in Holoenzyme Architectures 173
4. Extensive Conformational Variability 180
5. Perspectives 187
Acknowledgments 188
References 189

6. Enzymology of Microbial Dimethylsulfoniopropionate Catabolism 195

Mishtu Dey

1. Introduction 196
2. DMSP Demethylation Pathway and the Demethylase, DmdA 198
3. DMSP Cleavage Enzymes 202
4. The CoA-Transferase Pathway of DMS and 3-Hydroxypropionate Formation Using DddD 202
5. The Lyase Pathway of DMS and Acrylate Generation 203
6. Summary and Perspective 218
Acknowledgments 219
References 219

7. Sortase Transpeptidases: Structural Biology and Catalytic Mechanism **223**

Alex W. Jacobitz, Michele D. Kattke, Jeff Wereszczynski, and Robert T. Clubb

1. Introduction 224
2. Structural Biology: Enzyme Structure and Class-Specific Variations 228
3. Structural Biology: Molecular Basis of Substrate Recognition 241
4. Computational Studies 248
5. Catalytic Mechanism 251
6. In Vitro Transpeptidation Activity 253
7. Conclusions and Future Directions 258

References 259

8. Biology, Mechanism, and Structure of Enzymes in the α-D-Phosphohexomutase Superfamily **265**

Kyle M. Stiers, Andrew G. Muenks, and Lesa J. Beamer

1. Introduction 266
2. Biology 267
3. Catalytic Mechanism 280
4. Three-Dimensional Structure 284
5. Conclusions and Future Directions 295

Acknowledgment 296

References 296

9. Mechanistic Insights Into Catalytic RNA–Protein Complexes Involved in Translation of the Genetic Code **305**

Satya B. Routh and Rajan Sankaranarayanan

1. Introduction 306
2. RNPCs in Translation 307
3. Conclusion 338

Acknowledgments 339

References 339

CONTRIBUTORS

Lesa J. Beamer
University of Missouri, Columbia, MO, United States

Stephen L. Bearne
Dalhousie University, Halifax, NS, Canada

Christo Z. Christov
Faculty of Health and Life Sciences, Northumbria University, Newcastle upon Tyne, United Kingdom

Robert T. Clubb
The Molecular Biology Institute and the UCLA-DOE Institute of Genomics and Proteomics, University of California, Los Angeles, CA, United States

Martin Culka
Computational Biochemistry, University of Bayreuth, Bayreuth, Germany

Mishtu Dey
The University of Iowa, Iowa City, IA, United States

Gregg B. Fields
Florida Atlantic University; The Scripps Research Institute/Scripps Florida, Jupiter, FL, United States

Florian J. Gisdon
Computational Biochemistry, University of Bayreuth, Bayreuth, Germany

Alex W. Jacobitz
The Molecular Biology Institute and the UCLA-DOE Institute of Genomics and Proteomics, University of California, Los Angeles, CA, United States

Tatyana G. Karabencheva-Christova
Faculty of Health and Life Sciences, Northumbria University, Newcastle upon Tyne, United Kingdom

Michele D. Kattke
The Molecular Biology Institute and the UCLA-DOE Institute of Genomics and Proteomics, University of California, Los Angeles, CA, United States

Andrew P. Montgomery
School of Chemistry, Faculty of Science, Medicine and Health, University of Wollongong, Wollongong, NSW, Australia

Andrew G. Muenks
University of Missouri, Columbia, MO, United States

Satya B. Routh
CSIR–Centre for Cellular and Molecular Biology, Hyderabad, India

Rajan Sankaranarayanan
CSIR–Centre for Cellular and Molecular Biology, Hyderabad, India

Danielle Skropeta
School of Chemistry, Faculty of Science, Medicine and Health; Centre for Medical and Molecular Bioscience, University of Wollongong, Wollongong, NSW, Australia

Martin St. Maurice
Marquette University, Milwaukee, WI, United States

Kyle M. Stiers
University of Missouri, Columbia, MO, United States

Liang Tong
Columbia University, New York, NY, United States

G. Matthias Ullmann
Computational Biochemistry, University of Bayreuth, Bayreuth, Germany

Xingyong Wang
School of Chemistry, Faculty of Science, Medicine and Health; Centre for Medical and Molecular Bioscience, University of Wollongong, Wollongong, NSW, Australia

Jeff Wereszczynski
Center for Molecular Study of Condensed Soft Matter, Illinois Institute of Technology, Chicago, IL, United States

Kela Xiao
School of Chemistry, Faculty of Science, Medicine and Health, University of Wollongong, Wollongong, NSW, Australia

Haibo Yu
School of Chemistry, Faculty of Science, Medicine and Health; Centre for Medical and Molecular Bioscience, University of Wollongong; Illawarra Health and Medical Research Institute, Wollongong, NSW, Australia

PREFACE

The current thematic volume "Structural and Mechanistic Enzymology" presents state-of-the-art contributions, providing insights into the relationship between enzyme structure, catalysis, and function. Understanding of enzymatic molecular mechanisms requires a combined application of a broad range of experimental and computational techniques. Experimental methods such as NMR spectroscopy, X-ray crystallography, molecular biology, and kinetic studies, combined with a range of molecular modeling techniques such as molecular dynamics (MD), quantum chemistry, and docking, provide us a synergistic approach for gaining insights into enzyme catalysis, structure, and function.

Chapter 1 "Collagenolytic matrix metalloproteinase structure–function relationships: insights from molecular dynamics studies" provides a comprehensive review of a range of MD methods used in matrix metalloproteinases (MMPs) modeling. Insights into the detailed process of collagenolysis, performed by the MMP family, have been obtained by means of combining experimental (X-ray crystallography and NMR spectroscopy) and computational methods, ultimately relating MMPs structural data to their functional implications.

Chapter 2 "Computational glycobiology: mechanistic studies of carbohydrate-active enzymes and implication for inhibitor design" offers understanding from computational modeling perspective into the catalytic mechanisms of carbohydrate-active enzymes integrating structural, kinetic, and mutagenesis data.

Chapter 3 "Computational biochemistry—enzyme mechanisms explored" reviews a variety of simulation methods for enzyme mechanism modeling. The review asserts that a synergistic approach using biomolecular experimental studies complemented by computational modeling methods helps in elucidating and explaining enzyme mechanisms and experimental data.

Chapter 4 "A paradigm for C–H bond cleavage: structural and functional aspects of transition state stabilization by mandelate racemase" reviews structural and mechanistic studies revealing mechanistic strategies of mandelate racemase an enzyme that catalyzes proton transfer reaction, and further gives insights into homologous enzymes in the enolase family.

Chapter 5 "Striking diversity in holoenzyme architecture and extensive conformational variability in biotin-dependent carboxylases" presents structural studies of biotin-dependent carboxylases, involved in the metabolism of amino acids, fatty acids, and carbohydrates, and focuses on a characteristic common feature of structural flexibility that is a key property for function and regulation of these enzymes.

Chapter 6 "Enzymology of microbial dimethylsulfoniopropionate catabolism" focuses on the biochemistry of dimethylsulfoniopropionate catabolism—one of the most abundant organosulfur compounds on earth, synthesized by marine organisms that play a role in the global sulfur cycle and global climate.

Chapter 7 "Sortase transpeptidases: structural biology and catalytic mechanism" reviews structural, biochemical, and computational results on the sortase family of enzymes, with a particular focus on reaction mechanisms. One of the functions of sortase enzymes, potential drug targets, is attaching proteins to the microbial cell wall of Gram-positive bacteria.

Chapter 8 "Biology, mechanism, and structure of enzymes in the α-D-phosphohexomutase superfamily" presents recent research on the superfamily of α-D-phosphohexomutases enzymes, that catalyze reversible conversion of phosphosugars, key to primary metabolism in living organisms, and important for processes such as protein glycosylation.

Chapter 9 "Mechanistic insights into catalytic RNA–protein complexes involved in translation of the genetic code" reviews structure–function relationships of RNA–protein complexes, that are involved in protein biosynthesis, and further reveals mechanistic aspects of these complexes.

Tatyana Karabencheva-Christova
University of Northumbria at Newcastle,
Newcastle upon Tyne,
United Kingdom

CHAPTER ONE

Collagenolytic Matrix Metalloproteinase Structure–Function Relationships: Insights From Molecular Dynamics Studies

Tatyana G. Karabencheva-Christova*, Christo Z. Christov*, Gregg B. Fields†,‡,1

*Faculty of Health and Life Sciences, Northumbria University, Newcastle upon Tyne, United Kingdom
†Florida Atlantic University, Jupiter, FL, United States
‡The Scripps Research Institute/Scripps Florida, Jupiter, FL, United States
[1]Corresponding author: e-mail address: fieldsg@fau.edu

Contents

1. Matrix Metalloproteinases 2
2. MD Simulation Approaches 2
3. MD Simulation of the Collagen Triple Helix 5
4. MD Simulation of the MMP Cleavage Site in Collagen 6
5. MD Consideration of Water Surrounding Collagen 9
6. MD of MMP-1 or MMP-8 Interaction With Collagen 10
7. MD of MMP-2 Interaction With Collagen 17
8. Mechanism of Collagenolysis 19
Acknowledgments 20
References 21

Abstract

Several members of the zinc-dependent matrix metalloproteinase (MMP) family catalyze collagen degradation. Experimental data reveal a collaboration between different MMP domains in order to achieve efficient collagenolysis. Molecular dynamics (MD) simulations have been utilized to provide atomistic details of the collagenolytic process. The triple-helical structure of collagen exhibits local regions of flexibility, with modulation of interchain salt bridges and water bridges contributing to accessibility of individual chains by the enzyme. In turn, the hemopexin-like (HPX) domain of the MMP initially binds the triple helix and facilitates the presentation of individual strands to active site in the catalytic (CAT) domain. Extensive positive and negative correlated motions are

Advances in Protein Chemistry and Structural Biology, Volume 109
ISSN 1876-1623
http://dx.doi.org/10.1016/bs.apcsb.2017.04.001

observed between the CAT and HPX domains when collagen is bound. Ultimately, the MD simulation studies have complemented structural (NMR spectroscopy, X-ray crystallography) and kinetic analyses to provide a more detailed mechanistic view of MMP-catalyzed collagenolysis.

1. MATRIX METALLOPROTEINASES

Matrix metalloproteinases (MMPs) are a family of zinc-dependent proteolytic enzymes. It has long been recognized select MMPs cleave interstitial (types I–III) collagen into 1/4 and 3/4 length fragments (Amar, Smith, & Fields, 2017). The collagenolytic MMPs (MMP-1, MMP-8, MMP-13, membrane-type 1 (MT1)-MMP, and MT2-MMP) all possess a catalytic (CAT) domain, a hemopexin-like (HPX) domain, and a linker connecting the two domains. MMP-2 and MMP-9, which have been attributed collagenolytic activity in select circumstances (Amar et al., 2017), contain three fibronectin type II (FNII) inserts within their CAT domain, as well as a HPX domain and linker. The sequence homologies of the MMPs and the involvement of multiple subsites in the cleavage mechanism preclude the use of simple chemical techniques to concretely elucidate enzyme mechanism and collagen selectivity. The plethora of MMP X-ray crystallographic structures lacks insight into the dynamic aspect of enzyme–substrate interaction. Collagenolysis is a dynamic process with many simultaneous events taking place, some complementary, some antagonistic. Molecular dynamics (MD) simulations have demonstrated its utility in furthering the understanding of enzyme behavior and, in conjunction with X-ray crystallography and NMR data, show great promise in contributing to a comprehensive understanding of MMP collagenolytic mechanisms.

2. MD SIMULATION APPROACHES

A range of MD simulation techniques have been developed and are broadly applied in studying protein conformational dynamics and flexibility (Adcock & McCammon, 2006). MD techniques differ in the level of resolution used to describe the system and its respective interactions (Orozco, 2014). In terms of resolution, MD simulations can be done at the atomistic (AT) level of accuracy, where all atoms are explicitly represented (e.g., the simulated protein, the ligands, and the surrounding solvent water molecules are all treated at AT level), this way allowing for the movement of every single atom of the system to be taken into account explicitly when accessing the overall flexibility of the simulated system. Alternatively, the so-called coarse-grained (CG) MD simulations can be done, which are lower resolution methods, where solvents are not explicitly represented and also groups of protein atoms are treated as one unit with the aim of reducing the degrees of freedom of the system (Rinker, Allison, & van Gunsteren, 2012). CG MD simulations are usually applied to very large systems and/or in order to accelerate the simulation time, but CG MD simulations do not represent interactions in AT detail.

AT MD simulations also differ in respect to the applied energy function used to describe the molecular interactions. In this respect there are two types of MD approaches applied in protein modeling: classical molecular mechanics (MM) MD and combined quantum mechanics/molecular mechanics (QM/MM) MD (Vidossich & Magistrato, 2014). Modeling reaction mechanisms and reactivity in enzymes where bond breaking and creation has to be modeled cannot be done by applying classical MD simulations, and thus hybrid QM/MM coupled to MD techniques are widely used for this aim. In this approach the system is "divided" into two regions: QM region treated by QM method and the rest of the system treated by the classical MM force field. MM force fields have been parameterized using high-quality QM calculations and experimental data for model systems. First-generation force fields have been a subject of dramatic development and thus, nowadays, force fields are very refined. Broadly used force fields are AMBER, GROMOS, CHARMM, and OPLS-AA. MD simulations start with constructing a structural model of the system of interest, which in the present case is a collagen or MMP or both in a complex. For this aim ideally an experimentally derived three-dimensional structure is used or alternatively a collagen model or MMP model structure can be constructed using widely available computational tools. The MD simulations can vary on how the initial coordinates of the system are generated. For example, by using X-ray crystallographic or NMR-derived structures and also whether the X-ray crystallographic and NMR-derived waters are kept or not.

A number of MD programs such as GROMACS (Hess, Kutzner, van der Spoel, & Lindahl, 2008), AMBER (Case et al., 2005), CHARMM (Brooks et al., 2009), NAMD (Phillips, Zheng, Kumar, & Kalé, 2002), and DESMOND (Cappel et al., 2016) are widely applied in MD simulations of proteins. Improvements in AT MD simulations involve both extension of system size and simulation time. Nowadays, the growth in supercomputer power and the development of graphical processor units allow for MD simulations to be performed in the microsecond timescale.

The standard MD approach typically includes an unrestrained simulation of solvated proteins. The forces acting on each atom are calculated from the MM potential energy function of the system, and the Newton's equations of motion are solved numerically for all atoms in the system (Adcock & McCammon, 2006; Orozco, 2014). As an output the MD simulations provide trajectories describing the time evolution of the system. Due to the development of software and hardware, classical MD simulations can describe motions in the range of tens of microseconds. Some functionally important conformational transitions, however, have high-energy barriers and can hardly be explored by classical unrestrained MD simulations. Instead biased (or directed) MD techniques have been developed such as umbrella sampling simulations (potential of mean force) (Kästner, 2011). In this method the system is driven from one conformational state to another (in small steps called windows) usually separated by a high-energy barrier, along a predetermined reaction coordinate restrained by applying an umbrella potential (usually a harmonic biasing potential). This type of simulation allows for the free energy profile of the system transition between the two states to be estimated by applying the weighted histogram analysis method. In order to explore conformational states that occur beyond the limitations of unrestrained MD simulations accelerated methods such as metadynamics were developed. The method can explore larger numbers of conformational states by forbidding the system to visit already passed conformational states.

In order to calculate the free energy of ligand-binding Molecular Mechanics/Poisson Boltzmann Surface Area (MM/PBSA) and Molecular Mechanics Generalized Born Surface Area (MM/GBSA) methods have received broad application (Genheden & Ryde, 2015). These approaches have intermediate accuracy between the highly accurate free energy perturbations and thermodynamic integration methods and the lower in accuracy empirical scoring methods. The free energy is estimated as a sum of the standard MM terms for bonds, angles, dihedrals, Van der Waals, and electrostatic interactions plus the polar (derived by solving the PB equation or by using the GB model) and nonpolar contributions (obtained from the solvent accessible surface area) to the free energies of solvation minus the absolute temperature multiplied by the entropy. The MM/PBSA and MM/GBSA results depend on the particular settings of the calculation, on the sampling method, on the entropies, and on the continuum solvation method.

The method ignores conformational changes upon binding, and there is no information about the number of water molecules bound to the active site and their entropy before and after ligand binding.

3. MD SIMULATION OF THE COLLAGEN TRIPLE HELIX

In order to "visualize" the mechanism of collagenolysis, one would like to use the information obtained from enzymatic and structural studies to model the interaction of the triple-helical substrate with the enzyme. While there exists an established history of computational calculations for triple helices, MD simulations of these structures containing diverse sequences are more recent. MD simulations to analyze fluctuations in the triple-helical backbone and side chains based on single-site residue mutations were initially described using the AMBER 94 molecular mechanical force field (Klein & Huang, 1999; Mooney, Huang, Kollman, & Klein, 2001; Mooney, Kollman, & Klein, 2002). Simulations were performed up to 1 ns. Prior to the molecular mechanical calculations, a triple-helical structural model was created using AMBER software (Bhatnagar, Pattabiramam, Sorensen, Langridge, & MacElroy, 1988; Klein & Huang, 1999). MM parameters were subsequently improved for 4-hydroxyproline (Hyp) (Park, Radmer, Klein, & Pande, 2005). Later studies of mutation effects utilized the AMBER 7 or AMBER 8 molecular simulation package with the AMBER 99 force field (Bodian, Madhan, Brodsky, & Klein, 2008; Park, Klein, & Pande, 2007; Radmer & Klein, 2004). Eventually, the entire type I collagen sequence underwent MD simulation for 10 ns utilizing GROMACS version 3.3 with the AMBER 99 force field (Bodian, Radmer, Holbert, & Klein, 2011). The simulation required 11 months. Triple helicity was maintained throughout, but triple-helical structure was not uniform, as specific regions of microunfolding were observed (Fig. 1). The results were consistent with one- and two-dimensional NMR experiments using ^{2}H-, ^{13}C-, and/or ^{15}N-labeled collagens and triple-helical peptides (THP) which detected triple-helical backbone mobilities, even in fibrils (Fan, Li, Brodsky, & Baum, 1993; Jelinski, Sullivan, & Torchia, 1980; Jelinski & Torchia, 1979; Liu, Kim, Dai, Brodsky, & Baum, 1998; Yu et al., 1999).

MD simulations were later utilized to evaluate the triple-helical stability offered by interchain salt bridges (Gurry, Nerenberg, & Stultz, 2010). MD simulations of (Gly-Pro-Hyp)$_3$-Gly-Pro-Lys-Gly-Glu-Hyp-(Gly-Pro-Hyp)$_3$ and (Gly-Pro-Hyp)$_3$-Gly-Pro-Lys-Gly-Asp-Hyp-(Gly-Pro-Hyp)$_3$

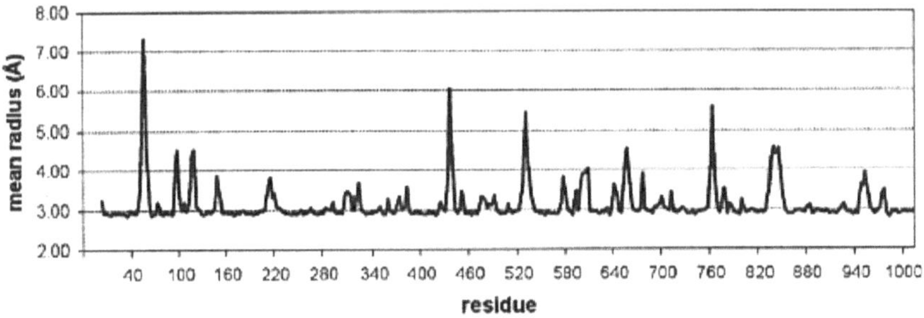

Fig. 1 Helical radius at each Gly position in the type I collagen triple helix. The average helical radius ranged from 2.8 to 7.3 Å. The sharp peaks correspond to regions of unwinding. *Reproduced by permission from Bodian, D. L., Radmer, R. J., Holbert, S., & Klein, T. E. (2011). Molecular dynamics simulations of the full triple helical region of collagen type I provide an atomic scale view of the protein's regional heterogeneity.* Pacific Symposium on Biocomputing, 16, *193–204.*

were performed using CHARMM 35b2 with the CHARMM22/CMAP all-atom parameters. Free energies of interchain salt bridge formation were calculated using umbrella sampling. The most stabilizing salt bridges between the Lys and Asp residues in the $(\text{Gly-Pro-Hyp})_3$-Gly-Pro-Lys-Gly-Glu-Hyp-$(\text{Gly-Pro-Hyp})_3$ THP occurred between the middle and trailing strand ($\Delta\Delta G = -2.4$ kcal/mol), while the most stabilizing salt bridges between the Lys and Glu residues in the $(\text{Gly-Pro-Hyp})_3$-Gly-Pro-Lys-Gly-Asp-Hyp-$(\text{Gly-Pro-Hyp})_3$ THP occurred between the leading and middle strand ($\Delta\Delta G = -3.6$ kcal/mol) (Fig. 2; Gurry et al., 2010).

4. MD SIMULATION OF THE MMP CLEAVAGE SITE IN COLLAGEN

Peptide substrate studies have shown that collagen primary structure was not the basis for discriminatory MMP collagenolytic behavior (Nagase & Fields, 1996). A model of the cleavage sites in interstitial collagens was thus developed based on the combination of collagen primary and supersecondary structures. This model suggested that all of the information necessary for efficient hydrolysis of collagen was contained in a 24-residue stretch (Fields, 1991). Cleavage site regions were distinguished by a low content (<10%) of charged residues in addition to being "tightly" triple helical (high Pro and Hyp content) prior to the cleavage site and "loosely" triple helical (low Pro and Hyp content) following the cleavage site (Fields, 1991). One of the implications of this model was that low Pro and Hyp

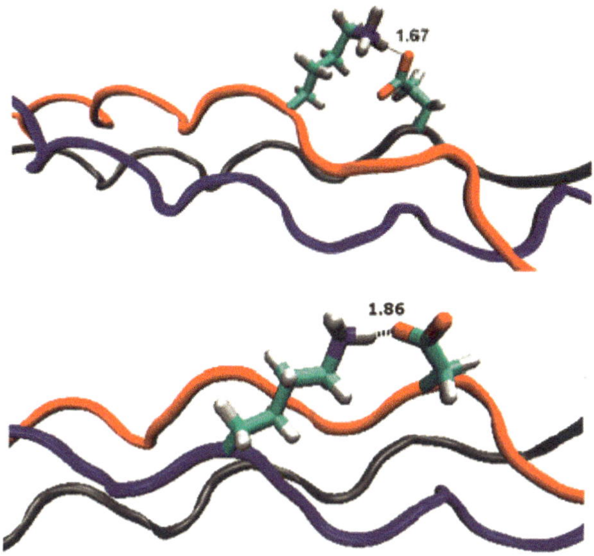

Fig. 2 Snapshots inside umbrella sampling simulation windows for (*top*) $(Gly\text{-}Pro\text{-}Hyp)_3$-Gly-Pro-Lys-Gly-Glu-Hyp-$(Gly\text{-}Pro\text{-}Hyp)_3$ middle+trailing strand interaction and (*bottom*) $(Gly\text{-}Pro\text{-}Hyp)_3$-Gly-Pro-Lys-Gly-Asp-Hyp-$(Gly\text{-}Pro\text{-}Hyp)_3$ leading+middle strand interaction, corresponding to the energy minimum for each. *Reproduced by permission from Gurry, T., Nerenberg, P. S., & Stultz, C. M. (2010). The contribution of interchain salt bridges to triple-helical stability in collagen.* Biophysical Journal, 98, *2634–2643.*

content creates a triple-helical region that is distinct, and more flexible, than a high Pro and Hyp content region. X-ray crystallographic studies of a type III collagen-model THP (T3-785, sequence $(Pro\text{-}Hyp\text{-}Gly)_3$-Ile-Thr-Gly-Ala-Arg-Gly-Leu-Ala-Gly-Pro-Hyp-Gly-$(Pro\text{-}Hyp\text{-}Gly)_3$; pdb IBKV) showed that Pro- and Hyp-rich regions had an average helical twist of 52.4 degree, corresponding to a 7/2 helix (51.4 degree unit twist), while Pro and Hyp poor regions had an average helical twist of 41.8 degree, corresponding to a 10/3 helix (36 degree unit twist) (Kramer, Bella, Mayville, Brodsky, & Berman, 1999; Okuyama, Wu, Jiravanichanun, Hongo, & Noguchi, 2006). Although the average molecular structure for collagen has been proposed to be the 7/2 helical symmetry (Okuyama, Wu, et al., 2006; Okuyama, Xu, Iguchi, & Noguchi, 2006), there are structural differences between Pro- and Hyp-rich and poor regions within triple helices.

MD simulations have explored regions close to or within the MMP cleavage region in collagen. MD simulations of T3-785 were performed using CHARMM with umbrella sampling to study localized unfolding, and involved increasing the radius of gyration of all Cα atoms in the reaction region of one chain (Stultz, 2002). Local unfolding occurred in the region where bridging water molecules associated with the triple helix, N-terminal to the Arg residue (Stultz, 2002). MD simulations of T3-785 using CHARMM version 31b1 with the param22 all-atom force field (with adjusted parameters for Hyp) found that the region containing Gly-Ile was an initiation site for local unwinding at 300 K within 1 ns (Ravikumar, Humphrey, & Hwang, 2007). This MD approach followed the conformational fluctuations of the triple helix. The results suggested that the MMP HPX domain might stabilize the unwound conformation at the MMP cleavage site (Ravikumar et al., 2007).

MD simulations were performed on the type I collagen cleavage site using the CHARMM19 extended-atom force field with CHARMM version 33b2 (Nerenberg & Stultz, 2008). The minimized structure was overlaid with a 30-Å sphere of equilibrated TIP3P water molecules. Unfolding of the α2(I) chain was found to be more favorable than unfolding of either of the two α1(I) chains. It was proposed that the lack of hydroxylation of the Pro residue in the α1(I) chain P_3 subsite allowed for greater flexibility of the α2(I) chain due to decreased hydrogen bonding between the chains (Nerenberg & Stultz, 2008). Kinetic and NMR spectroscopic studies of a series of THPs, which modeled the natural MMP cleavage site and a potential but noncleavable site just 2-triplets away from the natural site in type III collagen, suggested a correlation between the local conformation and dynamics of Ile/Leu at the collagenase cleavage site to the collagenolytic activities of the peptides (Xiao, Addabbo, Lauer, Fields, & Baum, 2010). The NMR studies showed that only a single Ile, located in the leading strand at the natural cleavage site, is more exposed to solvent and less locally stable than the Ile in the middle and trailing strands. The unique local instability of Ile at the cleavage site was proposed to arise from the conserved Pro at the P_3 subsite, which modulated the local dynamics of the three chains. Thus, a single locally unstable chain at the cleavage site, rather than two or three labile chains, may be the determining factor for collagen to be cleaved by MMP-1 (Xiao et al., 2010).

It has been recognized that interstitial collagens contain multiple potential cleavage sites (Fields, 1991). MD simulations were performed on the five potential cleavage sites in type III collagen using CHARMM, where structures were solvated with a sphere of TIP3P water molecules of 30 Å radius (Salsas-Escat & Stultz, 2010). All five potential cleavage sites indicated some local unfolding, but with different vulnerabilities (average exposure of the scissile bond). Structures were subsequently docked with the MMP-8

CAT domain. The sequences that were the most vulnerable fit best into the active site of MMP-8. The most vulnerable sequence of the five encompassed residues 757–807, which included the native MMP cleavage site (Gly ∼ Ile at 775–776). It was proposed that electrostatic interactions between Arg side chains and backbone moieties on adjacent chains stabilized the triple helix, and disruption of such interactions (perhaps by the HPX domain) would facilitate destabilization of the local triple-helical structure (Salsas-Escat & Stultz, 2010).

5. MD CONSIDERATION OF WATER SURROUNDING COLLAGEN

Solvent plays an important role in collagen hydrolysis, as indicated by deuterium isotope effect and activation energy studies of MMP-1 action on type I collagen (Welgus, Jeffrey, & Eisen, 1981). MD simulations of $[(\text{Pro-Hyp-Gly})_{10}]_3$ and T3-785 were performed in explicit solvent using the PMEMD module of the AMBER 9 package and the AMBER 03 force field to model the systems (Suárez, Díaz, & Suárez, 2008). The study primarily evaluated the free energy changes for the two structures, finding $\Delta G = -6.2$ and -1.5 kcal/mol for the folding of $[(\text{Pro-Hyp-Gly})_{10}]_3$ and T3-785, respectively. The balance of solute–solvent interactions largely determined the global thermodynamic stability of the triple-helical state, while the entropic penalty due to the loss of conformational freedom was reduced by Pro and Hyp (Suárez et al., 2008).

X-ray crystallographic analysis of T3-785 has indicated that hydration patterns are different in the Pro and Hyp poor region of the type III collagen cleavage site compared with Pro- and Hyp-rich regions, in that more ordered water is found in the Pro and Hyp poor region (Kramer & Berman, 1998). An MD study found that unwinding was controlled by dynamic water bridges between backbone atoms (Ravikumar & Hwang, 2008). Nine water bridging sites were identified in the Pro and Hyp poor region of T3-785 and were highly dynamic (lifetimes of only a few ps) (Ravikumar & Hwang, 2008). The weakest water bridges were found in the Gly-Ile region. Thus, local unwinding could be modulated by the enzyme breaking water bridges. MD simulations of T3-785, which contained 30 residues in each chain and 9 single-water bridge sites between them in the Pro and Hyp poor region, were performed using NAMD with the CHARMM all-atom force field param22 (with adjusted parameters for Hyp) (Tourell & Momot, 2016). The results indicated that water molecules closest to the bridge sites have long residence times and constantly break and reform water bridges (Tourell & Momot, 2016). The lack of tightly bound water seems at odds with experimental data, indicating that the bridge bonds are ice-like. However, additional intra- and interchain water bridge types have been identified (Bella, Brodsky, & Berman, 1995). Future MD studies may need to consider all water bridge types at the triple helix (Tourell & Momot, 2016).

6. MD OF MMP-1 OR MMP-8 INTERACTION WITH COLLAGEN

To investigate the ability of MMP-1 to unwind triple-helical collagen, collagen was incubated with the catalytically inactive, but correctly folded MMP-1($Glu_{200} \rightarrow$ Ala) in the presence of MMP-1 CAT (Chung et al., 2004). MMP-1 CAT alone did not cleave collagen nor did MMP-1 ($Glu_{200} \rightarrow$ Ala). However, MMP-1 CAT did cleave type I collagen when incubated with MMP-1($Glu_{200} \rightarrow$ Ala). These results suggested that MMP-1($Glu_{200} \rightarrow$ Ala), which has an intact HPX domain, oriented and helped unwind the collagen triple helix, while MMP-1 CAT cleaved the triple helix. If collagen hydrolysis was entirely due to enzyme conformational rearrangement, MMP-1($Glu_{200} \rightarrow$ Ala) would have had no effect on the ability of MMP-1 CAT to cleave collagen. It was suggested that MMP-1 preferentially bound the α2(I) chain (Chung et al., 2004).

At room temperature the MMP-8 CAT domain could cleave type I collagen at high enzyme concentrations (1.2 μM) and using long incubation times (up to 48 h) (Salsas-Escat, Nerenberg, & Stultz, 2010). This was attributed to the MMP-8 CAT domain being less efficient then full-length MMP-8 at trapping the vulnerable state. Although comparison with gelatin hydrolysis was used to demonstrate that the MMP-8 CAT domain was not simply cleaving denatured collagen (Salsas-Escat et al., 2010), it was still possible that the triple-helical collagen was in equilibrium with an unfolded form (Leikina, Mertts, Kuznetsova, & Leikin, 2002), especially given the time frame of the reaction.

Mathematical modeling has suggested that collagen may exist in a "vulnerable" state and that MMPs do not actively unwind or destabilize collagen but instead trap the vulnerable state (Nerenberg, Salsas-Escat, & Stultz, 2008; Salsas-Escat et al., 2010). The vulnerable state would be sampled frequently and with strong binding affinity by MMPs (Fig. 3; Nerenberg et al., 2008). The MMP HPX domain would contribute to the affinity for the

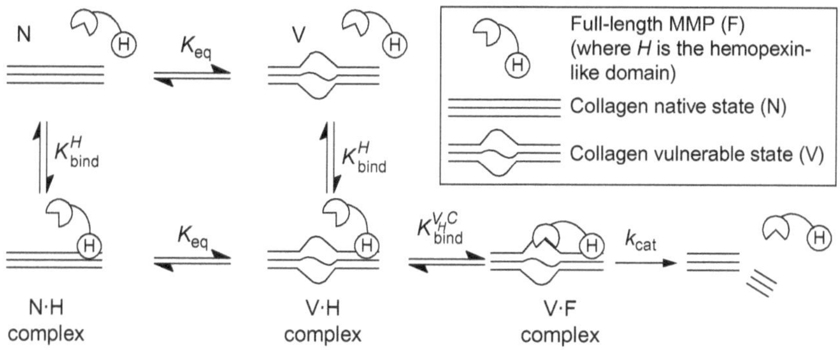

Fig. 3 A conformational selection mechanism for collagenolysis with full-length MMPs. Collagen exists in an equilibrium between native (N) and vulnerable (V) states determined by the equilibrium constant K_{eq}. Full-length MMP (F) can bind to the native state of collagen via the HPX domain (H), forming the N•H complex. The N•H complex can transition to the V•H complex via the equilibrium between native and vulnerable states, K_{eq}. The full-length enzyme (F) can bind directly to the vulnerable state (V) via the HPX domain (H) forming the V•H complex. Once a V•H complex is formed, the CAT domain of the full-length MMP can bind to the vulnerable state forming the V•F complex. The V•F complex is then degraded with catalytic rate k_{cat}. *Reproduced by permission from Salsas-Escat, R., Nerenberg, P. S., & Stultz, C. M. (2010). Cleavage site specificity and conformational selection in type I collagen degradation.* Biochemistry, 49, *4147–4158.*

vulnerable state. However, the modeling utilized catalytic rate constants from MMP-1 hydrolysis of denatured collagen chains (Nerenberg et al., 2008; Salsas-Escat et al., 2010), and the enzymatic processing of collagen chains *within a triple-helical context* vs denatured can be very different (Minond et al., 2007). Nonetheless, the vulnerable state hypothesis provided an intriguing, energetically reasonable collagenolysis mechanism.

In order to provide insight into MMP-1-catalyzed collagenolysis, we performed a set of long-range AT MD simulations on MMP-1 bound to a type II collagen-model THP (Singh, Fields, Christov, & Karabencheva-Christova, 2016b). The X-ray crystallographic structure of the MMP-1•THP complex had been solved (pdb 4AUO) (Manka et al., 2012). Effects of the linker region on interdomain interactions and interactions with the substrate were studied by MD simulations of in silico engineered MMP-1 with the linker of MT1-MMP or MMP-13. In order to explore the dynamic properties of MMP-1, we performed an extensive set of MD simulations for 300 ns (wild-type MMP-1) and 100 ns for the modified linker systems by using Gromacs 4.5.5 package with GROMOS96 43a1 force field (Singh et al., 2016b).

The study confirmed the closed or collapsed state of the MMP-1 X-ray crystallographic structure (Manka et al., 2012) characterized by close orientation of the MMP-1 HPX and CAT domains. The MD simulation did not cause the MMP-1 conformation to open but allowed the effects of the flexibility on the closed conformation to be explored. The simulation provided insight into the role of the linker for modulating conformational dynamics between the HPX and CAT domains and revealed the dynamic nature of interactions between the THP, HPX domain, and CAT domain (Fig. 4; Singh et al., 2016b). Correlated motions observed between the residues from the CAT domain and locally destabilized THP region would be important for the formation of the productive E•S complex. The correlated motions would also support a more favorable orientation of the enzyme electrostatic environment that would facilitate the catalyzed hydrolysis reaction in the active site (e.g., stabilizing the transition state).

The linker played a key role in modulating the interactions between the domains and formation of the active complex (Singh et al., 2016b). The MD simulations with exchanged linker regions from MT1-MMP and MMP-13 revealed the important role of the linker region to influence overall enzyme flexibility, the pattern of the correlated motions between the HPX and CAT domains, and the interactions between the triple-helical substrate and the enzyme.

Experimental studies showed that a variety of mutations in the MMP-1 HPX and CAT domains have significant effects on the collagenolytic activity (Arnold et al., 2011; Lauer-Fields et al., 2009; Manka et al., 2012). Computational studies provided AT explanation of these effects and revealed that the mutations have not only local structural effects but also long-range impact on the protein structure, dynamics, and the interactions with a triple-helical substrate (Singh, Fields, Christov, & Karabencheva-Christova, 2016a). The mutations change the flexibility around the local site but also influence distant regions from the HPX domain, the linker region, and the CAT domain. In addition, the mutations modulate the intensity and the nature of the correlated and anticorrelated motions. An increased collagenase activity in the MMP-1 L338A/H339A mutant corresponded to the appearance of a unique anticorrelated motion and decreased correlated motions (Fig. 5), while decreased collagenase activity in the other mutations corresponded both to increased and to decreased anticorrelated motions (Singh et al., 2016a). The MD studies provided structural and dynamic information which helped to explain the experimentally measured effects

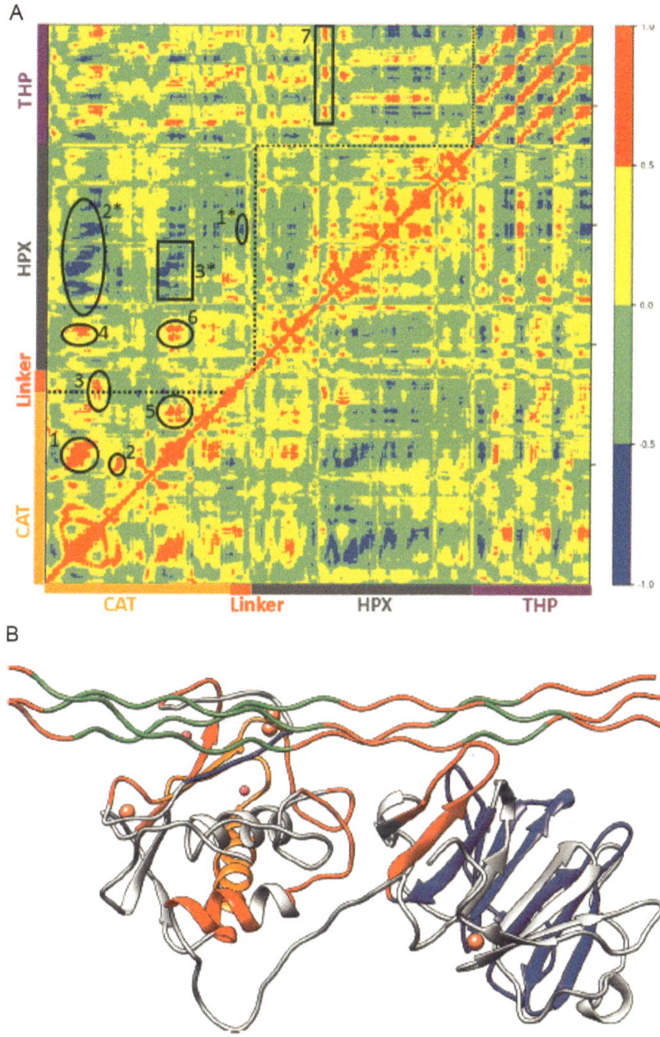

Fig. 4 Dynamic crosscorrelation analysis of MMP-1•THP. (A) Intensities are indicated by the *bar* on the right. Numbering of correlations corresponds to specific interactions as follows: (1) The CAT domain residues 105–115 show correlation toward residues 175–195. (2) Residues 140–150 show correlation toward residues 170–180. (3) Residues 117–125 show correlation to residues 235–240. (4) CAT domain residues 105–115 show correlation toward HPX domain residues 287–294. (5) CAT domain residues 180–195 show correlation toward CAT domain residues 220–230. (6) Residues 180–195 show correlation toward HPX domain residues 280–290. Residues 180–195 also show correlation to THP residues 780–785. (7) HPX domain residues 315–320 are correlated to the ends of the THP strands. (1*) Residues 160–165 show negative correlation to HPX domain *(Continued)*

of the mutations. With the exception of the E200A mutant, there are no X-ray crystallographic structures of MMP-1•THP with mutated residues, and thus the MD study provided this missing structural information and asserted that the mutations have delicate, distinct, and specific effects on the structure, interactions, and dynamics in MMP-1•THP.

MD simulations were utilized to dock MMP-1 with the cleavage site region in type III collagen (Subramanian, Singam, Berinski, Subramanian, & Wade, 2016). The X-ray crystallographic structure of the active form of MMP-1 (pdb 1FBL) was used, and a model triple helix was constructed using the Gencollagen Web server. MD simulations were performed using AMBER 10 with the FF99SB force field. Although the X-ray crystallographic structure represents a closed form of MMP-1, more open structures were generated by manually changing the dihedral angles of the residues in the linker region. The HPX domain was found to bind two Arg residues, located in the P_8' and P_{20}' subsites of the type III triple helix, via Asp338 and/or Glu313 (Subramanian et al., 2016). The side chains of Hyp at P_{14}' and Met at P_{16}' made favorable nonpolar interactions with the loop containing Phe308 and Tyr309. After binding the triple helix, the CAT domain could reorient to properly face the cleavage site. The dual Arg "electrostatic ruler" motif was modeled to be exposed in the type III collagen fibril. The two Arg residues found in type III collagen are conserved in the P_5' and P_{17}' subsites of types I and II collagen.

MD has provided insight into the degradation of type I collagen fibrils by MMP-8 (Lu & Stultz, 2013). Mathematical modeling and MD simulations (using CHARMM version 36b2 with the CHARMM19 extended atom force field) indicated a two-step process, whereby the α2(I) chain "unfolds" from the triple helix, and then the MMP-8 CAT domain reorients the chain to fit into the active site (Fig. 6; Lu & Stultz, 2013).

Fig. 4—Cont'd residues 275–280. (2*) CAT domain residues 100–120 show negative correlation toward HPX domain residues 310–380. (3*) CAT domain residues 175–185 show negative correlation toward HPX domain residues 330–350. (B) The correlated and anticorrelated motions between the residues of MMP-1•THP are projected on to the 3-D structure, where *red* and *blue* follow the scheme above and *orange* shows residues participating in both correlated and anticorrelated motion. The Zn^{2+} and Ca^{2+} are shown in *purple* and *pink*, respectively. The THP and MMP-1 are shown in *green* and *gray*, respectively. *Reproduced by permission from Singh, W., Fields, G. B., Christov, C. Z., & Karabencheva-Christova, T. G. (2016b). Importance of the linker region in matrix metalloproteinase-1 domain interactions.* RSC Advances, 6, *23223–23232.*

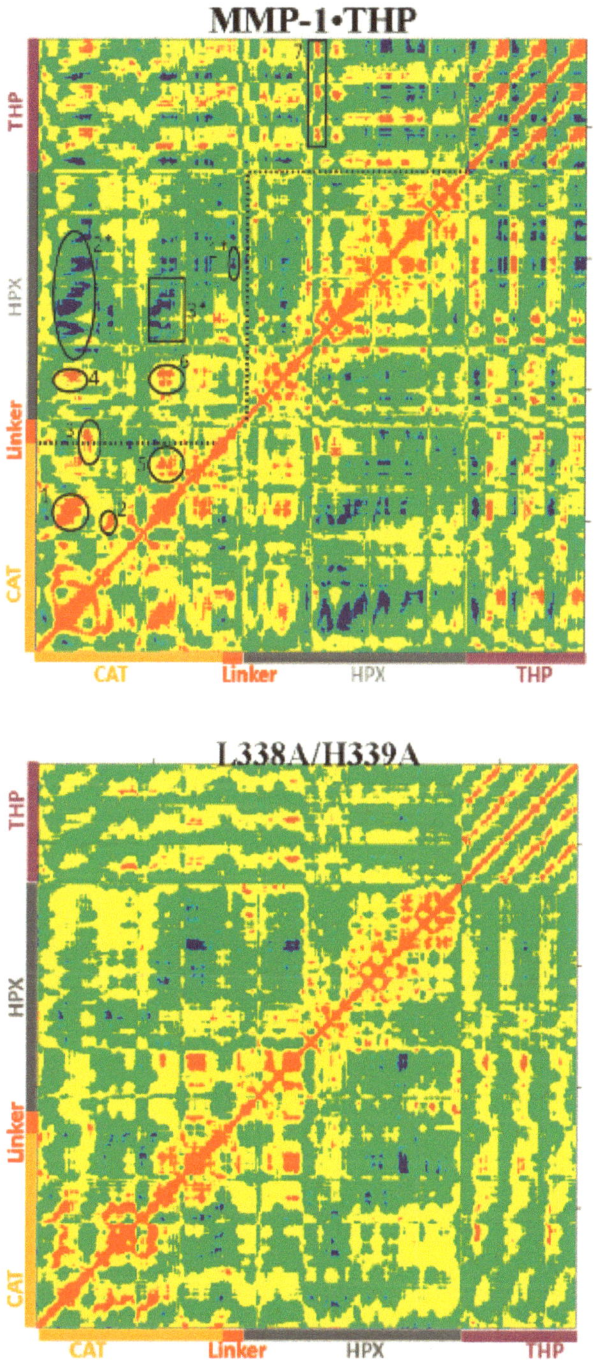

Fig. 5 See legend on next page.

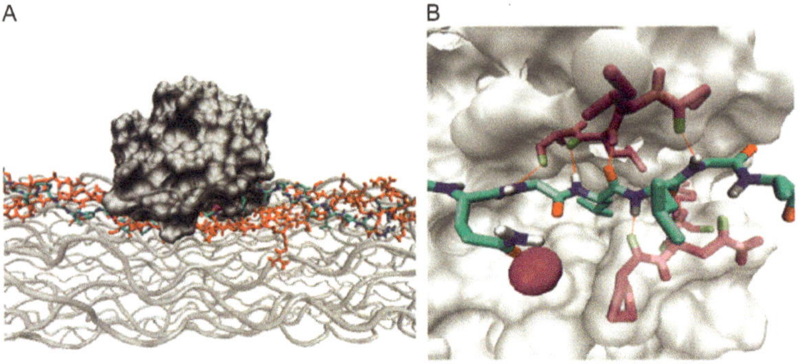

Fig. 6 A representative docked structure of MMP-8 with the type I collagen fibril corresponding to the lowest energy state. (A) Side view of the fibril structure docked into the catalytic site of MMP-8. The central triple helix of the fibril is shown in *stick* representation, and the rest of the fibril surface is shown as *gray* transparent tubes. The $\alpha1(I)$ chains are colored *red*, and the $\alpha2(I)$ chain is colored *cyan*. MMP-8 is shown as a surface representation in *gray*. (B) View of the catalytic site, with the docked $\alpha2(I)$ chain. The Zn^{2+} ion in the catalytic site is shown in *purple*. The hydrogen-bonding pattern of the $\alpha2(I)$ chain backbone to the MMP-8 CAT domain is shown in *red lines* and closely follows that of one shown in the X-ray crystallographic structure of MMP-8 with a bound peptide (pdb 2OY4). *Reproduced by permission from Lu, K. G., & Stultz, C. M. (2013). Insight into the degradation of type-I collagen fibrils by MMP-8.* Journal of Molecular Biology, 425, *1815–1825.*

MD simulations were utilized to examine conserved water molecules in the MMP-1 CAT domain (Chakrabarti, Bairagya, Mukhopadhyay, & Sekar, 2017). Five X-ray crystallographic structures of MMP-1 were utilized (pdb 1CGE, 1CGF, 1CGL, 1HFC, and 2TCL) in conjunction with the Nanoscale Molecular Dynamics (NAMD v.2.9) program and CHARMM27 force field. During the 50 ns MD simulation, three conserved water molecules were found to be coordinated to the catalytic Zn^{2+} and one conserved water molecule with the structural Zn^{2+} (Chakrabarti et al., 2017). The two Zn^{2+} ions are bridged by two conserved water molecules, which may ultimately impact the hydrolytic mechanism of the enzyme.

Fig. 5 The dynamic crosscorrelation analysis (DCCA) of MMP-1•THP and the L338A/H339A MMP-1 mutant•THP. The scale of correlated motion ranges from +1 to −1 and represents positive (*red*) and negative (*blue*) motions of Cα atoms. The areas labeled 1–7 indicate positive correlated motions, while the areas labeled 1*–3* indicate negative correlated motions. *Reproduced by permission from Singh, W., Fields, G. B., Christov, C. Z., & Karabencheva-Christova, T. G. (2016a). Effects of mutations on structure-function relationships of matrix metalloproteinase-1.* International Journal of Molecular Sciences, 17, *E1727.*

7. MD OF MMP-2 INTERACTION WITH COLLAGEN

Diaz and coworkers observed a conformational shift from a closed X-ray crystallographic MMP-2 structure to an elongated structure using AT MD simulations on a 100-ns timescale (Díaz, Suárez, & Valdés, 2008). MMP-2 could adopt extended conformations in solution prior to collagen hydrolysis (Díaz & Suárez, 2012). The interaction of MMP-2 with a triple-helical substrate, fTHP-5, was subsequently analyzed by MD simulations (Díaz, Suárez, & Valdés, 2013). The coordinates for the triple-helical substrate were built using the TCL/TK triple-helical collagen building script (The BuScr) (Rainey & Goh, 2004). MD simulations of fTHP-5 were performed using the SANDER module of AMBER 10 (Díaz et al., 2013). Subsequently, fTHP-5 was surrounded by a 25-Å water box using the TIP3P potential. Following MD simulations, fTHP-5 structures were docked within the active site of MMP-2. The most favored docked structures then underwent further MD simulation. The most energetically favorable binding event occurred between the MMP-2 active site and the B (middle) chain of fTHP-5 (Díaz et al., 2013). Local unwinding of the triple helix was required prior to binding to the active site, but the energy penalty was similar for all three chains (10 kcal/mol) and thus none had an intrinsic preference to be the first one hydrolyzed (Díaz et al., 2013). Further MD simulation analysis was performed for the prereactive complex between MMP-2 and fTHP-5 (Díaz & Suárez, 2015). Three different models were obtained based on molecular interactions and positioning of the FNII repeats and the CAT domain. Ultimately two models, MMP-2(X)/fTHP-5 and MMP-2(D-1)/fTHP-5, were deemed the most energetically favorable, where in the former model the HPX domain was close to active site cleft, while in the latter model the HPX domain was at the rear of the CAT domain and distal to the active site cleft (Fig. 7; Díaz & Suárez, 2015). This indicated flexibility in domain positioning prior to catalysis.

MD simulations have primarily been limited to the interaction of MMPs with collagen and binding of the initial strand. The manipulations of the triple helix for hydrolysis, and the release of product, have been less studied. QM/MM and MD simulations have been utilized to study the interaction of the peptide Ace-Gln-Gly ~ Ile-Ala-Gly-Nme with MMP-2 (Díaz & Suárez, 2008a, 2008b; Díaz, Suárez, & Suárez, 2010; Vasilevskaya, Khrenova, Nemukhin, & Thiel, 2015, 2016). Examination of hydrolysis of Ace-Gln-Gly ~ Ile-Ala-Gly-Nme by MMP-2 indicated that release of Ace-Gln-Gly was the rate-limiting step for catalysis, and release of

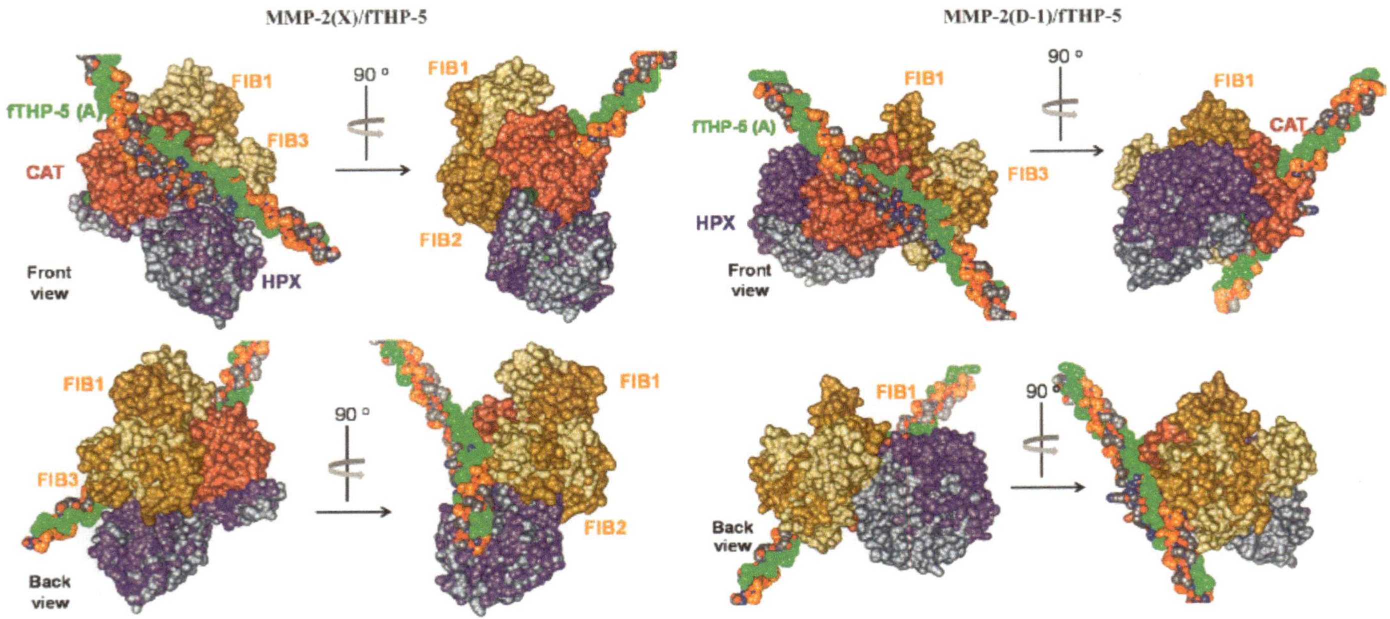

Fig. 7 Different views (90 degree rotated) for a surface representation of the average structures obtained from the last 10 ns of the MMP-2/fTHP-5 MD simulations. For comparison, the average structures of the HPX and FNII (FIB) domains obtained from the last 10 ns of the corresponding MMP-2 simulations are also displayed in *lighter colors*. The MMP-2 structures were superposed over the MMP-2/fTHP-5 structures using the backbone coordinates of the CAT domain. *Reproduced by permission from Díaz, N., & Suárez, D. (2015). Extensive simulations of the full-length matrix metalloproteinase-2 enzyme in a prereactive complex with a collagen triple-helical peptide.* Biochemistry, 54, *1243–1258.*

Ace-Gln-Gly occurred before release of Ile-Ala-Gly-Nme (Vasilevskaya et al., 2015). The difficulty of releasing the Ace-Gln-Gly fragment was due to the negatively charged C-terminus being coordinated in a bidentate fashion to the active site Zn^{2+} (Vasilevskaya et al., 2015).

QM/MM examination of the initial stages of hydrolysis of Ace-Gln-Gly ~ Ile-Ala-Gly-Nme by MMP-2 indicated that a 12-Å water layer was sufficient to provide electrostatic stabilization of the reactive species (enzyme•substrate complex, transition state, and tetrahedral intermediate) (Vasilevskaya et al., 2016).

8. MECHANISM OF COLLAGENOLYSIS

The highest MO conformations sampled by MMP-1 when free in solution appeared to be much more poised for interaction with collagen than the compact X-ray crystallographic structures (Cerofolini et al., 2013). In the conformations belonging to this cluster (i) the collagen-binding residues of the HPX domain were solvent exposed and (ii) the CAT domain was already correctly positioned for its subsequent interaction with the collagen. A structural rearrangement involving a ~50 degree rotation around a single axis of the CAT domain with respect to the HPX domain was sufficient to position the CAT domain right in front of the preferred cleavage site in triple-helical collagen. The transition seemed to be feasible at physiological temperature as the difference in free energy between these steps in the pathway was favorable (−0.133 kcal/mol). The reorientation of the CAT domain was observed in MD simulation studies (Subramanian et al., 2016).

Hydrogen–deuterium exchange mass spectrometry, NMR spectroscopy, X-ray crystallography, and enzymatic assay analysis of mutants have led to the identification of MMP-1 HPX domain residues Ile290, Arg291, Phe301, Val319, and Asp338 as important for collagenolysis (Arnold et al., 2011; Lauer-Fields et al., 2009; Manka et al., 2012). Based on the X-ray crystallographic structure of MMP-1•THP (pdb 4AUO), Phe320 was found to be an important contributor, along with Ile290 and Arg291, to the S_{10}'-binding pocket (Manka et al., 2012). The S_{10}'-binding pocket bound the P_{10}' subsite of the THP, which possessed a Leu residue important for interaction of triple helices with MMP-1 (Arnold et al., 2011; Manka et al., 2012; Robichaud, Steffensen, & Fields, 2011). In MMP-13 the S_{10}' pocket is shifted, with Arg297 at one edge of the pocket moved 7.3 Å relative to MMP-1 (Stura, Visse, Cuniasse, Dive, & Nagase, 2013).

The interaction of MMP-1 with a THP has been investigated utilizing NMR spectroscopy, leading to a plausible multistep mechanism for collagenolysis (Bertini et al., 2012). In this mechanism, the initial binding of the HPX domain to the collagen triple helix through specific residues in blades I and II of the HPX domain is followed by the interaction of the CAT domain with the triple helix in front of the cleavage site, and by a subsequent back rotation of the CAT and HPX domains toward the closed conformation observed by X-ray crystallography that drives the unwinding/perturbation of the triple helix and causes the displacement of one peptide chain into the active site. The MMP-1•THP complex in which one strand of the triple helix is displaced from the other two was strongly supported by changes in intensity of the interhelical NOEs upon THP binding (Bertini et al., 2012). MD simulation studies support the reorientation of the CAT and HPX domains, and the movement of the domains (as predicted based on NMR spectroscopic studies) may better orient the destabilized strand of the triple helix into the active site (Lu & Stultz, 2013; Subramanian et al., 2016).

Models for collagenolysis have also been proposed based on X-ray crystallographic structures (Manka et al., 2012; Stura et al., 2013). In these models, significant movement of the triple helix (3–6 Å), rotation/sliding of the triple helix, and/or significant movement of a single strand from the triple helix (~4 Å) is needed to accommodate a chain in the active site. It is not clear if these models are energetically favorable. A MD study was performed using the MMP-1•THP X-ray crystallographic structure (pdb 4AUO) with the GROMACS simulation package, version 5.0.5 (Nash, Birch, & de Leeuw, 2017). Metadynamics calculations indicated unwinding at the cleavage site and insertion of a single chain into the active site (Nash et al., 2017). However, free energy calculations from umbrella sampling of the transition from noninserted collagen to inserted collagen were energetically unfavorable (Nash et al., 2017). This suggested that there could be a thermal fluctuation in the triple helix to promote insertion into the active site.

ACKNOWLEDGMENTS

T.K. and C.C. acknowledge support by Marie Curie International Career Development Fellowships, NSCCS Grants, and HEC-Biosim Grants, while GBF acknowledges support by the National Institutes of Health (CA98799, AR063795, and NHLBI contract 268201000036C-0-0-1).

REFERENCES

Adcock, S. A., & McCammon, J. A. (2006). Molecular dynamics: Survey of methods for simulating the activity of proteins. *Chemical Reviews*, *106*, 1589–1615.

Amar, S., Smith, L., & Fields, G. B. (2017). Matrix metalloproteinase collagenolysis in health and disease. *Biochimica et Biophysica Acta (BBA)—Molecular Cell Research*. in press.

Arnold, L. H., Butt, L., Prior, S. H., Read, C., Fields, G. B., & Pickford, A. R. (2011). The interface between catalytic and hemopexin domains in matrix metalloproteinase 1 conceals a collagen binding exosite. *The Journal of Biological Chemistry*, *286*, 45073–45082.

Bella, J., Brodsky, B., & Berman, H. M. (1995). Hydration structure of a collagen peptide. *Structure*, *3*, 893–906.

Bertini, I., Fragai, F., Luchinat, C., Melikian, M., Toccafondi, M., Lauer, J. L., et al. (2012). Structural basis for matrix metalloproteinase 1 catalyzed collagenolysis. *Journal of the American Chemical Society*, *134*, 2100–2110.

Bhatnagar, R. S., Pattabiramam, N., Sorensen, K. R., Langridge, R., & MacElroy, R. D. (1988). Inter-chain proline:proline contacts contribute to the stability of the triple helical conformation. *Journal of Biomolecular Structure & Dynamics*, *6*, 223–233.

Bodian, D. L., Madhan, B., Brodsky, B., & Klein, T. E. (2008). Predicting the clinical lethality of osteogenesis imperfecta from collagen glycine mutations. *Biochemistry*, *47*, 5424–5432.

Bodian, D. L., Radmer, R. J., Holbert, S., & Klein, T. E. (2011). Molecular dynamics simulations of the full triple helical region of collagen type I provide an atomic scale view of the protein's regional heterogeneity. *Pacific Symposium on Biocomputing*, *16*, 193–204.

Brooks, B. R., Brooks, C. L., 3rd, Mackerell, A. D., Jr., Nilsson, L., Petrella, R. J., Roux, B., et al. (2009). CHARMM: The biomolecular simulation program. *Journal of Computational Chemistry*, *30*, 1545–1614.

Cappel, D., Hall, M. L., Lenselink, E. B., Beuming, T., Qi, J., Bradner, J., et al. (2016). Relative binding free energy calculations applied to protein homology models. *Journal of Chemical Information and Modeling*, *56*, 2388–2400.

Case, D. A., Cheatham, T. E., Darden, T., Gohlke, H., Luo, R., Merz, K. M., et al. (2005). The AMBER biomolecular simulation programs. *Journal of Computational Chemistry*, *26*, 1668–1688.

Cerofolini, L., Fields, G. B., Fragai, M., Geraldes, C. F., Luchinat, C., Parigi, G., et al. (2013). Examination of matrix metalloproteinase-1 (MMP-1) in solution: A preference for the pre-collagenolysis state. *The Journal of Biological Chemistry*, *288*, 30659–30671.

Chakrabarti, B., Bairagya, H. R., Mukhopadhyay, B. P., & Sekar, K. (2017). New biochemical insight of conserved water molecules at catalytic and structural Zn2+ ions in human matrix metalloproteinase-I: A study by MD-simulation. *Journal of Molecular Modeling*, *23*, 57.

Chung, L., Dinakarpandian, D., Yoshida, N., Lauer-Fields, J. L., Fields, G. B., Visse, R., et al. (2004). Collagenase unwinds triple helical collagen prior to peptide bond hydrolysis. *The EMBO Journal*, *23*, 3020–3030.

Díaz, N., & Suárez, D. (2008a). Molecular dynamics simulations of the active matrix metalloproteinase-2: Positioning of the N-terminal fragment and binding of a small peptide substrate. *Proteins*, *72*, 50–61.

Díaz, N., & Suárez, D. (2008b). Peptide hydrolysis catalyzed by matrix metalloproteinase 2: A computational study. *The Journal of Physical Chemistry. B*, *112*, 8412–8424.

Díaz, N., & Suárez, D. (2012). Alternative interdomain configurations of the full-length MMP-2 enzyme explored by molecular dynamics simulations. *The Journal of Physical Chemistry. B*, *116*, 2677–2686.

Díaz, N., & Suárez, D. (2015). Extensive simulations of the full-length matrix metalloproteinase-2 enzyme in a prereactive complex with a collagen triple-helical peptide. *Biochemistry*, *54*, 1243–1258.
Díaz, N., Suárez, D., & Suárez, E. (2010). Kinetic and binding effects in peptide substrate selectivity of matrix metalloproteinase-2: Molecular dynamics and QM/MM calculations. *Proteins*, *78*, 1–11.
Díaz, N., Suárez, D., & Valdés, H. (2008). From the X-ray compact structure to the elongated form of the full-length MMP-2 enzyme in solution: A molecular dynamics study. *Journal of the American Chemical Society*, *130*, 14070–14071.
Díaz, N., Suárez, D., & Valdés, H. (2013). Unraveling the molecular structure of the catalytic domain of matrix metalloproteinase-2 in complex with a triple-helical peptide by means of molecular dynamics simulations. *Biochemistry*, *52*, 8556–8569.
Fan, P., Li, M. H., Brodsky, B., & Baum, J. (1993). Backbone dynamics of $(Pro-Hyp-Gly)_{10}$ and a designed collagen-like triple-helical peptide by ^{15}N NMR relaxation and hydrogen-exchange measurements. *Biochemistry*, *32*, 13299–13309.
Fields, G. B. (1991). A model for interstitial collagen catabolism by mammalian collagenases. *Journal of Theoretical Biology*, *153*, 585–602.
Genheden, S., & Ryde, U. (2015). The MM/PBSA and MM/GBSA methods to estimate ligand-binding affinities. *Expert Opinion on Drug Discovery*, *10*, 449–461.
Gurry, T., Nerenberg, P. S., & Stultz, C. M. (2010). The contribution of interchain salt bridges to triple-helical stability in collagen. *Biophysical Journal*, *98*, 2634–2643.
Hess, B., Kutzner, C., van der Spoel, D., & Lindahl, E. (2008). GROMACS 4: Algorithms for highly efficient, load-balanced, and scalable molecular simulation. *Journal of Chemical Theory and Computation*, *4*, 435–447.
Jelinski, L. W., Sullivan, C. E., & Torchia, D. A. (1980). 2H NMR study of molecular motion in collagen fibrils. *Nature*, *284*, 531–534.
Jelinski, L. W., & Torchia, D. A. (1979). 13C/1H high power double magnetic resonance investigation of collagen backbone motion in fibrils and in solution. *Journal of Molecular Biology*, *133*, 45–65.
Kästner, J. (2011). Umbrella sampling. *WIREs Computational Molecular Science*, *1*, 932–942.
Klein, T. E., & Huang, C. C. (1999). Computational investigations of structural changes resulting from point mutations in a collagen-like peptide. *Biopolymers*, *49*, 167–183.
Kramer, R. Z., Bella, J., Mayville, P., Brodsky, B., & Berman, H. M. (1999). Sequence dependent conformational variations of collagen triple-helical structure. *Nature Structural Biology*, *6*, 454–457.
Kramer, R. Z., & Berman, H. M. (1998). Patterns of hydration in crystalline collagen peptides. *Journal of Biomolecular Structure and Dynamics*, *16*, 367–380.
Lauer-Fields, J. L., Chalmers, M. J., Busby, S. A., Minond, D., Griffin, P. R., & Fields, G. B. (2009). Identification of specific hemopexin-like domain residues that facilitate matrix metalloproteinase collagenolytic activity. *The Journal of Biological Chemistry*, *284*, 24017–24024.
Leikina, E., Mertts, M. V., Kuznetsova, N., & Leikin, S. (2002). Type I collagen is thermally unstable at body temperature. *Proceedings of the National Academy of Sciences of the United States of America*, *99*, 1314–1318.
Liu, X., Kim, S., Dai, Q.-H., Brodsky, B., & Baum, J. (1998). Nuclear magnetic resonance shows asymmetric loss of triple helix in peptides modeling a collagen mutation in brittle bone disease. *Biochemistry*, *37*, 15528–15533.
Lu, K. G., & Stultz, C. M. (2013). Insight into the degradation of type-I collagen fibrils by MMP-8. *Journal of Molecular Biology*, *425*, 1815–1825.
Manka, S. W., Carafoli, F., Visse, R., Bihan, D., Raynal, N., Farndale, R. W., et al. (2012). Structural insights into triple-helical collagen cleavage by matrix metalloproteinase 1. *Proceedings of the National Academy of Sciences of the United States of America*, *109*, 12461–12466.

Minond, D., Lauer-Fields, J. L., Cudic, M., Overall, C. M., Pei, D., Brew, K., et al. (2007). Differentiation of secreted and membrane-type matrix metalloproteinase activities based on substitutions and interruptions of triple-helical sequences. *Biochemistry*, *46*, 3724–3733.

Mooney, S. D., Huang, C. C., Kollman, P. A., & Klein, T. E. (2001). Computed free energy differences between point mutations in a collagen-like peptide. *Biopolymers*, *58*, 347–353.

Mooney, S. D., Kollman, P. A., & Klein, T. E. (2002). Conformational preferences of substituted prolines in the collagen triple helix. *Biopolymers*, *64*, 63–71.

Nagase, H., & Fields, G. B. (1996). Human matrix metalloproteinase specificity studies using collagen sequence-based synthetic peptides. *Biopolymers*, *40*, 399–416.

Nash, A., Birch, H. L., & de Leeuw, N. H. (2017). Mapping intermolecular interactions and active site conformations: From human MMP-1 crystal structure to molecular dynamics free energy calculations. *Journal of Biomolecular Structure and Dynamics*, *35*, 564–573.

Nerenberg, P. S., Salsas-Escat, R., & Stultz, C. M. (2008). Do collagenases unwind triple-helical collagen before peptide bond hydrolysis? Reinterpreting experimental observations with mathematical models. *Proteins: Structure, Function, and Bioinformatics*, *70*, 1154–1161.

Nerenberg, P. S., & Stultz, C. M. (2008). Differential unfolding of the α1 and α2 chains in type I collagen and collagenolysis. *Journal of Molecular Biology*, *382*, 246–256.

Okuyama, K., Wu, G., Jiravanichanun, N., Hongo, C., & Noguchi, K. (2006a). Helical twists of collagen model peptides. *Biopolymers (Peptide Science)*, *84*, 421–432.

Okuyama, K., Xu, X., Iguchi, M., & Noguchi, K. (2006b). Revision of collagen molecular structure. *Biopolymers (Peptide Science)*, *84*, 181–191.

Orozco, M. (2014). A theoretical view of protein dynamics. *Chemical Society Reviews*, *43*, 5051–5066.

Park, S., Klein, T. E., & Pande, V. S. (2007). Folding and misfolding of the collagen triple helix: Markov analysis of molecular dynamics simulations. *Biophysical Journal*, *93*, 4108–4115.

Park, S., Radmer, R. J., Klein, T. E., & Pande, V. S. (2005). A new set of molecular mechanics parameters for hydroxyproline and its use in molecular dynamics simulations of collagen-like peptides. *Journal of Computational Chemistry*, *26*, 1612–1616.

Phillips, J. C., Zheng, G., Kumar, S., & Kalé, L. V. (2002). NAMD: Biomolecular simulation on thousands of processors, Proceedings of the IEEE/ACM SC2002 conference (technical paper 277): IEEE Press.

Radmer, R. J., & Klein, T. E. (2004). The severity of osteogenesis imperfecta and the structure of a collagen-like peptide modeling a lethal mutation site. *Biochemistry*, *43*, 5314–5323.

Rainey, J. K., & Goh, M. C. (2004). In interactive triple-helical collagen builder. *Bioinformatics*, *20*, 2458–2459.

Ravikumar, K. M., Humphrey, J. D., & Hwang, W. (2007). Spontaneous unwinding of a labile domain in a collagen triple-helix. *Journal of Mechanics of Materials and Structures*, *2*, 999–1010.

Ravikumar, K. M., & Hwang, W. (2008). Region-specific role of water in collagen unwinding and assembly. *Proteins*, *72*, 1320–1332.

Rinker, S., Allison, J. R., & van Gunsteren, W. F. (2012). On developing coarse-grained models for biomolecular simulation: A review. *Physical Chemistry Chemical Physics*, *14*, 12423–12430.

Robichaud, T. K., Steffensen, B., & Fields, G. B. (2011). Exosite interactions impact matrix metalloproteinase collagen specificities. *The Journal of Biological Chemistry*, *286*, 37535–37542.

Salsas-Escat, R., Nerenberg, P. S., & Stultz, C. M. (2010). Cleavage site specificity and conformational selection in type I collagen degradation. *Biochemistry*, *49*, 4147–4158.

Salsas-Escat, R., & Stultz, C. M. (2010). Conformational selection and collagenolysis in type III collagen. *Proteins: Structure, Function, and Bioinformatics*, *78*, 325–335.

Singh, W., Fields, G. B., Christov, C. Z., & Karabencheva-Christova, T. G. (2016a). Effects of mutations on structure-function relationships of matrix metalloproteinase-1. *International Journal of Molecular Sciences*, *17*, E1727.

Singh, W., Fields, G. B., Christov, C. Z., & Karabencheva-Christova, T. G. (2016b). Importance of the linker region in matrix metalloproteinase-1 domain interactions. *RSC Advances*, *6*, 23223–23232.

Stultz, C. M. (2002). Localized unfolding of collagen explains collagenase cleavage near imino-poor sites. *Journal of Molecular Biology*, *319*, 997–1003.

Stura, E. A., Visse, R., Cuniasse, P., Dive, V., & Nagase, H. (2013). Crystal structure of full-length human collagenase 3 (MMP-13) with peptides in the active site defines exosites in the catalytic domain. *The FASEB Journal*, *27*, 4395–4405.

Suárez, E., Díaz, N., & Suárez, D. (2008). Entropic control of the relative stability of triple-helical collagen peptide models. *The Journal of Physical Chemistry. B*, *112*, 15248–15255.

Subramanian, S. R., Singam, E. R., Berinski, M., Subramanian, V., & Wade, R. C. (2016). Identification of an electrostatic ruler motif for sequence-specific binding of collagenase to collagen. *The Journal of Physical Chemistry. B*, *120*, 8580–8589.

Tourell, M. C., & Momot, K. I. (2016). Molecular dynamics of a hydrated collagen peptide: Insights into rotational motion and residence times of single-water bridges in collagen. *The Journal of Physical Chemistry. B*, *120*, 12432–12443.

Vasilevskaya, T., Khrenova, M. G., Nemukhin, A. V., & Thiel, W. (2015). Mechanism of proteolysis in matrix metalloproteinase-2 revealed by QM/MM modeling. *Journal of Computational Chemistry*, *36*, 1621–1630.

Vasilevskaya, T., Khrenova, M. G., Nemukhin, A. V., & Thiel, W. (2016). Methodological aspects of QM/MM calculations: A case study on matrix metalloproteinase-2. *Journal of Computational Chemistry*, *37*, 1801–1809.

Vidossich, P., & Magistrato, A. (2014). QM/MM molecular dynamics studies of metal binding proteins. *Biomolecules*, *4*, 616–645.

Welgus, H. G., Jeffrey, J. J., & Eisen, A. Z. (1981). Human skin fibroblast collagenase: Assessment of activation energy and deuterium isotope effect with collagenous substrates. *The Journal of Biological Chemistry*, *256*, 9516–9521.

Xiao, J., Addabbo, R. M., Lauer, J. L., Fields, G. B., & Baum, J. (2010). Local conformation and dynamics of isoleucine in the collagenase cleavage site provide a recognition signal for matrix metalloproteinases. *The Journal of Biological Chemistry*, *285*, 34181–34190.

Yu, Y.-C., Roontga, V., Daragan, V. A., Mayo, K. H., Tirrell, M., & Fields, G. B. (1999). Structure and dynamics of peptide-amphiphiles incorporating triple-helical proteinlike molecular architecture. *Biochemistry*, *38*, 1659–1668.

CHAPTER TWO

Computational Glycobiology: Mechanistic Studies of Carbohydrate-Active Enzymes and Implication for Inhibitor Design

Andrew P. Montgomery*[,1], Kela Xiao*[,1], Xingyong Wang*[,†], Danielle Skropeta*[,†], Haibo Yu*[,†,‡,2]

*School of Chemistry, Faculty of Science, Medicine and Health, University of Wollongong, Wollongong, NSW, Australia

†Centre for Medical and Molecular Bioscience, University of Wollongong, Wollongong, NSW, Australia

‡Illawarra Health and Medical Research Institute, Wollongong, NSW, Australia

[2]Corresponding author: e-mail address: hyu@uow.edu.au

Contents

1. Introduction 26
2. Computational Methods 31
   2.1 Combined Quantum Mechanics/Molecular Mechanics Methods 31
   2.2 pK_a and pK_a Shift Calculations in Enzymes 32
   2.3 Minimum Energy Pathway and Potential of Mean Force 33
   2.4 Binding Free Energy Calculations 34
3. Puckering of Carbohydrate Structures 36
   3.1 Carbohydrate Structures 36
   3.2 Cremer–Pople Puckering Coordinates 37
   3.3 The Conformational Catalytic Itinerary in GHs 38
4. Identification of Catalytically Competent Protonation States of Key Residues 40
   4.1 pK_a Cycling of the General Acid/Base in GH33 and GH34 Families 41
   4.2 Rationalizing pK_a Shifts in *Bacillus circulans* Xylanase With Computational Studies 45
5. Clarification of Catalytic Mechanisms 48
   5.1 Catalytic Mechanisms of Hen Egg-White Lysozyme 48
   5.2 Reaction Mechanisms of Inverting *O*-GlcNAc Glycosyltransferase 49
6. Transition-State Analogues as Inhibitors of CAZymes 54
   6.1 Transition-State Analogue Inhibitors 54
   6.2 Computational Insight Into the Design of Transition-State Analogue Inhibitors of CAZymes 55

[1] These authors contributed equally to this work.

Advances in Protein Chemistry and Structural Biology, Volume 109
ISSN 1876-1623
http://dx.doi.org/10.1016/bs.apcsb.2017.04.003

7. Conclusions and Future Directions 65
Acknowledgments 66
References 66

Abstract

Carbohydrate-active enzymes (CAZymes) are families of essential and structurally related enzymes, which catalyze the creation, modification, and degradation of glycosidic bonds in carbohydrates to maintain essentially all kingdoms of life. CAZymes play a key role in many biological processes underpinning human health and diseases (e.g., cancer, diabetes, Alzheimer's diseases, AIDS) and have thus emerged as important drug targets in the fight against pathogenesis. The realization of the full potential of CAZymes remains a significant challenge, relying on a deeper understanding of the molecular mechanisms of catalysis. Considering numerous unsettled questions in the literature, while with a large amount of structural, kinetic, and mutagenesis data available for CAZymes, there is a pressing need and an abundant opportunity for collaborative computational and experimental investigations with the aim to unlock the secrets of CAZyme catalysis at an atomic level. In this review, we briefly survey key methodology development in computational studies of CAZyme catalysis. This is complemented by selected case studies highlighting mechanistic insights provided by computational glycobiology. Implication for inhibitor design by mimicking the transition state is also illustrated for both glycoside hydrolases and glycosyltransferases. The challenges for such studies will be noted and finally an outlook for future directions will be provided.

1. INTRODUCTION

Carbohydrate-active enzymes (CAZymes) are a large group of structurally related enzymes that catalyze the formation, modification, and degradation of glycosidic bonds in carbohydrates, which are of vital importance for life on earth (Lombard, Golaconda Ramulu, Drula, Coutinho, & Henrissat, 2014). Two of the most notable families of CAZymes are glycoside hydrolases (GHs) and glycosyltransferases (GTs), which are responsible for the breakage and synthesis of glycosidic bonds, respectively (Lairson, Henrissat, Davies, & Withers, 2008; Rye & Withers, 2000). CAZymes play a key role in a variety of biological processes closely related to both human health and diseases (e.g., cancer, diabetes, Alzheimer's diseases, and AIDS) and have been utilized as important drug targets in combating pathogenesis (Gloster & Vocadlo, 2012; Williams & Davies, 2001).

Structure and sequence-based classification categorizes GHs into more than 135 families (Lombard et al., 2014). Alternatively, based on their reaction mechanisms, GHs can be separated into two distinct classes proposed by

Koshland in 1953 (Koshland, 1953; Rye & Withers, 2000): inverting and retaining GHs that catalyze hydrolysis with inversion or retention of the anomeric configuration, respectively. Inverting GHs (Fig. 1A) operates via a direct replacement of the leaving group by a water molecule. This mechanism requires a general acid residue and a general base residue. In retaining GHs (Fig. 1B), one catalytic residue functions both as a general acid by protonating the aglycone during the glycosylation step and as a general base by deprotonating the attacking water during the deglycosylation step. The other catalytic residue acts as a nucleophile during the glycosylation step

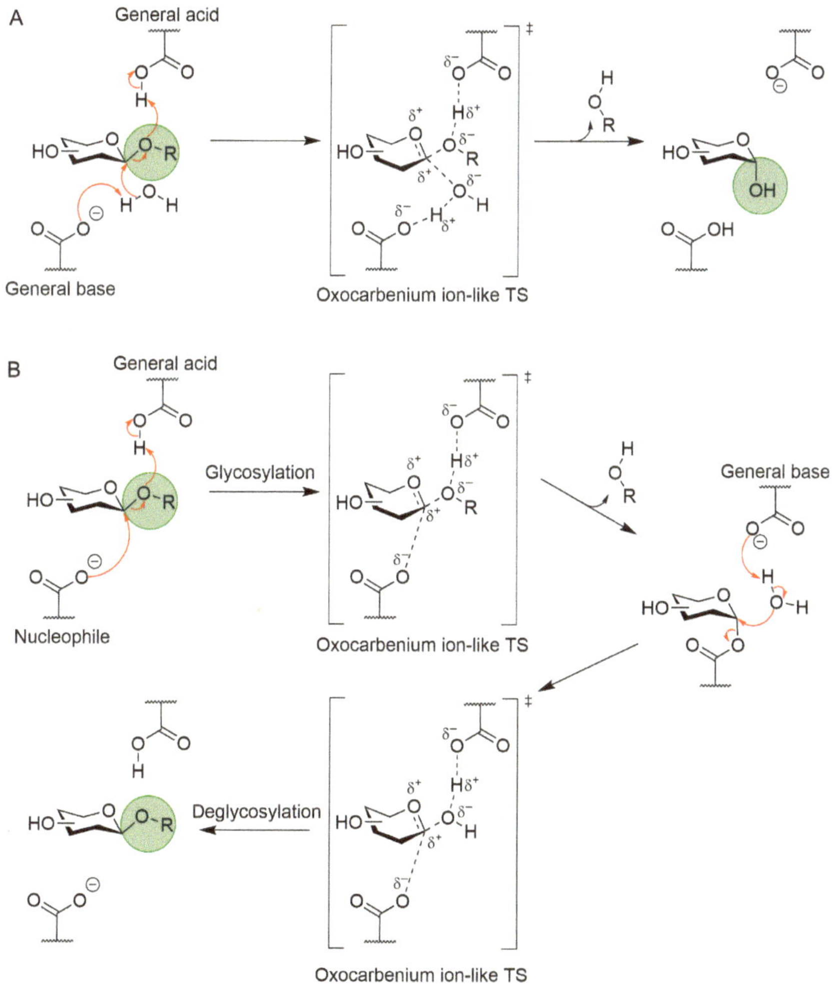

Fig. 1 Mechanisms of inverting (A) and retaining (B) glycoside hydrolases.

to form a glycosyl-enzyme intermediate and acts as a leaving group during the deglycosylation step. Each reaction step for these two classes has been shown to involve an oxocarbenium ion-like transition-state (TS) (Berti & Tanaka, 2002; Lee, Bain, & Berti, 2004; Stubbs & Marx, 2003; Vocadlo, Wicki, Rupitz, & Withers, 2002; Zechel & Withers, 2000) (Fig. 1). The anomeric carbon (C1) in the TS is sp^2 hybridized and positively charged due to the cleavage of C1–O1 bond, along with the shrinkage in C1–O5 distance and the formation of a double bond (Biarnés, Nieto, Planas, & Rovira, 2006). The glycopyranosyl ring exhibits a distorted conformation away from a relaxed chair conformation (Biarnés et al., 2007; Spiwok, Králová, & Tvaroška, 2010), which is attributed to the planarity of C5, O5, C1, and C2 atoms imposed by the sp^2-hybridized TS. This distorted conformation is also observed in Michaelis complex structures (Davies, Ducros, Varrot, & Zechel, 2003; Davies, Planas, & Rovira, 2012; Fushinobu et al., 2008; Guérin et al., 2002; Rye & Withers, 2000; Sabini et al., 1999; Tailford et al., 2008; Vasella, Davies, & Böhm, 2002). It is worth noting that though most GHs follow the classical Koshland mechanisms, a number of completely distinct mechanisms for enzymatic cleavage of glycosides have been discovered (Jongkees & Withers, 2014).

According to the CAZy database, there are more than 100 families of GTs (Lombard et al., 2014). GTs are responsible for the transfer of a sugar unit to, most commonly, the nucleophilic oxygen of a hydroxyl group of the acceptor. To achieve this transfer, GTs utilize an activated sugar substrate that contains a (substituted) phosphate leaving group. Donor sugar substrates are generally activated in the form of nucleoside diphosphate sugars (e.g., UDP-Gal); however, nucleoside monophosphate sugars (e.g., CMP-Neu5Ac), lipid phosphates (e.g., dolichol phosphate oligosaccharides), and unsubstituted phosphates are also used. The acceptor substrates used by GTs are most commonly other sugars but can also be a lipid, protein, nucleic acid, or other small molecules (Lairson et al., 2008). Analogous to GHs, GTs catalyze the synthesis of the glycosidic bond with inversion or retention of the anomeric stereochemistry with respect to the donor sugar (Ardèvol & Rovira, 2015). The mechanism of inverting GTs (Fig. 2A) is well established as a single-displacement S_N2 reaction with a general base catalyst increasing the nucleophilicity of the attacking group. In contrast, the mechanism of retaining GTs remains controversial (Lairson et al., 2008). For those GTs with a properly positioned nucleophile within the active site (aspartate or glutamate), a double-displacement mechanism (Fig. 2B) was proposed,

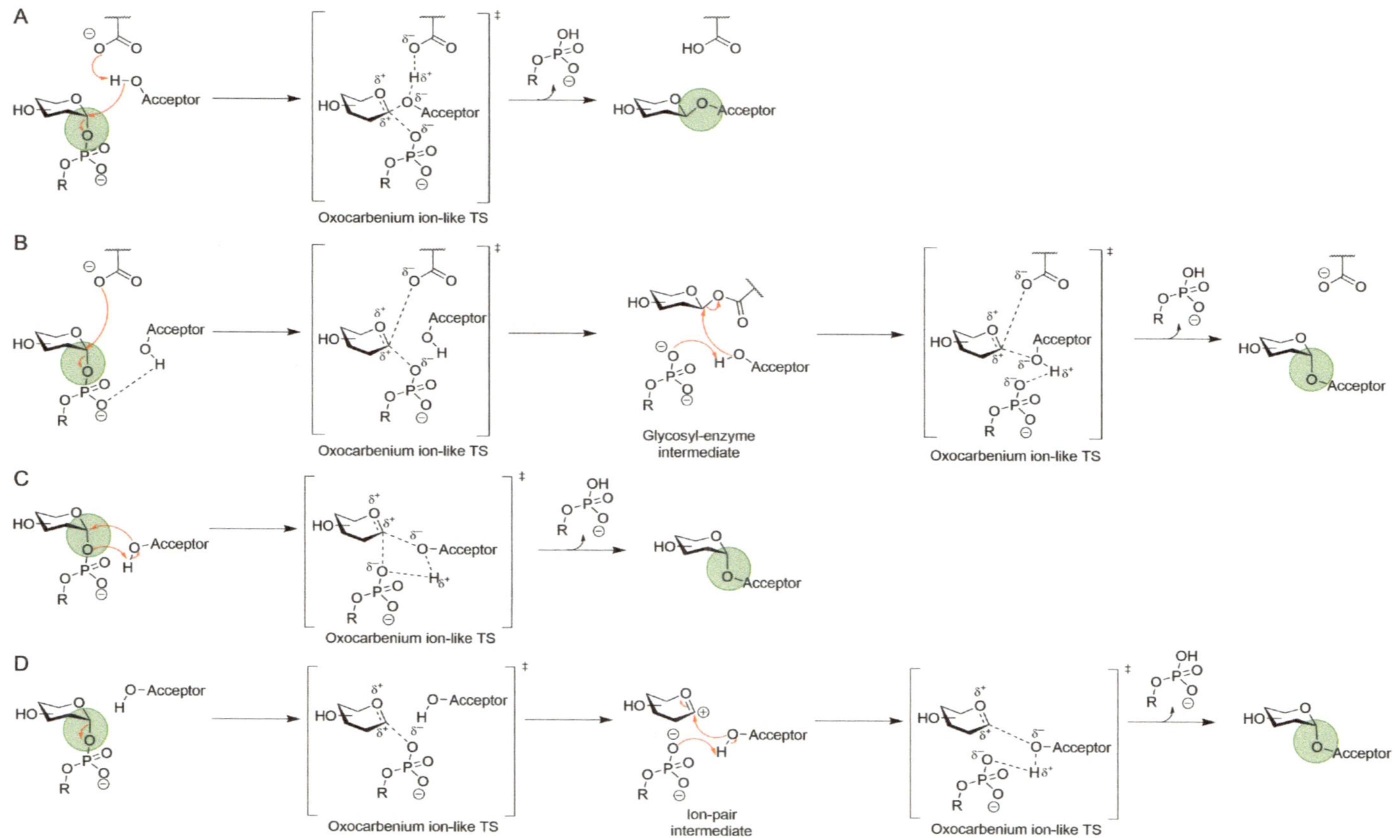

Fig. 2 Reaction mechanisms of glycosyltransferases. (A) S_N2 mechanism for inverting glycosyltransferases. (B) Double-displacement mechanism of retaining glycosyltransferases. (C) Concerted front-face mechanism of retaining glycosyltransferases. (D) Stepwise front-face mechanism of retaining glycosyltransferases.

which involves the formation of a covalent glycosyl-enzyme intermediate (Monegal & Planas, 2006; Rojas-Cervellera, Ardèvol, Boero, Planas, & Rovira, 2013; Soya, Fang, Palcic, & Klassen, 2011). However, up to now only two GTs from family 6, α-1,3-GT and blood-group A and B α-1,3-GT, both adopting fold-type A, have been reported to have such active-site nucleophiles (Ardèvol & Rovira, 2015). It should also be noted that there is not enough direct experimental evidence supporting this mechanism. For those GTs that lack a nucleophile within the reaction center, a front-face mechanism, also known as the internal return (S_Ni-like) mechanism, was suggested (Ardèvol & Rovira, 2011; Errey et al., 2010; Gómez, Polyak, Thiel, Lluch, & Masgrau, 2012; Lairson et al., 2008; Lee et al., 2011; Lira-Navarrete et al., 2014). Within this scenario, the reaction could take place either in a concerted (Fig. 2C) or in a stepwise style (Fig. 2D), depending on whether a short-lived ion-pair intermediate is formed.

In the last decade, glycobiology has attracted considerable research interest and great effort has been made to the study of the structure, mechanism, and function of CAZymes (Varki et al., 2009), leading to a boom in CAZyme-based synthetic methodologies and therapeutics. However, a thorough understanding of the catalytic mechanisms still remains a significant challenge, which influences the realization of the full potential of CAZymes. As a perfect complement to experimental studies, computational glycobiology has proved a powerful and robust tool in probing the reaction pathway and providing structural, energetic, and electronic details of CAZymatic reactions (Ardèvol & Rovira, 2015; Fadda & Woods, 2010; Woods & Tessier, 2010). A major challenge in computational glycobiology is to develop algorithms and models that can accurately mimic the highly heterogeneous electrostatic environment and capture the complex energetic landscapes of carbohydrate structures along the reaction pathways. It can also faithfully describe the molecular factors that modulate the precise pK_a values of ionizable groups in enzymes, which is crucial to the identification of the catalytically competent protonation states. Ultimately, these mechanistic investigations are driven by the goal to rationally design selective and potent inhibitors targeting these key enzymes for therapeutic applications, where computer-aided drug design is increasingly playing an important role.

This review will briefly survey key methodology used in computational glycobiology and computer-aided drug design accompanied by selected case studies providing mechanistic insights into CAZymes.

2. COMPUTATIONAL METHODS

2.1 Combined Quantum Mechanics/Molecular Mechanics Methods

Combined quantum mechanics and molecular mechanics (QM/MM) methods have been proven to be remarkably useful in revealing the mechanistic details of localized enzymatic reactions (Senn & Thiel, 2009; Warshel & Levitt, 1976). As recognized by the 2013 Nobel Prize in Chemistry, QM/MM approaches have made a broad impact in the field of enzymology and are now widely used for interpreting experimental data, testing mechanistic hypotheses, and inspiring new experiments. First introduced by Warshel and Levitt, significant development in QM/MM methodology has been made (Warshel, 1991; Warshel & Levitt, 1976). QM/MM studies of CAZymes can be carried out with ab initio methods often based on minimal energy pathway (MEP) calculations, where the thermal fluctuations are either approximated or neglected (Roston & Cui, 2016). This, however, may in some cases provide misleading conclusions for flexible biomolecular systems (Klahn, Braun-Sand, Rosta, & Warshel, 2005; Riccardi et al., 2006; Roux & Yu, 2010; van Gunsteren et al., 2006; Yang, Yu, & Cui, 2008; Yu & Roux, 2009; Zhu, Yang, & Yu, 2014). In order to properly take into account the thermal fluctuations and the entropic contributions, potential of mean force (PMF)-based free energy calculations have been carried out often with more efficient semiempirical QM models (Christensen, Kubař, Cui, & Elstner, 2016; Thiel, 2014). Such PMF calculations can be performed for different proposed mechanisms, providing the most unique mechanistic definition for an enzymatic reaction. The predicted energetic properties can be directly compared to experimental thermodynamics and kinetics data when available (Riccardi et al., 2006; Yang et al., 2008). Among them, DFTB3 (Gaus, Cui, & Elstner, 2011; Yang, Yu, York, Cui, & Elstner, 2007) is a popular choice as the QM model in computational enzymology. It is a recent extension of the self-consistent-charge density-functional tight-binding method (SCC-DFTB) (Elstner et al., 1998) and derived from a third-order expansion of the density functional theory (DFT) total energy around a given reference density. It has been shown to substantially improve the description of charged systems, especially regarding hydrogen-binding energies and proton affinities (Gaus et al., 2011). These improvements make DFTB3 particularly applicable to biomolecular systems. It has been shown that DFTB3/MM achieves a balance between the accuracy and the efficiency and enables relatively long molecular dynamics (MD) sampling (Riccardi et al., 2006; Yu & Griffiths, 2014).

To properly model the structure and dynamics of biomolecular systems, electrostatic interactions have to be treated in a reliable and efficient way. This can be achieved via the Ewald summation method together with periodic boundary conditions. Although this naturally avoids the artifacts from truncating the long-range electrostatics interactions, it results in a high computational cost, which severely limits the size of the systems and the affordable sampling time, in addition to concerns regarding the artificial periodicity (Reif & Oostenbrink, 2014; Rocklin, Mobley, Dill, & Hünenberger, 2013). Alternatively coupled with spherical boundary conditions, two different robust and efficient schemes have been introduced to treat long-range electrostatics interactions, namely the generalized solvent boundary potential (GSBP) method (Im, Bernèche, & Roux, 2001; Schaefer, Riccardi, & Cui, 2005) and the solvent macromolecule boundary potential (SMBP) method (Benighaus & Thiel, 2009; Zienau & Cui, 2012). The GSBP-based QM/MM protocol enables relatively long conformational sampling with semiempirical methods, while the SMBP-based QM/MM protocol with ab initio or DFT as the QM model provides an affordable approach for multiple MEP calculations.

2.2 pK_a and pK_a Shift Calculations in Enzymes

A significant challenge for computational enzymology is to reproduce pK_a values and assign catalytically competent protonation states for key residues in a highly heterogeneous environment. A variety of different approaches have been proposed to predict pK_a shifts. Three popular choices are thermodynamic integration (TI) based on explicit solvent QM/MM simulations (Kamerlin, Haranczyk, & Warshel, 2009; Li & Cui, 2003), multiconformation continuum electrostatics (MCCE) (Song, Mao, & Gunner, 2009), and PROPKA (Olsson, Søndergaard, Rostkowski, & Jensen, 2011).

The QM/MM TI-based method computes pK_a shifts via a thermodynamic cycle according to the following equation:

$$\Delta pK_a = pK_a(E-A) - pK_a(\text{model}-A) = \frac{1}{2.303kT}\Delta\Delta G$$

with

$$\Delta\Delta G = \Delta G(E-A(H/D)) - \Delta G(\text{model}-A(H/D))$$

where ΔG(E-A(H/D)) and ΔG(model-A(H/D)) represent the free energy difference between protonated and deprotonated states for a specific residue in protein and for a specific model compound in aqueous solution, respectively. Both terms can be obtained from a QM/MM TI simulations (Li & Cui, 2003). pK_a can be computed by summing up ΔpK_a and the known experimental pK_a value (model-A).

MCCE (Song et al., 2009) is a method that combines continuum electrostatics (CE) and molecular mechanics sampling of important rotamers. Different dielectric constants of protein and water are defined in Poisson–Boltzmann equations of CE. The Boltzmann distribution of side-chain ionization states and conformers is defined as a function of pH. With Monte Carlo (MC) sampling of coupling conformations and ionization, an explicit heterogeneous dielectric response that provides coupled ionization and position changes accurately is introduced, significantly improving the traditional pK_a calculations based on a single static structure.

As a highly efficient empirical method, PROPKA (Olsson et al., 2011) is designed to evaluate the pK_a values for both surface and buried residues and the effects from substrate binding. The desolvation effects and intraprotein interactions, along with (non)covalent intraligand interactions, are taken into account in PROPKA.

The three methods have their own strengths and limitations; thus, it is a better choice to perform a comparative study with different approaches benchmarked by experimental data to obtain reliable pK_a predictions.

2.3 Minimum Energy Pathway and Potential of Mean Force

To fully understand an enzymatic reaction, insights on the energetic and structural information along the entire reaction pathway are required. Among the different simulation techniques, two of the most popular methods to model the reaction pathway of an enzymatic reaction and to obtain the properties of TSs are the minimum energy pathway (MEP) and potential of mean force (PMF).

MEP on the potential energy surface (PES) can be obtained through adiabatic mapping by minimizing the energy of the system at intervals along predefined reaction coordinates that connect the reactant state and the product state (van der Kamp & Mulholland, 2013). A TS structure is represented by the structure highest in energy on this MEP. It is crucial that MEPs are evaluated with multiple enzyme–substrate conformations as the barriers and TS structures can vary significantly based on the starting structure

(Klahn et al., 2005; Riccardi et al., 2006; Zhu et al., 2014). The calculated reaction barrier can be directly compared to an (apparent) activation energy obtained from the experimental apparent first-order reaction rate constant using TS theory (Garcia-Viloca, Gao, Karplus, & Truhlar, 2004).

PMF is another method used to calculate the free energy differences that can be very useful for probing chemical reactions. In contrast to MEP, PMF relies on MD at a finite temperature, properly taking into account thermal fluctuations and flexibility of both the enzyme and the substrate. It can be determined as:

$$W(\xi) = W(\xi^*) - k_B T \ln\left[\frac{\langle \rho(\xi) \rangle}{\langle \rho(\xi^*) \rangle}\right],$$

where ξ is the predefined reaction coordinate, $\langle \rho(\xi^*) \rangle$ and $W(\xi^*)$ are arbitrary reference values. $\langle \rho(\xi) \rangle$ represents the average distribution along the reaction coordinate ξ and is defined as follows:

$$\langle \rho(\xi) \rangle = \frac{\int dR\delta(\xi'[R] - \xi)e^{-U(R)/k_B T}}{\int dRe^{-U(R)/k_B T}},$$

where $U(R)$ is the function of coordinates R for the total system energy (McQuarrie, 2000). In practice, it is difficult to obtain an accurate PMF along a reaction coordinate from free (unbiased) MD simulations due to the presence of low probability regions, as high energy regions are generally sampled less frequently. Enhanced sampling can be achieved through advanced sampling techniques including umbrella sampling (Torrie & Valleau, 1977), local elevation (Huber, Torda, & van Gunsteren, 1994)/metadynamics simulations (Barducci, Bonomi, & Parrinello, 2011), and the string method (Pan, Sezer, & Roux, 2008).

2.4 Binding Free Energy Calculations

Protein–ligand binding is of key importance to both biological function and pharmaceutical activity. Achieving the desired therapeutic effect of a compound is dependent on the magnitude of binding affinity of the ligand to the target receptor. Thus, designing tight-binding ligands while maintaining biological efficacy and safety is the key objective in drug discovery projects. The accurate prediction of protein–ligand binding free energies is a principal target of computer-aided drug design. A variety of free energy simulation methods employ an analysis of all-atom MD or MC simulations to determine the relative free energy difference between two similar ligands via either a chemical or an alchemical path (Chipot & Pohorille, 2007; Hadden, Tessier, Fadda, & Woods, 2015). In most cases, the calculation of relative binding free energies is sufficient and is favored as a significant reduction in computational effort is required when compared to the requirements of calculating the absolute binding free energy (Wang et al., 2015).

Methods based on the calculation of the free energy along a thermodynamic pathway employing alchemical transformations are widely used in the calculation of relative binding free energies between different ligands. The alchemical transformations are usually performed using the double-decoupling method in which two separate calculations are carried out for decoupling the ligand from the solution and receptor environments (Hansen & van Gunsteren, 2014). The calculations are performed based on a thermodynamic cycle, such as that shown in Fig. 3. In this cycle, $\Delta\Delta G$ between different compounds (ΔG_1 and ΔG_3) is calculated by subtraction of ΔG_2 and ΔG_4, which is the free energy (ΔG) associated with the alchemical transformation of one inhibitor to another in complex with the protein or in aqueous solution.

Alternatively, the standard binding free energy of a ligand to a protein can be calculated through a combination of an alchemical transformation and a geometric transformation (Deng & Roux, 2006; Jin, Yu, & Huang, 2016; Wang, Deng & Roux, 2006). Even with modern computing facilities, this still presents a daunting challenge to achieve sufficient conformational sampling. FEP coupled with replica-exchange molecular dynamics (FEP/λ-REMD) has been developed to improve the sampling and

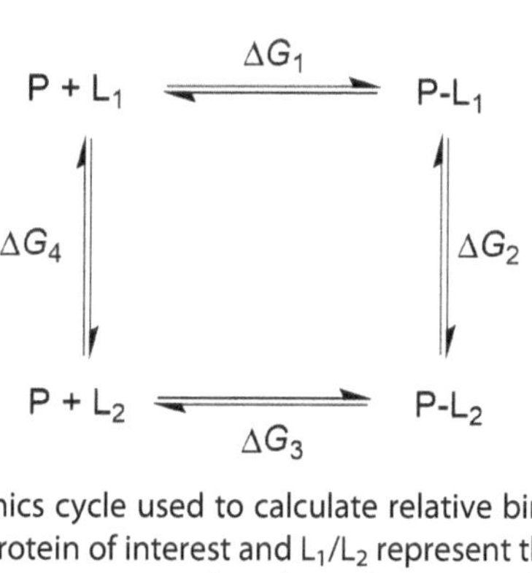

Fig. 3 The thermodynamics cycle used to calculate relative binding free energy $\Delta\Delta G$, where P represents the protein of interest and L_1/L_2 represent the different inhibitors of interest. The *horizontal legs* correspond to the ΔG associated with binding affinities, whereas the *vertical legs* correspond to the ΔG for the alchemical transformation of L_1 to L_2. The binding free energy difference for L_1 and L_2 can be calculated according to equation $\Delta\Delta G = \Delta G_2 - \Delta G_4$.

accelerate the convergence of the free energy calculations (Jiang, Hodoscek, & Roux, 2009; Woods, Essex, & King, 2003).

Binding free energy calculations performed with implicit solvent models can also be considered. By regarding solvent effects implicitly, using continuum solvent methods, the computationally intensive evaluation of individual contributions of a large number of solvent molecules can be avoided. The computational cost can be further reduced by considering only the end point states in the free energy calculations. End point, implicit solvent approaches therefore have the advantage of a relatively low computational cost, while important solvent effects are still regarded (Chen et al., 2016; Genheden & Ryde, 2015; Homeyer & Gohlke, 2012; Kleinjung & Fraternali, 2014). The most well-known method of this style is the molecular mechanics/Poisson–Boltzmann surface area (MM-PBSA) approach or the molecular mechanics/generalized born surface area (MM-GBSA) approach (Kollman et al., 2000; Massova & Kollman, 2000). However, their limitations from crude and questionable approximations and their system-dependent performance have been documented in the literature (Genheden & Ryde, 2015).

3. PUCKERING OF CARBOHYDRATE STRUCTURES

3.1 Carbohydrate Structures

The complexity of carbohydrate structures is evident from their diverse ring sizes, derivatization, ring substituents, and their stereochemistry. The changes in the stereochemistry of the ring substituents can strongly influence the properties of carbohydrate structures and thus make them highly complex. As an example, Fig. 4 shows the structures of D-glucopyranose that has two different stereochemistries, α (**1**) and β (**2**). Additionally, variations in conformations of the sugar ring are important in stabilizing the structures of carbohydrates, with the orientation of exocyclic groups closely related to the conformers of the sugar ring. Thus, this flexibility makes carbohydrate structures more sophisticated. As there is a myriad of carbohydrate-based

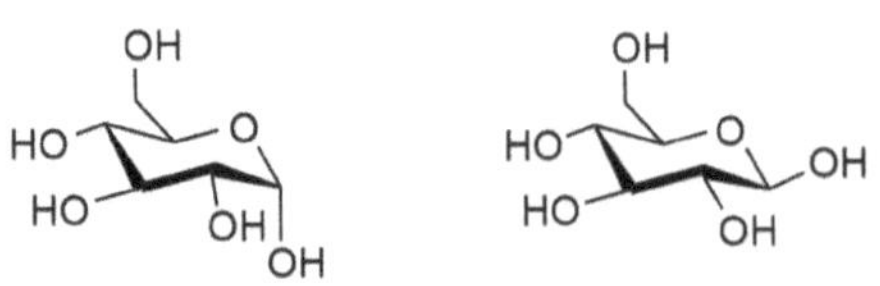

1 α-D-Glucopyranose **2** β-D-Glucopyranose

Fig. 4 α (**1**) and β (**2**) conformations of D-glucopyranose.

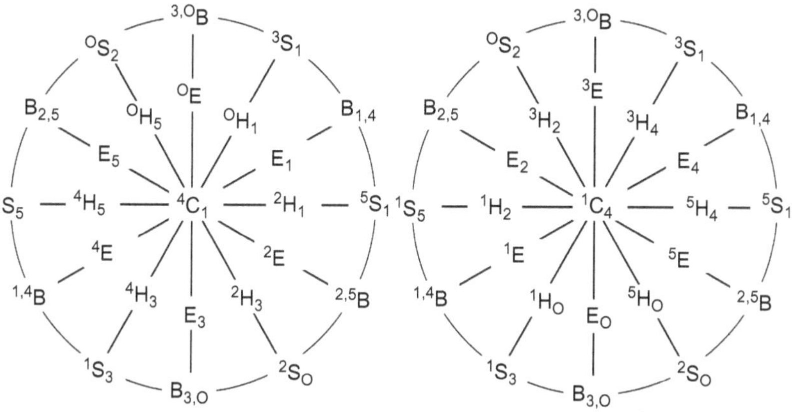

Fig. 5 Stoddart diagrams of pyranose ring in north and south poles (Raich, Nin-Hill, Ardèvol, & Rovira, 2016).

structures in nature, here we will focus on discussing the conformations of pyranose (six-membered ring sugars).

The pyranose ring nomenclature described by Stoddart (1971) first and adopted by IUPAC-IUB (1980) is summarized in Fig. 5, which includes two diagrams, introducing all the accessible conformers, together with their connectivity. The 38 canonical conformations shown in Fig. 5 represent distinct ways in which the pyranose ring can be classified into chair, half-chair, envelop, boat, and skew-boat (represented by the letters C, H, E, B, and S, respectively) (Raich et al., 2016).

3.2 Cremer–Pople Puckering Coordinates

Cremer and Pople mapped the conformations of pyranose onto a spherical representation using three polar coordinates: a radius Q, two phase angles ϕ and θ (Fig. 6) (Cremer & Pople, 1975). These three coordinates are defined by the following equations:

$$\begin{cases} Q\sin\theta\cos\phi = \sqrt{\dfrac{1}{3}\sum_{j=1}^{6} z_j \cos\left[\dfrac{2\pi}{6}2(j-1)\right]} \\ Q\sin\theta\cos\theta = \sqrt{\dfrac{1}{3}\sum_{j=1}^{6} z_j \cos\left[\dfrac{2\pi}{6}2(j-1)\right]} \\ Q\sin\theta\cos\theta = \sqrt{\dfrac{1}{3}\sum_{j=1}^{6} (-1)^{j-1} z_j} \end{cases}.$$

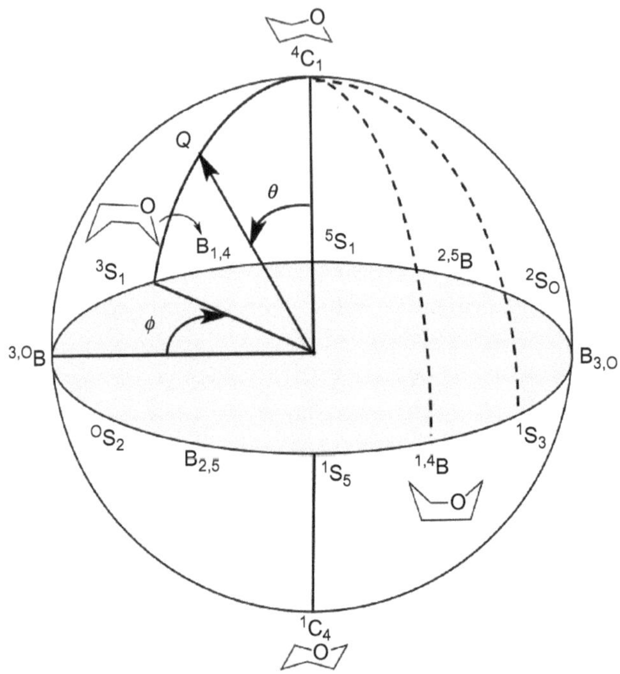

Fig. 6 Cremer–Pople puckering coordinates of a pyranose ring (Ardèvol & Rovira, 2015).

All the possible pyranose ring conformations would fall within the ellipsoid of puckering radius Q (Fig. 6). Stoddart's representation (Stoddart, 1971) corresponds to the projection of Cremer–Pople sphere-like representation and can be computed according to the Cartesian coordinates (Biarnés et al., 2007). Cremer and Pople's puckering coordinates offer a convenient approach to study the conformational free energy of the pyranose ring in vacuo (Mayes, Broadbelt, & Beckham, 2014), in aqueous solution (Barnett & Naidoo, 2010), or along reaction pathways (Davies et al., 2012). On the other hand, the complexity of carbohydrate puckering also presents a challenge for MD simulations. Deficiencies have been identified in adopted models (Sattelle & Almond, 2010), and improvements have been introduced for better descriptions of puckering (Govender, Gao, & Naidoo, 2014; Pol-Fachin, Rusu, Verli, & Lins, 2012).

3.3 The Conformational Catalytic Itinerary in GHs

Conformational changes along the reaction coordinates are referred to as the catalytic itinerary (Davies et al., 2003; Vocadlo & Davies, 2008). Knowing the conformational catalytic itinerary for a specific GH can be very helpful in the design potent inhibitors, such as TS analogues, whose shapes mimic the TS of a substrate (Caines et al., 2007). TS structures can be interpolated in accordance with the Cremer–Pople puckering coordinates if the distorted conformations of both the Michaelis complex and the covalent intermediate of retaining GH (or product state of inverting GH) are known.

X-ray crystallography is the most commonly used tool to infer the conformational catalytic itinerary harnessed by GHs. For example, an X-ray study trapped the three-dimensional structures of the Michaelis complex and covalent intermediate of a retaining GH26 β-mannanase, showing a distorted β-$^{1}S_5$ and α-$^{O}S_2$ conformation, respectively (Ducros et al., 2002). These data suggested that the TS may adopt a $B_{2,5}$ boat conformation, which is flanked by Michaelis complex $^{1}S_5$ and intermediate $^{O}S_2$ conformations on the Cremer–Pople itinerary, predicting that a $^{1}S_5 \rightarrow B_{2,5} \rightarrow {}^{O}S_2$ itinerary occurs for β-mannanase. This prediction was supported in the later work on a large number of putative mannosidase inhibitors bound to the β-mannosidase BtMan2A (GH2) (Tailford et al., 2007) with kinetic analysis (Tailford et al., 2008). Offen et al. reported a distorted $^{1}S_5$ conformation for the Michaelis complex for the β-mannosidase Man2A (GH2) arguing strongly in support of the same itinerary of $^{1}S_5 \rightarrow B_{2,5} \rightarrow {}^{O}S_2$ for GH2 and 26 (Offen, Zechel, Withers, Gilbert, & Davies, 2009).

Although X-ray crystallography is a very powerful tool to investigate GH itineraries, it has its own limitations. The mutated enzymes and the nonnatural substrates often required may affect the nearby active sites and result in distortion of the sugar ring in subsite −1. In addition, the TS has a short life and is unstable; as a result, it cannot be observed by this type of direct measurement. Complementarily, computational simulations can overcome these limitations, providing equilibrium data for TS conformation and complete conformational itinerary gaps between experimentally captured states (Ardèvol & Rovira, 2015). Recently, a joint experimental and computational study was carried out to probe the transition of a GH76 α-1,6-mannanase, a key enzyme in bacterial and fungal mannoprotein metabolism (Thompson et al., 2015). Experimentally, a Michaelis complex state in $^{O}S_2$ conformation and an inhibitor complex in $B_{2,5}$ conformation suggested a boat conformation for the TS. The complete interpretation was achieved through Car–Parrinello MD metadynamics simulations of the Michaelis complex and the inhibitor complex. The free energy landscape of the mannose natural donor and inhibitor isofagomine (IFG) in complex with GH76 clearly demonstrated that GH76 dramatically reshapes the energetic accessible conformational space. The $^4C_1$ conformer is no longer the global minima, being 6 kcal mol^{-1} higher compared to the $^1S_5/B_{2,5}$ conformer of IFG and >15 kcal mol^{-1} higher compared to the $^OS_2/B_{2,5}$ conformer of the α-mannoside. Taken together, computational studies mapped out the full itinerary for α-1,6-mannanase and suggested that it operates through a canonical $^OS_2 \rightarrow B_{2,5} \rightarrow {}^1S_5$ itinerary.

Consistent with X-ray crystallographic studies, all the computational studies have noted that the substrates in the Michaelis complex of different GHs adopt different distorted structures, concurring with the concept that different GHs follow unique catalytic conformational itineraries (Biarnés et al., 2006; Iglesias-Fernández, Raich, Ardèvol, & Rovira, 2015; Knott et al., 2013; Thompson et al., 2012). All these examples illustrate that calculation of the free energy landscape could be a powerful tool to capture the conformations of both the sugar donor and TS (Biarnés, Ardevol, Iglesias-Fernández, Planas, & Rovira, 2011).

In general, one of the common features coming out of these computational studies of GHs is that both the nucleophile and the leaving group are conformationally restrained so that the anomeric substitution performed by the anomeric migration toward the nucleophile needs to break the glycosidic bond in subsite −1 through the conformational changes. It has been proposed that ring distortion provides electronic assistance and structural changes to preactivate the enzymatic reaction and thus facilitate the TS formation (Ardèvol & Rovira, 2015; Biarnés et al., 2006).

4. IDENTIFICATION OF CATALYTICALLY COMPETENT PROTONATION STATES OF KEY RESIDUES

The identification of catalytic residues and determination of their catalytically competent protonation states are essential steps in the mechanistic studies of CAZymes. The protonation states and their apparent pK_a values are determined by the intrinsic pK_a values of these functional groups and the specific electrostatic environment they sit in. There are numerous examples in the literature where the pK_a values of catalytic residues are substantially altered (Harris & Turner, 2002). Such pK_a shifts (the difference between the intrinsic and apparent pK_a values) are modulated by interactions between the catalytic groups with other charged or polar groups, and by the polarity or dielectric constant of the medium that surrounds it. Understanding such pK_a shifts will provide a basis for understanding the electrostatics, entropic, and potential allosteric aspect of the machinery in the catalytic process. Additionally, molecular simulations and electrostatics calculations can be applied to delineate the factors modulating the often observed pK_a shifts in enzymes at atomistic levels, which are often challenging to probe directly with experimental studies.

4.1 pK_a Cycling of the General Acid/Base in GH33 and GH34 Families

The GH33 and 34 families catalyze the hydrolysis of sialosides (Fig. 7). Asp59 in *Trypanosoma cruzi trans*-sialidase (TcTS, GH33) or Asp151 in NA (influenza neuraminidase, GH34) is believed to act as a general acid/base in the catalytic mechanism for the GH33 and 34 families (Fig. 7). From the apo state to the enzyme–substrate complex state (holo) and then to the intermediate state (int), the general acid/base will go through so-called pK_a cycling (Ludwiczek et al., 2013; McIntosh et al., 1996; Vocadlo, Davies, Laine, & Withers, 2001). It has been postulated that such pK_a cycling is

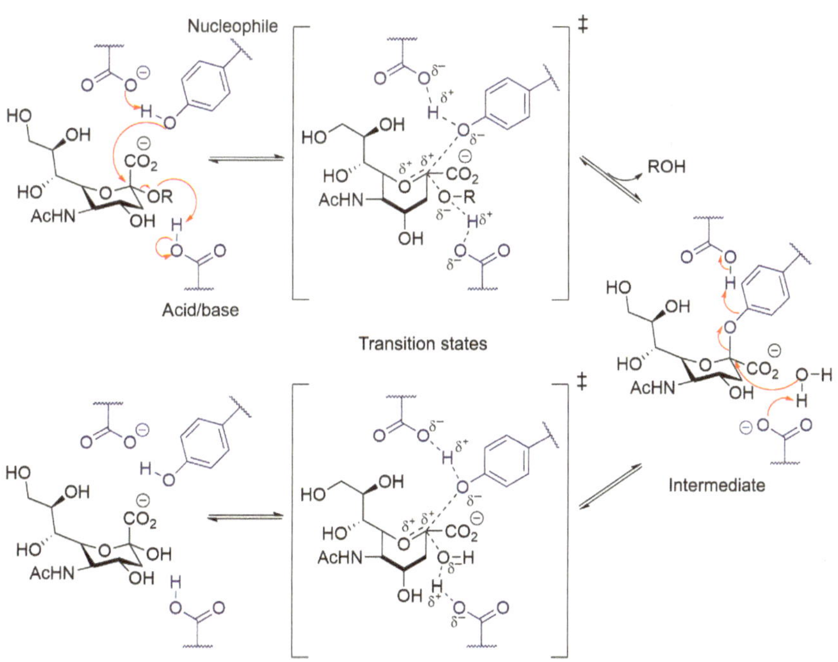

Fig. 7 The proposed catalytic mechanism for glycoside hydrolase families GH33 and 34 (Yu & Griffiths, 2014).

likely to be a common phenomenon in retaining GHs. To obtain a better understanding of the molecular basis for such pK_a cycling, we carried out pK_a shift calculations of the general acid/base, Asp59 in TcTS of GH33 and Asp151 in NA of GH34, based on different computational methods (Yu & Griffiths, 2014).

4.1.1 pK_a Calculations Based on Explicit Solvent Simulations

pK_a values of Asp59 in TcTS and Asp151 in NA were obtained with explicit QM/MM solvent simulations based on three different snapshots from the equilibrium simulations (Table 1). We can clearly see that these proposed general acid residues have a pK_a value well below 7 at the apo state and then have a substantial pK_a upshift, becoming protonated upon substrate binding. This is consistent with their catalytic role as a general acid during the glycosylation step (Fig. 7). The pK_a values obtained from QM/MM simulation are in good agreement with available experimental data, while PROPKA failed to capture the corresponding pK_a shifts. This illustrated the importance of explicit solvation and thermal fluctuations in pK_a calculations. TcTS has been shown to have an optimal pH of 7.9 (Scudder, Doom, Chuenkova, Manger, & Pereira, 1993), under which Asp59 (ISOI-holo) acts as the

Table 1 The pK_a Values of Asp59 in TcTS or Asp151 in NA Based on Three Independent Explicit Solvent Simulations and PROPKA

System	QM/MM[a]	PROPKA	Protonation States
Asp59 in *Trypanosoma cruzi trans*-sialidase			
1MS3-apo	0.8 ± 0.4	2.2	Deprotonated
1SOI-apo	1.0 ± 0.5	2.1	Deprotonated
1SOI-holo	8.5 ± 0.7	2.4	Protonated
Asp151 in influenza neuraminidase			
2QWB-apo	4.1 ± 0.1	3.6	Deprotonated
2QWB-holo	8.7 ± 0.4	3.6	Protonated
3W09-int	6.3 ± 0.4	4.2	Deprotonated

[a]The average pK_as from three different sets of simulations. The error bars in pK_a calculations are estimated from the standard deviation of the three sets of calculated values.

Adapted from Yu, H., & Griffiths, T. M. (2014). pKa cycling of the general acid/base in glycoside hydrolase families 33 and 34. *Physical Chemistry Chemical Physics, 16*(12), 5785–5792. http://dx.doi.org/10.1039/C4CP00351A by permission of the PCCP Owner Societies.

Table 2 Perturbation Analysis of the Deprotonation Free Energy of Asp59 in TcTS or Asp151 in NA With Explicit Solvent Simulations[a]

System	Ref.[b]	ΔAll[c]	ΔWat[d]	ΔProt[e]	ΔLig[f]
Asp59 in *Trypanosoma cruzi trans*-sialidase					
1SOI-apo	117.2	202.2	118.5	199.4	—
1SOI-holo	127.3	203.3	174.7	220.1	78.9
Asp151 in influenza neuraminidase					
2QWB-apo	121.5	201.4	114.1	211.5	—
2QWB-holo	127.0	199.6	168.9	213.5	86.8

[a]The deprotonation free energy of Asp59 in TcTS or Asp151 in NA was calculated by integrating the free energy derivatives recalculated from the DFTB3/MM FEP trajectories under different QM/MM coupling perturbations. The values are in kcal mol^{-1}.
[b]Ref. refers to the perturbative analysis on the originally simulated systems while keeping all the parameters untouched.
[c]All the partial charges on the protein (inner and outer atoms), water, and the substrate were set to 0.
[d]All the partial charges on water were set to 0.
[e]All the partial charges on the protein (inner and outer atoms) were set to 0.
[f]All the partial charges on the substrate were set to 0.
Reproduced from Yu, H., & Griffiths, T. M. (2014). pKa cycling of the general acid/base in glycoside hydrolase families 33 and 34. *Physical Chemistry Chemical Physics, 16*(12), 5785–5792. http://dx.doi.org/10.1039/C4CP00351A by permission of the PCCP Owner Societies.

general acid (pK_a 8.5; Table 1). The optimal pH range for NA predicted by QM/MM pK_a calculations is between 6.3 and 8.7 (Table 2; Yu & Griffiths, 2014). This is consistent with the experimental value of around 6.0 (Deshairs, Kessler, Aymard, & Quash, 1986; Gubareva, Robinson, Bethell, & Webster, 1997; McKimm-Breschkin et al., 2013).

4.1.2 Perturbative Analysis of pK_a Shifts for the General Acid/Base

To rationalize the molecular basis for pK_a shifts in the GH33 and 34 families, we set up perturbative analyses, in which the free energy difference between the protonated state and the deprotonated state is recalculated for the structures of QM/MM trajectories by modifying relevant parameters in the QM/MM coupling model. Such perturbation can provide valuable information about how the environment establishes the pK_a values for the dual role of the general acid/base. The partial charges on groups of interest were scaled to zero and the free energy derivatives were reevaluated from the QM/MM FEP trajectories. The results are summarized in Table 2. We can clearly see that the contributions from water in 1SOI-apo and 2QWB-apo are relatively small, about −1.3 and −7.4 kcal mol^{-1}, while the ones in 1SOI-holo and 2QWB-holo are substantially large, about −47.3 and −41.9 kcal mol^{-1}, respectively. This may be attributed to the different net charge for the system between the apo state and the holo state. The coordination numbers of water molecules of Asp59 or Asp151 in 1SOI-apo and 2QWB-apo at both the protonated state and the deprotonated state are higher than the one in 1SOI-holo and 2QWB-holo, which suggests that the substrate binding makes Asp59 or Asp151 less exposed to water and contributes to their pK_a shifts. The net contribution from protein in all four cases is relatively large, more than −80 kcal mol^{-1} (Table 2). Notably, no significant difference was found for the contributions from key protein residues between the apo state and the holo state (Fig. 8). Combined with the perturbative analysis, we can conclude that the substrate binding, together with the glycosylation and the solvation environment, contributes to the pK_a of Asp59 in TcTS and Asp151 in NA for the dual role (Yu & Griffiths, 2014).

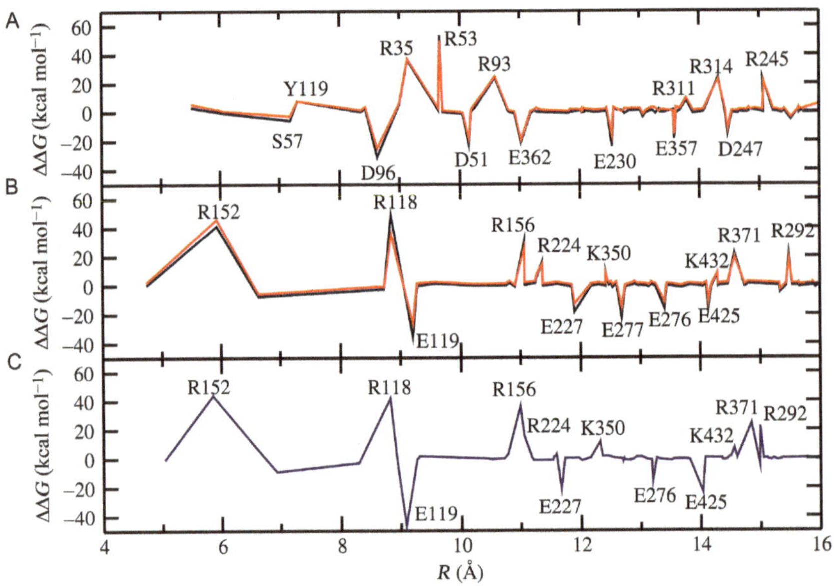

Fig. 8 Contribution to the electrostatic component of the free energy of deprotonation from protein atoms (only MM atoms are considered) in the TcTS and NA from the DFTB3/MM simulations. The contribution is plotted against the distance between its Cα from the Cγ of Asp59 in TcTS or Asp151 in NA. (A) IS0O-apo (*black*) and ISOI-holo (*red*); (B) 2QWB-apo (*black*) and 2QWB-holo (*red*); (C) 3W09-int (*blue*). *Adapted from Yu, H., & Griffiths, T. M. (2014). pKa cycling of the general acid/base in glycoside hydrolase families 33 and 34.* Physical Chemistry Chemical Physics, 16*(12), 5785–5792. http://dx.doi.org/10.1039/C4CP00351A by permission of the PCCP Owner Societies.*

4.2 Rationalizing pK_a Shifts in *Bacillus circulans* Xylanase With Computational Studies

B. circulans xylanase (BcX) is a 20kDa retaining *endo*-β-(1,4)-xylanase belonging to the GH11 family (Lombard et al., 2014), which has been well studied using experimental methods, such as NMR spectroscopy, X-ray crystallography, and mutagenesis studies (Joshi et al., 2001; Ludwiczek et al., 2013; McIntosh et al., 1996; Paes, Berrin, & Beaugrand, 2012; Pollet, Delcour, & Courtin, 2010). The double-displacement mechanism in BcX, together with the experimental pK_a values for Glu78 and Glu172 (McIntosh et al., 1996), is shown in Fig. 9. It is generally believed that Glu78 functions as a nucleophile (Miao, Ziser, Aebersold, & Withers, 1994; PDB code: 1BVV), while Glu172 acts as a general acid/base during the catalytic reaction (Fig. 9).

We examined the performance of three different computational methods (QM/MM TI, MCCE (Song et al., 2009), and PROPKA (Olsson et al., 2011)) in calculating the pK_a shifts for BcX, including wild type (WT), WT-2FXb, and the mutants N35H, N35H-2FXb, N35D, and N35D-2FXb. Additionally, a complementary understanding of the molecular basis for pK_a shifts in BcX has been provided (Xiao & Yu, 2016).

4.2.1 *Accuracy of Three Different Methods for pK_a Calculations*

The calculated pK_a values for Glu172 and other key residues of BcX based on different methods are shown in Tables 3 and 4. Encouragingly, all three methods correctly assigned proper protonation states for Glu172 at the apo state and the intermediate state at pH_{opt}. The error analyses suggest that MCCE DEAFAULT provided the best results in pK_a calculations. The QM/MM provided satisfactory predictions except overestimating the interactions between the positively charged His35 and deprotonated

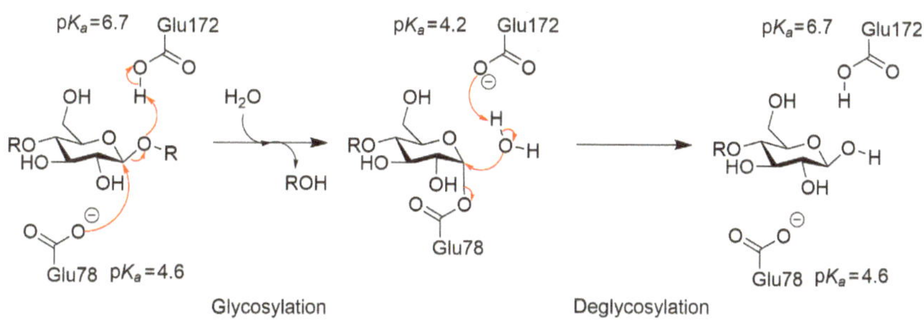

Fig. 9 The double-displacement mechanism of the GH11 retaining BcX and the measured pK_a values for Glu172 and Glu78 (Xiao & Yu, 2016).

Table 3 The pK_as for Glu172 Based on QM/MM Simulations, MCCE, and PROPKA Methods

Protein	Exp.	QM/MM	MCCE	PROPKA
WT (pH_{opt} 5.6)	6.7[a]	7.3	4.8	7.5
WT-2FXb	4.2[a]	4.3	4.9	6.8
N35H (pH_{opt} 4.4)	3.4[a]	0.7	2.6	4.2
N35H-2FXb	2.8[a]	2.8	3.0	4.4
N35D (pH_{opt} 4.6)	8.4[b]	12.7	7.1	10.3
N35D-2FXb	>9[a,c]	9.6	8.7	5.7

[a]Data from Ludwiczek et al. (2013).
[b]Data from Joshi et al. (2000).
[c]The exact experimental pK_a is not available due to Asp35 and Glu172 being strongly coupled in N35D-2FXb (Ludwiczek et al., 2013).
Reproduced from Xiao, K., & Yu, H. (2016). Rationalising pKa shifts in *Bacillus circulans* xylanase with computational studies. *Physical Chemistry Chemical Physics, 18*(44), 30305–30312. http://dx.doi.org/10.1039/C6CP02526A by permission of the PCCP Owner Societies.

Glu172 in N35H, and the one between the negatively charged Asp35 and deprotonated Glu172, which is likely due to the lack of explicit polarization in the MM potential (Yu & van Gunsteren, 2005). The decreased pK_a value of Glu172 in N35H is consistent with the one in V37R-BcX, which is attributed to the attractive electrostatic interaction between Arg37 and Glu172 (Pokhrel, Joo, & Yoo, 2013). Similarly, the elevated pK_a value of Glu172 in N35D is attributed to the substitution of Asn35 for Asp35 (Joshi et al., 2000; Ludwiczek et al., 2013). The results shown in Table 4 indicate that MCCE is more applicable than PROPKA for BcX (Xiao & Yu, 2016).

4.2.2 Perturbative Analyses of Glu172 pK_a

In silico mutagenesis was carried out to probe the individual effects from water molecules and nearby key residues on the pK_a shifts for Glu172 in BcX. Table 5 presents the pK_a shift results due to the contributions from residue 35, global charges (the charges of tyrosine and lysine on the surface), surface tyrosine, and surface lysine. We can clearly see that the pK_a shifts that result from residue 35 are substantially larger than that from the other three cases. The net contributions from global charges are significantly smaller in all systems. We can conclude that residue 35 in BcX makes a greater contribution to the (de)protonation process of Glu172,

Table 4 The pK_as for Other Residues Based on MCCE and PROPKA Methods

Protein		Exp.	MCCE	PROPKA
WT (pH_{opt} 5.6)	Asp4	3.0[a]	3.5	2.7
	Asp11	2.5[a]	2.4	3.0
	Asp83	<2.0[a]	3.2	4.0
	Asp101	<2.0[a]	2.9	2.9
	Asp106	2.7[a]	2.9	3.9
	Asp119	3.2[a]	3.3	3.3
	Asp121	3.6[a]	3.5	4.0
	Glu78	4.6[a]	4.5	6.0
	His149	<2.3[a]	3.8	4.1
	His156	~6.5[a]	7.3	6.5
N35H (pH_{opt} 4.4)	Glu78	5.5[b]	4.0	6.4
	His35	7.5[b]	8.4	7.3
N35H-2FXb	His35	5.6[b]	6.0	7.2
N35D (pH_{opt}4.6)	Glu78	5.7[c]	4.7	6.1
	Asp35	3.7[c]	4.0	6.5
N35D-2FXb	Asp35	1.9–3.4[c,d]	5.3	11.4

[a]Data from Joshi, Hedberg, and Mcintosh (1997).
[b]Data from Ludwiczek et al. (2013).
[c]Data from Joshi et al. (2000).
[d]The exact experimental pK_a is not available due to Asp35 and Glu172 being strongly coupled in N35D-2FXb (Ludwiczek et al., 2013).
Reproduced from Xiao, K., & Yu, H. (2016). Rationalising pKa shifts in *Bacillus circulans* xylanase with computational studies. *Physical Chemistry Chemical Physics, 18*(44), 30305–30312. http://dx.doi.org/10.1039/C6CP02526A by permission of the PCCP Owner Societies.

and changing the global charge of BcX through random succinylation does not significantly affect pK_a values of Glu172 consistent with experimental observations (Ludwiczek et al., 2013). It has been argued that the effects from the perturbation on the global charges are largely screened due to the solvated environment. Additionally, the surface tyrosine shows larger individual effects on pK_a values of Glu172 than the surface lysine (Table 5) (Xiao & Yu, 2016).

Table 5 pK_a Shifts Based on Perturbative Analyses of the Deprotonation Glu172 With QM/MM Simulations[a]

System	Δ35[b]	ΔTyr and Lys[c]	ΔTyr[d]	ΔLys[e]
WT	7.3	3.5	−1.8	5.3
WT-2FXb	7.8	0.1	−2.2	5.5
N35H	38.7	4.6	−1.3	6.0
N35H-2FXb	20.3	2.9	−2.3	5.2
N35D	−22.5	4.8	−1.0	5.7
N35D-2FXb	−26.0	3.4	−1.9	5.3

[a]The deprotonation free energy of Glu172 was computed by integrating the free energy derivatives that recalculated from the QM/MM trajectories for different l windows, and pK_a shifts were then obtained based on the recomputed free energy differences.
[b]All the partial charges on the residue 35 were set to 0.
[c]All the partial charges on surface tryosine and lysine were set to 0.
[d]All the partial charges on surface tryosine were set to 0.
[e]All the partial charges on surface lysine were set to 0.
Reproduced from Xiao, K., & Yu, H. (2016). Rationalising pKa shifts in *Bacillus circulans* xylanase with computational studies. *Physical Chemistry Chemical Physics, 18*(44), 30305–30312. http://dx.doi.org/10.1039/C6CP02526A by permission of the PCCP Owner Societies.

5. CLARIFICATION OF CATALYTIC MECHANISMS

Understanding the catalytic mechanisms of GHs and GTs has a major impact on the design of enzyme inhibitors, which are important for the treatment of diseases related to these enzymes (Gloster & Vocadlo, 2012; Williams & Davies, 2001). However, it is experimentally difficult to get direct information, especially about TSs and sometimes mutations, and chemical modifications are required as a surrogate to probe the catalytic mechanism of the WT enzymes. As CAZymes are flexible enzymes binding particularly flexible substrates, QM/MM is the technique of choice as it enables the dynamics of the system to be taken into account (Ardèvol & Rovira, 2015). These studies permit the mapping out of full reaction pathways with great structural and energetic details of all the species and have pushed forward our knowledge about CAZyme mechanisms over the past four decades (Warshel & Levitt, 1976).

5.1 Catalytic Mechanisms of Hen Egg-White Lysozyme

As the first enzyme to have its 3D structure determined by X-ray crystallography, hen egg-white lysozyme (HEWL) (Blake et al., 1965) has witnessed the success of combined experimental and computational efforts that lead to clarification of the catalytic itinerary. There is a long-standing debate on whether the oxocarbenium ion formed during the reaction is a long-lived intermediate (Phillips mechanism (Phillips, 1967)), or a TS toward the covalent intermediate (Koshland mechanism (Koshland, 1953)) (Fig. 10).

Electrospray ionization mass spectrometry, α-deuterium kinetic isotope effects, and crystallography (Vocadlo et al., 2001) have indicated a covalent intermediate mechanism for HEWL. However, the evidence was still not entirely convincing as these measurements were performed either on a mutant HEWL (E35Q) or using an unnatural substrate. Based on QM/MM umbrella sampling MD simulations running backward from product **A**, Bowman, Grant, and Mulholland (2008) investigated the free energy profile for the reaction pathway of WT HEWL with its natural substrate. After protonation of the glycosidic oxygen (O1) with a barrier of ~16 kcal mol^{-1}, the TS **B** was gradually formed as the highest energy point (Fig. 11) featuring an oxocarbenium ion character, with planar conformation (Fig. 12) and positive charges on C1 and O5, which was consistent with experimental kinetic isotope effects. The overwhelming stability of covalent intermediate **C** (Fig. 12) over TS **B** was demonstrated in Fig. 13. The endocyclic protonation mechanism was also excluded due to high-energy barrier. This computational study finally settled the long-standing debate in the literature about the chemistry in lysozyme.

5.2 Reaction Mechanisms of Inverting *O*-GlcNAc Glycosyltransferase

As an important post-translational CAZyme, *O*-linked 2-acetamido-2-deoxy-β-D-glucopyranosidase (*O*-GlcNAc) GT (OGT) catalyzes the transfer of *N*-acetylglucosamine from UDP-*N*-acetylglucosamine to the hydroxyl group of the Ser/Thr of cytoplasmic, nuclear, and mitochondrial proteins. Although OGT has all the features of a dissociative S_N2-type reaction through an oxocarbenium species, at least three catalytic mechanisms were proposed based on rather similar X-ray structures, differing in the choice of catalytic bases: His498 (Lazarus, Nam, Jiang, Sliz, & Walker, 2011), α-phosphate (Schimpl et al., 2012), and Asp554 (Lazarus et al., 2012). There is various experimental evidence arguing against each of these proposed mechanisms (Withers & Davies, 2012). For instance, it is generally believed that α-phosphate has a pK_a of 2–3 thus from an energetic point of view; it is unfavorable to act as a catalytic base to abstract a proton from Ser. However, recent NMR titration experiments (Jancan & Macnaughtan, 2012) and pK_a

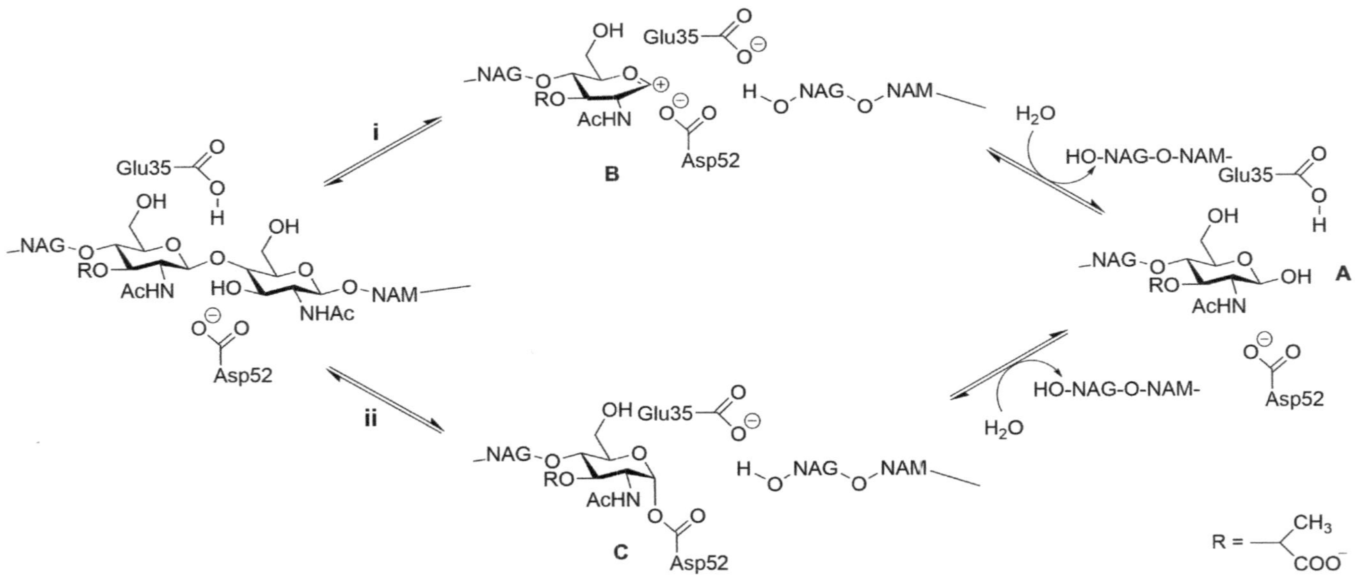

Fig. 10 Two possible catalytic mechanisms for hen egg-white lysozyme. Path (i) the Phillips mechanism; path (ii) the Koshland mechanism.

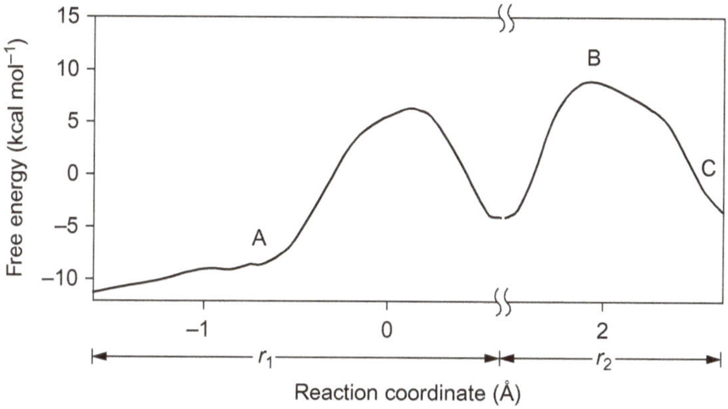

Fig. 11 Free energy profiles for the proton transfer from Glu35 to NAMD (the D site *N*-acetyl-muramic acid) O1 (r_1) and subsequent breaking of NAMD C1–O1 bond in HEWL (r_2). *Reproduced from Bowman, A. L., Grant, I. M., & Mulholland, A. J. (2008). QM/MM simulations predict a covalent intermediate in the hen egg white lysozyme reaction with its natural substrate.* Chemical Communications, *37(37), 4425–4427. http://dx.doi.org/10.1039/B810099C by permission of The Royal Society of Chemistry.*

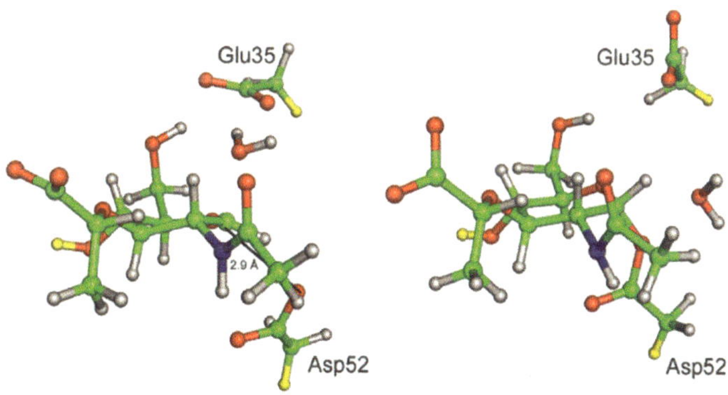

Fig. 12 Structures of TS **B** ($r_2 = 2.0$ Å, *right*) and covalent intermediate **C** ($r_2 = 2.5$ Å, left). *Adapted from Bowman, A. L., Grant, I. M., & Mulholland, A. J. (2008). QM/MM simulations predict a covalent intermediate in the hen egg white lysozyme reaction with its natural substrate.* Chemical Communications, *37(37), 4425–4427. http://dx.doi.org/10.1039/B810099C by permission of The Royal Society of Chemistry.*

calculations (Vipperla, Griffiths, Wang, & Yu, 2017) based on DFT/PCM models suggested that it has a much higher pK_a of 6–7 than previously thought. This demonstrates that more systematic comparative studies are required to conclusively pinpoint the exact catalytic mechanism for OGT. Based on hybrid QM/MM Car–Parrinello MD using the string method, Kumari et al. (2015) studied the reaction pathways using α-phosphate ($\mathbf{M_{PO4}}$) and Asp554 ($\mathbf{M_{Asp}}$) as the catalytic bases. An additional mechanism

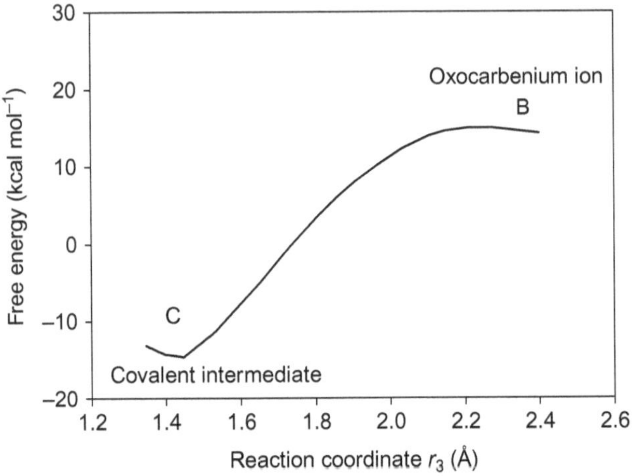

Fig. 13 Free energy profiles for the breaking of the bond between NAMD C1 and Asp52 Oδ2. Reaction coordinate: $r_3 = d$(NAMD C1–Asp52 Oδ2). *Reproduced from Bowman, A. L., Grant, I. M., & Mulholland, A. J. (2008). QM/MM simulations predict a covalent intermediate in the hen egg white lysozyme reaction with its natural substrate.* Chemical Communications, 37*(37), 4425–4427. http://dx.doi.org/10.1039/B810099C by permission of The Royal Society of Chemistry.*

($\mathbf{M_{wPO4}}$) was also found, in which a proton is transferred to the α-phosphate via a water molecule. The free energy profiles clearly show that the $\mathbf{M_{PO4}}$ mechanism is favorable with the lowest rate-limiting free energy barrier (23.5 kcal mol^{-1}), while those for $\mathbf{M_{Asp}}$ and $\mathbf{M_{wPO4}}$ mechanisms are significantly greater in energy (41.7 and 40.9 kcal mol^{-1}; Fig. 14). The TS for $\mathbf{M_{PO4}}$ exhibits a slightly dissociative character with similar bond distances (Fig. 15) for C1–O1 (2.52 Å) and C1–OG (2.15 Å), whereas this bond length difference is much larger in the other two mechanisms. However, there are still unsettled issues in terms of the configuration of Ser in the Michaelis complex state, as it has been suggested that its different configurations might favor one mechanism over the other.

Despite great success having been made with the computational modeling of GH and GT catalytic mechanisms, many uncertainties still remain and require further investigation. For instance, the conformational catalytic itinerary and the specific TS conformations for several GHs remain unclear, and it is still a mystery why different GT families have evolved to follow a specific reaction pathway. Moreover, more efforts need to be made to evaluate and eliminate the discrepancies caused by the choice of different computational approaches.

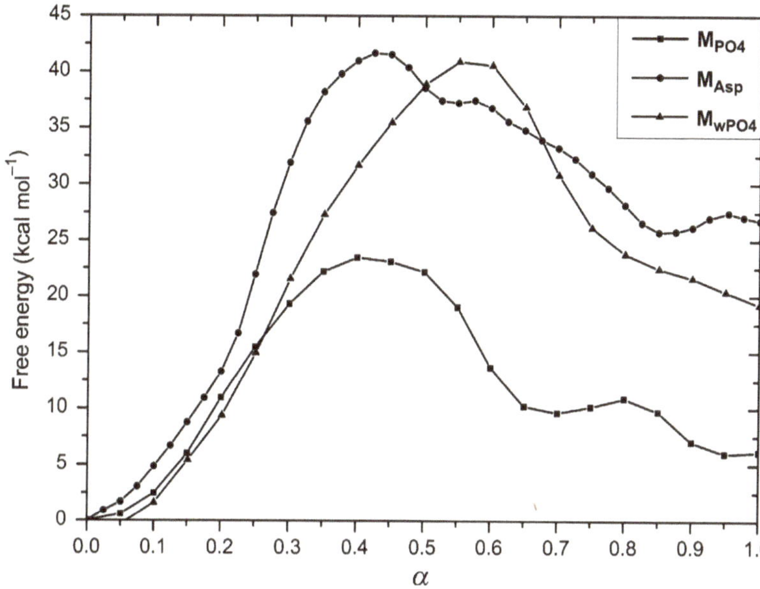

Fig. 14 Free energy profiles of three reaction pathways. Michaelis complex corresponds to $\alpha = 0$, while the product complex corresponds to $\alpha = 1$. *Adapted with permission from Kumari, M., Kozmon, S., Kulhanek, P., Stepan, J., Tvaroska, I., & Koca, J. (2015). Exploring reaction pathways for O-GlcNAc transferase catalysis. A string method study.* The Journal of Physical Chemistry B, 119*(12), 4371–4381. http://dx.doi.org/10.1021/jp511235f. Copyright © 2015 American Chemical Society.*

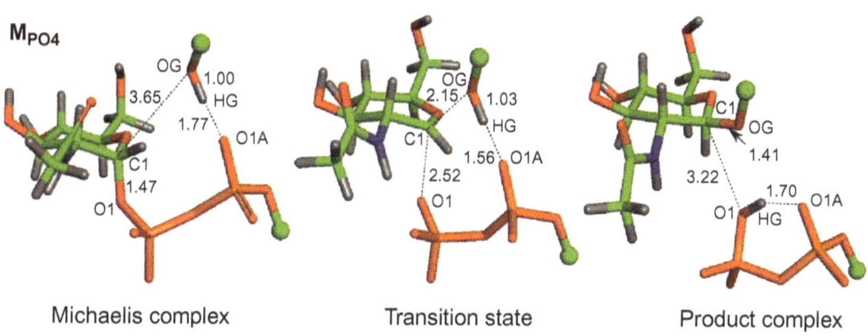

Fig. 15 Structures of Michaelis complex, TS, and product complex involved in the $\mathbf{M_{PO4}}$ mechanism. *Reproduced with permission from Kumari, M., Kozmon, S., Kulhanek, P., Stepan, J., Tvaroska, I., & Koca, J. (2015). Exploring reaction pathways for O-GlcNAc transferase catalysis. A string method study.* The Journal of Physical Chemistry B, 119*(12), 4371–4381. http://dx.doi.org/10.1021/jp511235f. Copyright © 2015 American Chemical Society.*

6. TRANSITION-STATE ANALOGUES AS INHIBITORS OF CAZYMES

6.1 Transition-State Analogue Inhibitors

Based on a statistical mechanics definition, the TS is the point where the energetics of bond making or breaking has equal probability of moving from the TS toward either the reactant or the product basins (Schramm, 2013). Knowledge of the TS allows for the design of analogues, which can be used as inhibitors of the enzyme function. TS analogues inhibit enzyme function as they fill the catalytic site and induce potential conformational changes that would normally occur to appropriately position the TS with the other reactants. However, the chemical stability of the analogue prevents the reaction from proceeding and instead forms stable thermodynamic interactions with the active site that resemble the geometry of the TS (Schramm, 2013).

The catalytic mechanisms of both GHs and GTs generally share similar oxocarbenium-like TSs (Lairson et al., 2008; Rye & Withers, 2000). For both classes of enzyme, some of the most potent inhibitors are those small molecules that are analogues of the TS. One of the most well-known and successful examples of TS analogue inhibitors of a CAZyme is the inhibitors of the GH known as neuraminidase. These inhibitors (Fig. 16), which are currently on the market as treatments for influenza, are known as zanamivir (**3**, marketed as Relenza by GlaxoSmithKline), oseltamivir (**4**, marketed as Tamiflu by Hoffmann La Roche and Gilead Sciences), and peramivir (**5**, marketed as Rapivab by BioCryst Pharmaceuticals). Using the natural substrate of neuraminidase, sialic acid (*N*-acetylneuraminic acid, Neu5Ac) (**6**, Fig. 16), these compounds were developed rationally using both crystallographic and computational techniques. These compounds have been reviewed in detail by von Itzstein (2007).

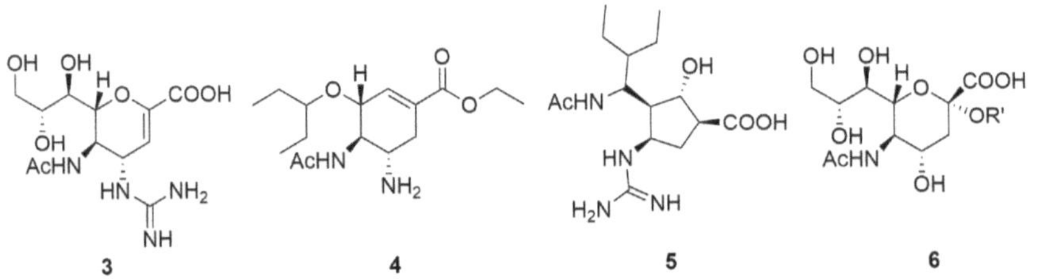

3 4 5 6

Fig. 16 Structures of TS analogue inhibitors of neuraminidase, zanamivir (**3**), oseltamivir (**4**), peramivir (**5**), and sialic acid (**6**).

6.2 Computational Insight Into the Design of Transition-State Analogue Inhibitors of CAZymes

6.2.1 *Understanding the Catalytic Mechanism of O-GlcNAcase and Relating This to the Design of Transition-State Analogue Inhibitors*

An example of a CAZyme whose mechanism has been studied using QM/MM techniques as a starting point to design novel inhibitors is *O*-glycoprotein 2-acetamino-2-deoxy-β-D-glucopyranosidase (*O*-GlcNAcase), a member of the GH84 family (Lombard et al., 2014). *O*-GlcNAcase hydrolyzes *O*-GlcNAc residues from post-translationally modified Ser or Thr residues of nuclear and cytoplasmic proteins (the reverse process of OGT). Given the potential in targeting protein *O*-GlcNAcylation for therapeutic benefit to chronic diseases such as neurodegeneration and cancer, small-molecule inhibitors of *O*-GlcNAcase have emerged as research tools for evaluating the phenotypic effects of increased *O*-GlcNAc levels in cultured cells, as well as in vivo (Cekic et al., 2016; Dorfmueller et al., 2006).

Mechanistic studies (Çetinbaş, Macauley, Stubbs, Drapala, & Vocadlo, 2006; Macauley, Whitworth, Debowski, Chin, & Vocadlo, 2005) and structural studies of bacterial homologues of human *O*-GlcNAcase (Dennis et al., 2006; Rao et al., 2006) have provided support for a two-step retaining mechanism (Fig. 17A). In this mechanism, two key general acid/base catalytic residues have been identified as Asp174 and Asp175 in human *O*-GlcNAcase (Çetinbaş et al., 2006) and Asp297 and Asp298 in the Gram-positive bacterium *Clostridium perfringens* (Rao et al., 2006). The first step of the reaction mechanism is a substrate-assisted catalytic process, where Asp174 directs and polarizes the 2-acetamido group of *O*-GlcNAc to act as a

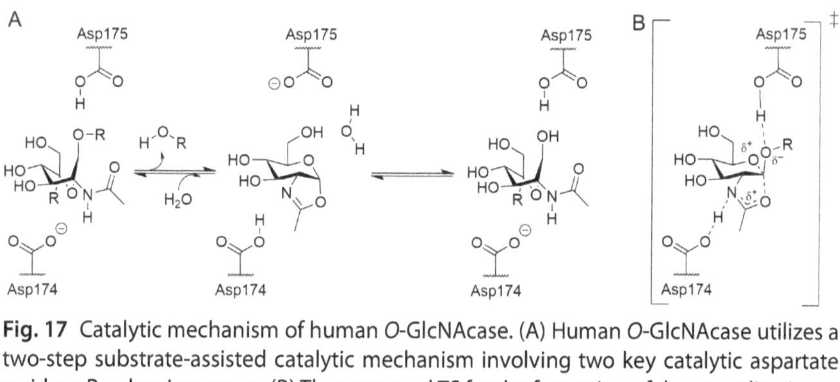

Fig. 17 Catalytic mechanism of human *O*-GlcNAcase. (A) Human *O*-GlcNAcase utilizes a two-step substrate-assisted catalytic mechanism involving two key catalytic aspartate residues R = leaving group. (B) The proposed TS for the formation of the oxazoline intermediate (Cekic et al., 2016).

nucleophile resulting in the formation of an oxazoline intermediate. Asp175 meanwhile acts as a general acid, donating a proton to the glycosidic oxygen during cleavage of the glycosidic bond. In the second step, Asp174 facilitates the departure of the 2-acetamido group, whereas Asp175 acts as a general base, promoting the attack of a water molecule to yield the β-hemiacetal product (Çetinbaş et al., 2006; Lameira, Alves, Tuñón, Martí, & Moliner, 2011). The hypothetical TS for the formation of the oxazoline intermediate (Fig. 17B) for O-GlcNAcase has been used as a model for the rational design of inhibitors for this enzyme.

Lameira et al. have performed QM/MM simulations to obtain both potential energy profiles and free energy profiles to study the catalytic mechanism of the formation of the oxazoline intermediate involving substrate-assisted catalysis for O-GlcNAcase of *C. perfringens* (Lameira et al., 2011). Using the starting reactive boat conformation (R_{boat}, Fig. 18), PESs were calculated for a stepwise and concerted mechanism (Fig. 19) for the formation of the oxazoline intermediate (P, Fig. 18). From these calculations, it could be seen that the rate-limiting energy barrier (between R and INT), calculated at the AM1/MM level, for the stepwise mechanism was 146.3 $kJ\,mol^{-1}$. The energy barrier between R and TS, calculated at the AM1/MM level, for the concerted mechanism was 246.6 $kJ\,mol^{-1}$. These results suggested that while the formation of the oxazoline intermediate can occur via a concerted mechanism, it is more favorable for the stepwise mechanism to proceed (Lameira et al., 2011).

The free energy profile of the stepwise mechanism was studied by calculating PMF that was calculated using an umbrella sampling approach. To study the first step of the mechanism (R_{boat} to INT), a 2D-PMF surface (Fig. 20A) was obtained using the R1–R2 and R3–R4 coordinates (Fig. 18). The 2D-PMF surface (Fig. 20A) was very similar to the corresponding PES (Fig. 19A). The second step's PMF (Fig. 20B) was traced using the starting point of the INT structure (Fig. 18) using the R5–R6 coordinate. The product found for this profile was the expected oxazoline/R-OH complex, P in Fig. 18. Corrected free energy surface (using B3LYP level) shows a barrier for the first step (R_{boat} to INT) to be 155 $kJ\,mol^{-1}$, which while too high for the kind of process in this study is significantly lower than the one obtained from the PES. The remaining error was attributed to the use of AM1 geometries in the high-level calculations (Lameira et al., 2011).

Lameira et al. compared the structures of three potent O-GlcNAcase inhibitors (Fig. 21A), PUGNAc (**7**, human O-GlcNAcase $K_i = 46\,nM$), GlcNAcstatin B (**8**, human O-GlcNAcase $K_i = 0.4\,nM$), and NAG-thiazoline

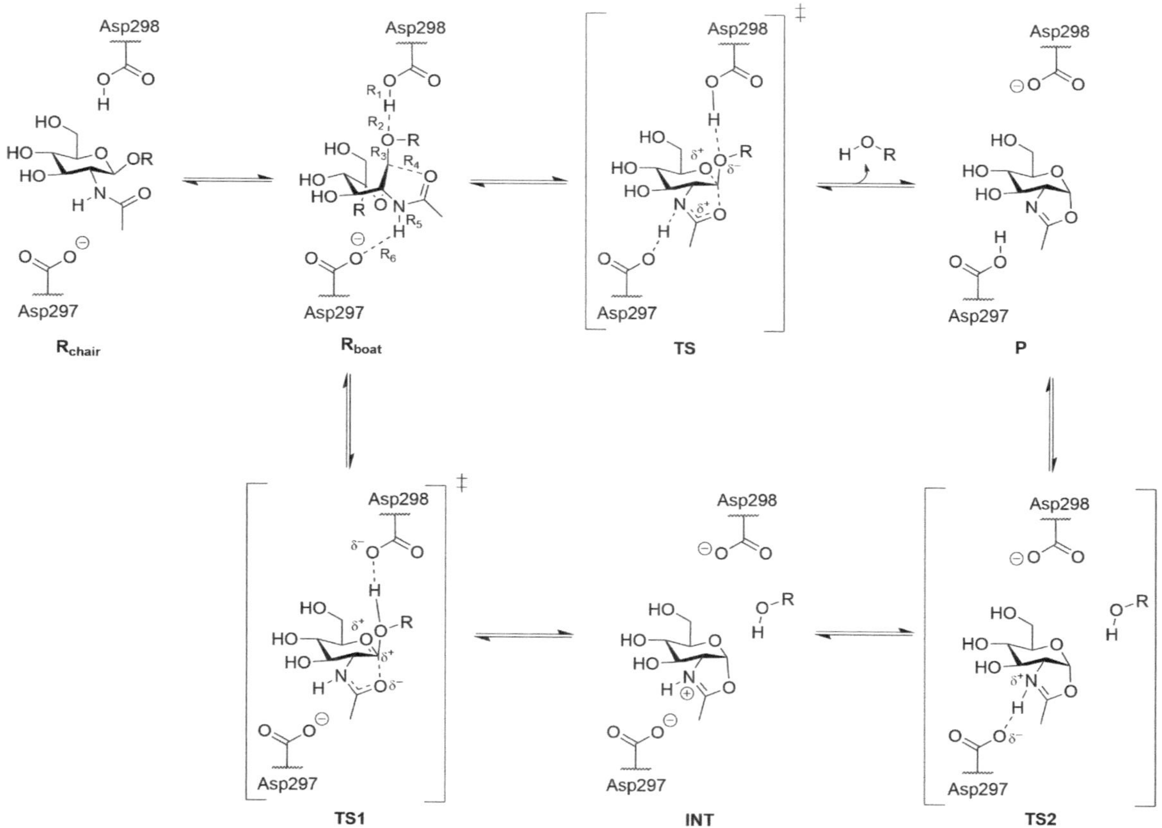

Fig. 18 Proposed mechanism of the first step of the reaction catalyzed by GlcNAcase of *Clostridium perfringens*. Key distances employed to explore the molecular mechanism are shown in R_{boat} (Lameira et al., 2011).

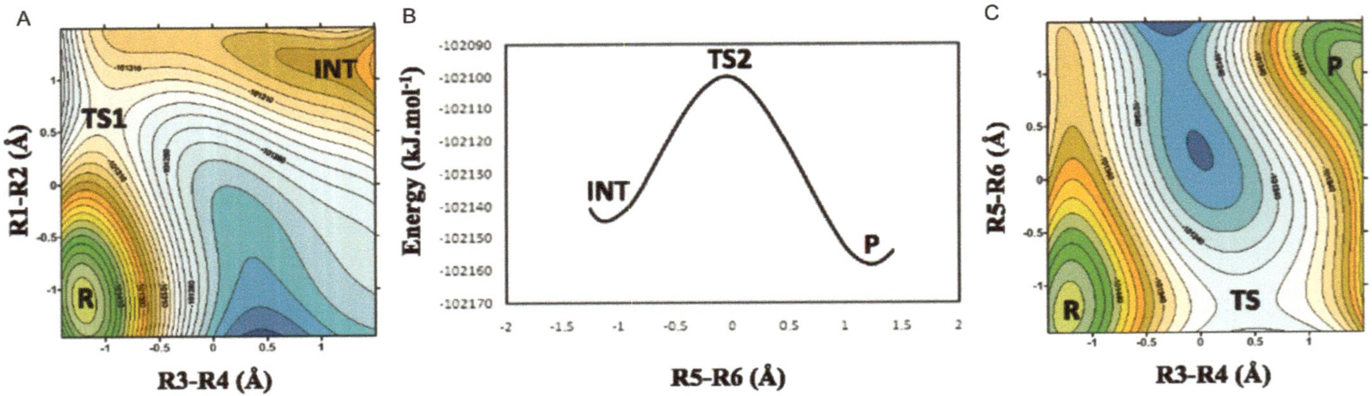

Fig. 19 Potential energy surfaces calculated at AM1/MM level obtained by means of (A) R1–R2 and R3–R4 antisymmetric combinations of distances, (B) R5–R6 antisymmetric combination, and (C) R3–R4 and R5–R6 antisymmetric combinations of distances. Values of contour potential energy lines are reported in $kJ\,mol^{-1}$ and coordinates in Å. *Reproduced with permission from Lameira, J., Alves, C. N., Tuñón, I., Martí, S., & Moliner, V. (2011). Enzyme molecular mechanism as a starting point to design new inhibitors: A theoretical study of O-GlcNAcase.* The Journal of Physical Chemistry B, 115*(20), 6764–6775. http://dx.doi.org/10.1021/jp202079e. Copyright © 2011 American Chemical Society.*

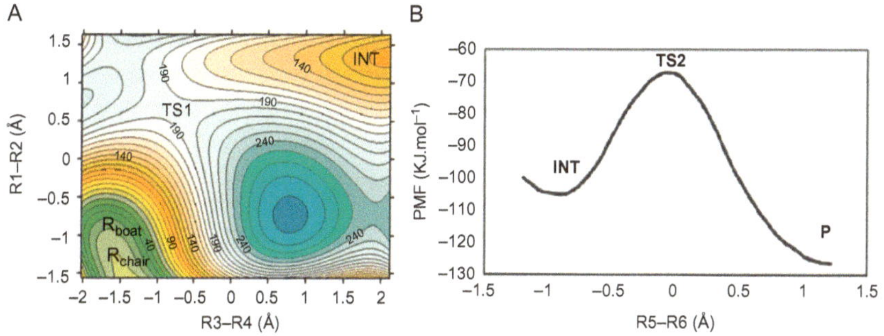

Fig. 20 (A) 2D-PMF (in kJ mol^{-1}) obtained for proton transfer from Asp298 to substrate (R1–R2, in Å) and attack of the 2-acetamido carbonyl oxygen on the anomeric center (R3–R4, in Å). (B) PMF corresponding to proton transfer from the substrate to Asp297 (R5–R6, in Å). *Values on isoenergetic lines* are reported each 10 kJ mol^{-1}. *Reproduced with permission from Lameira, J., Alves, C. N., Tuñón, I., Martí, S., & Moliner, V. (2011). Enzyme molecular mechanism as a starting point to design new inhibitors: A theoretical study of O-GlcNAcase.* The Journal of Physical Chemistry B, 115*(20), 6764–6775. http://dx.doi.org/10.1021/jp202079e. Copyright © 2011 American Chemical Society.*

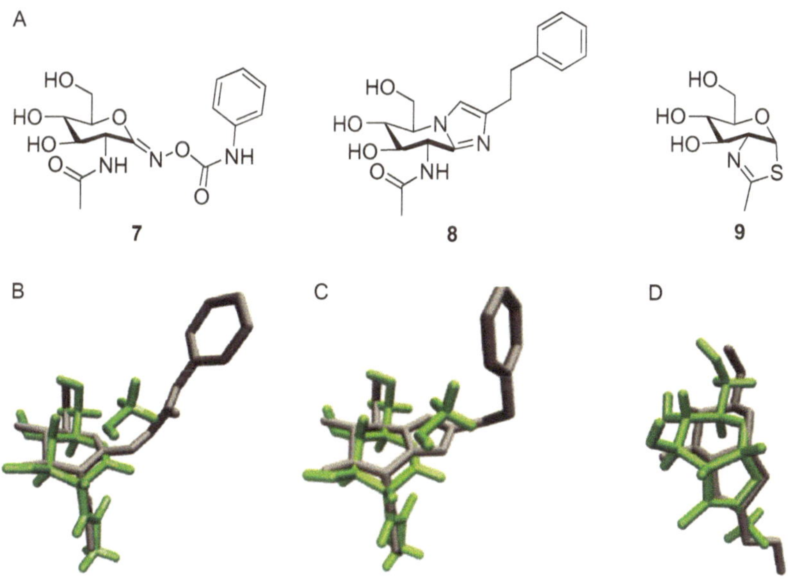

Fig. 21 (A) Structures of known inhibitors of *O*-GlcNAcase, PUGNAc (**7**), GlcNAcstatin B (**8**), and NAG-thiazoline (**9**). (B) Superposition of TS1 (*green*) and PUGNAc (*silver*) obtained from crystal structure of CpGH84H-PUGNAc complex at PDB code 2CBJ. (C) Superposition of TS1 (*green*) and GlcNAcstatin (*silver*) obtained from cpOGA in complex with GlcNAcstatin inhibitor, PDB code 2WB5. (D) Superposition of TS2 (*green*) and *N*-butyl-thiazoline (*silver*), obtained from structure of the *Bacteroides thetaiotaomicron* GH84 *O*-GlcNAcase, in complex with *N*-butyl-thiazoline. *Panels (B–D): Adapted with permission from Lameira, J., Alves, C. N., Tuñón, I., Martí, S., & Moliner, V. (2011). Enzyme molecular mechanism as a starting point to design new inhibitors: A theoretical study of O-GlcNAcase.* The Journal of Physical Chemistry B, 115*(20), 6764–6775. http://dx.doi.org/10.1021/jp202079e. Copyright © 2011 American Chemical Society.*

(**9**, human *O*-GlcNAcase $K_i = 70\,\text{nM}$) to the structures of the TS1- and TS2-protein complexes. It was found that NAG-thiazoline structure is close to the TS2-enzyme structure, whereas PUGNAc and GlcNAcstatin B inhibitors appear closer to the TS1-enzyme structure (Fig. 21B–D). This is also observed in MEP maps with NAG-thiazoline replicating TS2, and PUGNAc and GlcNAcstatin B inhibitors replicating the TS1 MEP surface. All these observations suggest that these three inhibitors can be considered TS analogues, which would explain their potency against *O*-GlcNAcase. Understanding of the relationship between the TS-enzyme structure and the inhibitor structure potency toward *O*-GlcNAcase has provided a platform for the further rational design of TS analogue inhibitors (Lameira et al., 2011).

6.2.2 Rational Design of Transition-State Analogue Inhibitors Targeting Human Sialyltransferases

Sialyltransferases (STs) are enzymes of the GT family that play an integral role in the biosynthesis of Neu5Ac (**6**, Fig. 16) containing oligosaccharides and glycoconjugates (Harduin-Lepers et al., 2001; Li & Chen, 2012). Generally in glycosylation reactions catalyzed by STs, the sugar nucleotide donor is cytidine 5′-monophosphate Neu5Ac (CMP-Neu5Ac), and the acceptor is an oligosaccharide or glycoconjugate terminated by a galactose (Gal), *N*-acetylgalactosamine (GalNAc), or other Neu5Ac residue. STs are classified based on the position of the glycosyl acceptor that Neu5Ac is transferred to. In humans, these are ST3, ST6, and ST8, which form an α-glycosidic bond between the C2 atom of Neu5Ac and the 3′-, 6′-, or 8′-hydroxyl group of the acceptor, respectively (Harduin-Lepers et al., 2001; Li & Chen, 2012). Hypersialylation of tumor cell surface proteins along with a marked upregulation of ST activity is a well-established hallmark of cancer. Due to the critical role of STs in tumor growth and progression, ST inhibition has emerged as a potential new antimetastatic strategy for a range of cancers including pancreatic and ovarian (Szabo & Skropeta, 2016).

It has been proposed that the mechanism of ST (Fig. 22) proceeds via nucleophilic attack on the Neu5Ac anomeric carbon by the deprotonated hydroxyl of an appropriate acceptor, facilitated by a catalytic histidine (Breton, Fournel-Gigleux, & Palcic, 2012). This generates an oxocarbenium-like TS, with the cytidine monophosphate (CMP) moiety acting as a leaving group. The β-configuration at the anomeric carbon of the donor is inverted in the final product (Breton et al., 2012). These aspects demonstrate that STs function as an inverting GT, utilizing an S_N2-like direct displacement mechanism (Audry et al., 2011; Lairson et al., 2008).

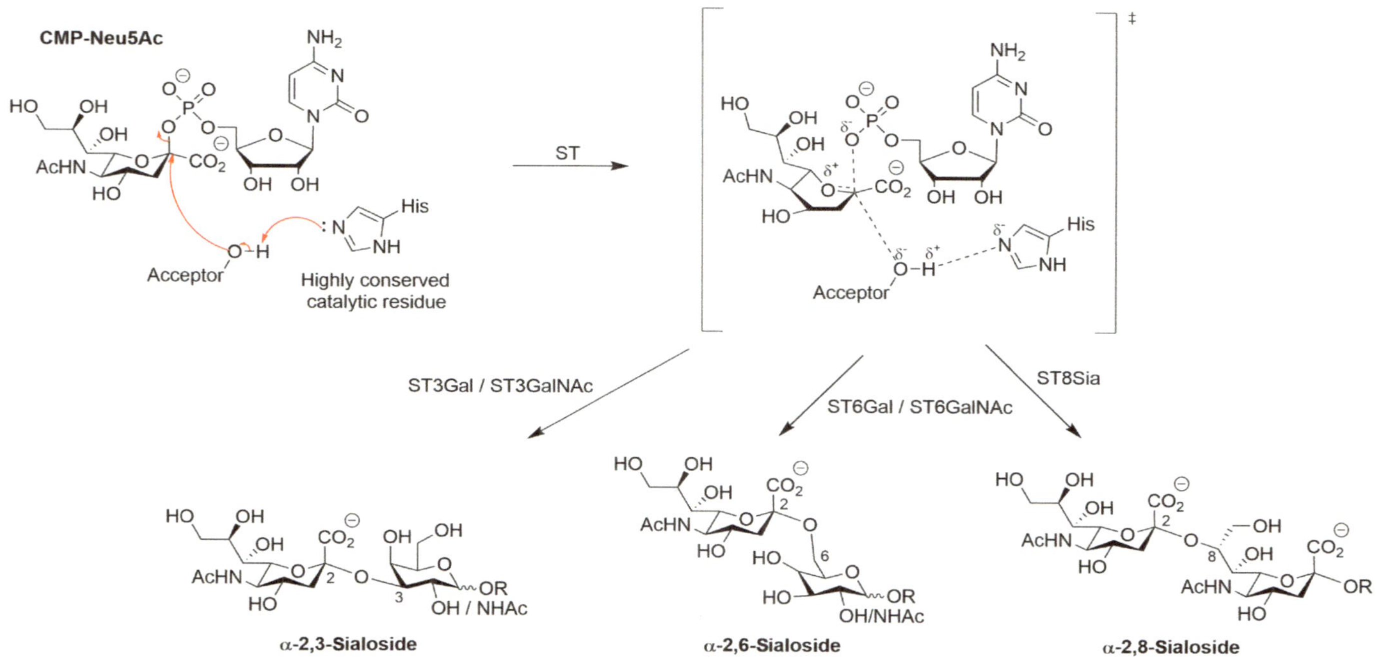

Fig. 22 Mechanism of the sialylation reaction and expected products by ST subtypes (Szabo & Skropeta, 2016).

TS analogue inhibitors originally designed by the Schmidt group (Schwörer & Schmidt, 2002; Skropeta, Schwörer, Haag, & Schmidt, 2004) (and derivatives of these compounds (Li, Niu, Xiong, Cao, & Ye, 2015)) have been shown to be some of the most potent inhibitors of STs. All these inhibitors have been designed with limited knowledge of the target ST structure. The recent characterization of the mammalian ST crystal structures (porcine ST3Gal I was the first in 2009) (Rao et al., 2009), in particular that of human ST6Gal I (Kuhn et al., 2013) and human ST8Sia III (Volkers et al., 2015), has provided starting points to begin the rational design of TS analogue inhibitors of ST.

One of the first examples of using computational modeling of TS analogue ST inhibitors was performed by our group (Montgomery, Szabo, Skropeta, & Yu, 2016). Using the human ST6Gal I structure (Kuhn et al., 2013), we investigated the potential binding effects of replacing the phosphodiester linker (e.g., **10**; Fig. 23) in derivatives with 1,2,3-triazoles (e.g., **11**; Fig. 23) and carbamates (e.g., **12**; Fig. 23), using molecular docking and MD simulations. These neutral derivatives were investigated as the charged phosphodiester linkage of TS analogue ST inhibitors, while contributing to potency, potentially may result in low bioavailability or instability in the ST active site due to the CMP leaving group moiety still being present within the inhibitor structure (Montgomery et al., 2016).

Molecular docking was performed for 11 reported inhibitors, and both the (*R*)- and (*S*)-configured diastereomers of 12 triazole- and carbamate-linked derivatives were examined using 6 different MD snapshots of human ST6Gal I. The mean binding affinities for all analogous compounds, where only the linker was altered, were observed to be not significantly different. For these triazole- and carbamate-linked compounds the position within the active site was also comparable to those obtained for their phosphodiester-linked counterparts (Fig. 24A), indicating that these linkers may be feasible replacements of the reported phosphodiester linker reported in the lead literature structure (Schwörer & Schmidt, 2002;

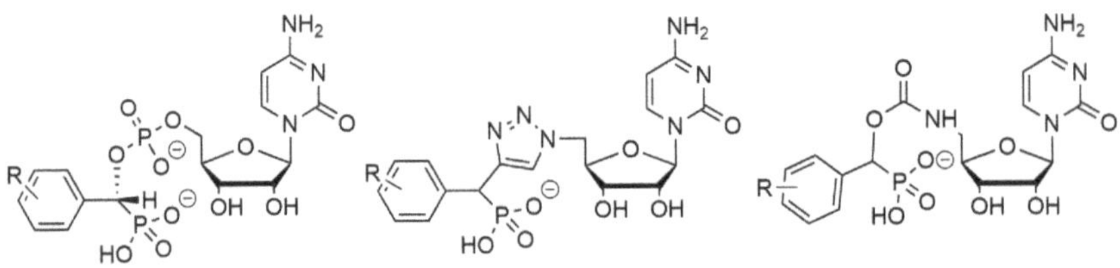

10 11 12

Fig. 23 Examples of the phosphodiester- (**10**), 1,2,3-triazole- (**11**), and carbamate-linked (**12**) derivatives investigated by Montgomery et al. (2016).

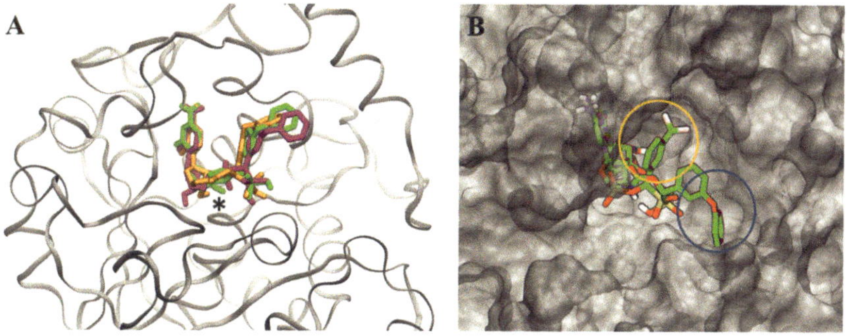

Fig. 24 (A) Comparison of binding modes obtained for *m*-phenoxy *(R)*-diastereomer of the phosphodiester- (*purple*), 1,2,3-triazole- (*green*), and carbamate-linked (*orange*) derivatives. The backbone of human ST6Gal I is shown as *gray ribbons*. (B) Surface model of human ST6Gal I highlighting the two possible binding pockets with the predicted sialyl acceptor site is *circled in blue* with the predicted Neu5Ac donor site *circled in orange*. The compounds shown are *m*-phenoxy carbamate-linked (*S*)-diastereomer and the *m*-trifluoromethyl phosphodiester-linked (*R*)-diastereomer. *Reproduced from Montgomery, A., Szabo, R., Skropeta, D., & Yu, H. (2016). Computational characterisation of the interactions between Human ST6Gal I and transition-state analogue inhibitors: Insights for inhibitor design.* Journal of Molecular Recognition, *29(5), 210–222. http://dx.doi.org/10.1002/jmr.2520 with permission from Wiley.*

Skropeta et al., 2004). The docking study also demonstrated generally no significant difference between analogous carbamate- and triazole-linked compounds. This suggests that in most cases there may be no preference for either a flexible carbamate- or a rigid triazole linker as an isostere for the phosphodiester linker (Montgomery et al., 2016).

From the binding modes obtained from the molecular docking study, it could be seen that the different functionalities ("R" in Fig. 23) of the aromatic ring were either located within or oriented toward one of two binding pockets (Fig. 24B), both of which could be exploited in the future design of inhibitors of human ST6Gal I. The first pocket was occupied by compounds such as the unsubstituted benzyls and those bearing substituents such as *m*-trifluoromethyl and *m*-methoxy. This pocket was identified as one that the Neu5Ac of the sialyl donor could be accommodated within, based on the docking results obtained for CMP-3-F-Neu5Ac and CMP-Neu5Ac. In compounds bearing larger substituents, such as a phenoxy, these groups were located within or oriented toward the predicted sialyl acceptor site. This provides a rationale for the exploration of a potential route to selectivity, as each ST subfamily would be expected to have differences in this region to accommodate the different sialyl acceptors (Montgomery et al., 2016).

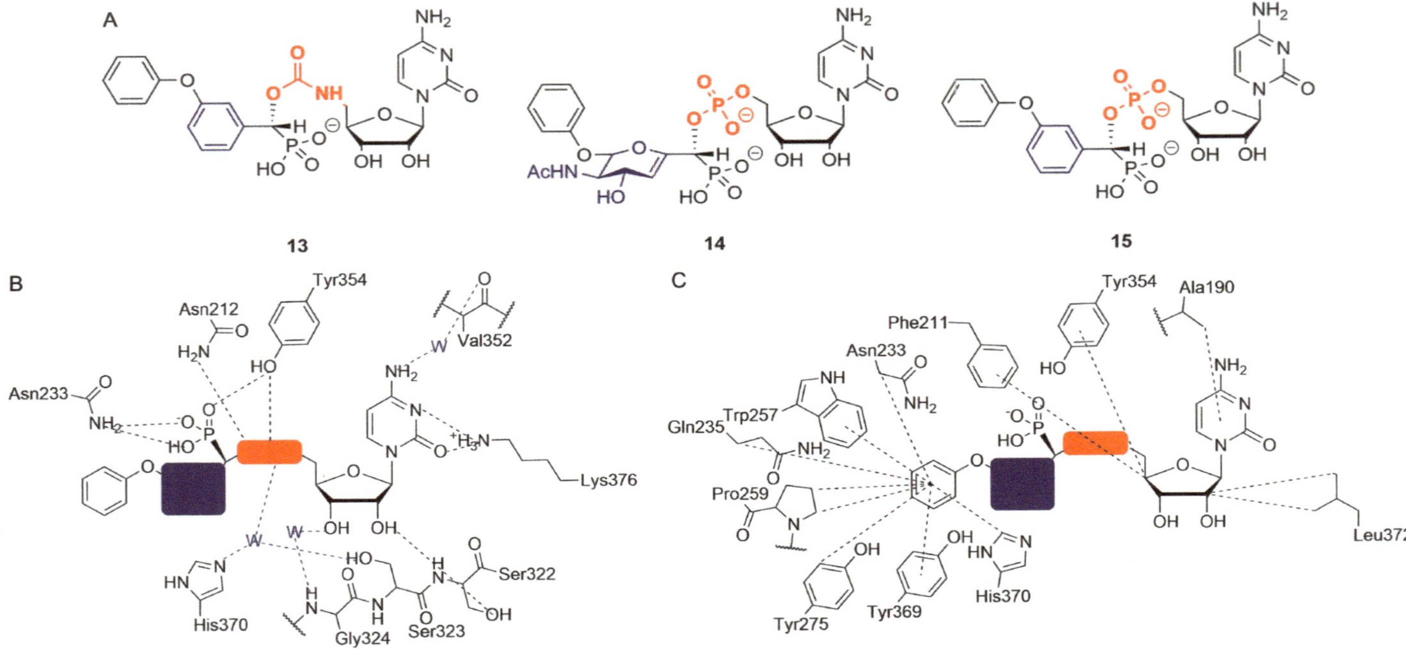

Fig. 25 (A) Structures of reported phosphodiester-linked TS analogues (**13** and **14**) and novel carbamate-linked derivative (**15**) used in inhibitor-ST6Gal I complex MD simulations. Schematic representation of conserved hydrogen bonds (B) and hydrophobic contacts (C) observed in the MD simulations. The consistent components of each inhibitor are shown as chemical structures. The *red box* represents the different linkers, and the *blue box* represents the different Neu5Ac mimic. *Adapted from Montgomery, A., Szabo, R., Skropeta, D., & Yu, H. (2016). Computational characterisation of the interactions between Human ST6Gal I and transition-state analogue inhibitors: insights for inhibitor design.* Journal of Molecular Recognition, *29(5), 210–222. http://dx.doi.org/10.1002/jmr.2520 with permission from Wiley.*

MD simulations of three different TS analogues (**13**, **14**, and **15**, Fig. 25A) complexed with human ST6Gal I were performed, and interactions in the complexes were characterized by monitoring hydrogen bonds and hydrophobic contacts over the course of the simulation (Fig. 25B and C, respectively). This analysis replicated all the hydrogen-bonding interactions observed in the human ST6Gal I–CMP complex crystal structure (Kuhn et al., 2013) and demonstrated that in a dynamic system the phenoxy substituent of these compounds is oriented toward the proposed sialyl acceptor site. This encourages the targeting of this site as a potential route to design selective human ST6Gal I inhibitors. Furthermore, these simulations demonstrated that the carbamate linker of **13** maintains the interactions with human ST6Gal I observed for the phosphodiester linker in the simulations of the reported inhibitors **14** and **15** (Montgomery et al., 2016).

Overall, the simulations we performed have demonstrated that 1,2,3-triazole- and carbamate-linked TS analogue ST inhibitors should be explored as phosphodiester isosteres. These conclusions have been further supported by our current work, in particular FEP calculations where each linker has been perturbed to the other. These calculations have indicated that replacing the charged phosphodiester with a neutral 1,2,3-triazole or carbamate does not have a negative impact on binding free energy to human ST6Gal I.

7. CONCLUSIONS AND FUTURE DIRECTIONS

Herein, we reviewed the mechanistic insights provided by computational studies of CAZymes. The molecular factors modulating the protonation and deprotonation processes of key catalytic resides in the active sites were highlighted with pK_a shift calculations and in silico mutagenesis-based electrostatics analyses. Comparative QM/MM simulations of different proposed reaction pathways have been shown to be powerful in settling mechanistic debates within the literature. More importantly from a practical point of view, the knowledge gained from such studies are shown to be valuable in guiding inhibitor design targeting these important enzymes. The limitations regarding various computational enzymology methods have also been discussed. With the active development in the field, we believe that such validated computational models will provide robust interpretation of experimental data for an unambiguous understanding of catalytic mechanisms at the atomic level for CAZymes while at the same time start to play more important roles in computer-aided drug design.

ACKNOWLEDGMENTS

We wish to acknowledge the Australian Government as H.Y. is the recipient of an Australian Research Council Future Fellowship (Project No. FT110100034), for an Australian Government Research Training Program Award scholarship for A.M., and for an Australian International Postgraduate Award scholarship for K.X. We also acknowledge the University of Wollongong for a Vice-Chancellor's Postdoctoral Research Fellowship to X.W. This research was in part supported under the Australian Research Council's Discovery Projects funding scheme (Project No. DP170101773).

REFERENCES

Ardèvol, A., & Rovira, C. (2011). The molecular mechanism of enzymatic glycosyl transfer with retention of configuration: Evidence for a short-lived oxocarbenium-like species. *Angewandte Chemie International Edition*, *50*(46), 10897–10901. http://dx.doi.org/10.1002/anie.201104623.

Ardèvol, A., & Rovira, C. (2015). Reaction mechanisms in carbohydrate-active enzymes: Glycoside hydrolases and glycosyltransferases. Insights from ab initio quantum mechanics/molecular mechanics dynamic simulations. *Journal of the American Chemical Society*, *137*(24), 7528–7547. http://dx.doi.org/10.1021/jacs.5b01156.

Audry, M., Jeanneau, C., Imberty, A., Harduin-Lepers, A., Delannoy, P., & Breton, C. (2011). Current trends in the structure-activity relationships of sialyltransferases. *Glycobiology*, *21*(6), 716–726. http://dx.doi.org/10.1093/glycob/cwq189.

Barducci, A., Bonomi, M., & Parrinello, M. (2011). Metadynamics. *Wiley Interdisciplinary Reviews: Computational Molecular Science*, *1*(5), 826–843. http://dx.doi.org/10.1002/wcms.31.

Barnett, C. B., & Naidoo, K. J. (2010). Ring puckering: A metric for evaluating the accuracy of AM1, PM3, PM3CARB-1, and SCC-DFTB carbohydrate QM/MM simulations. *The Journal of Physical Chemistry B*, *114*(51), 17142–17154. http://dx.doi.org/10.1021/jp107620h.

Benighaus, T., & Thiel, W. (2009). A general boundary potential for hybrid QM/MM simulations of solvated biomolecular systems. *Journal of Chemical Theory and Computation*, *5*(11), 3114–3128. http://dx.doi.org/10.1021/ct900437b.

Berti, P. J., & Tanaka, K. S. (2002). Transition state analysis using multiple kinetic isotope effects: Mechanisms of enzymatic and non-enzymatic glycoside hydrolysis and transfer. *Advances in Physical Organic Chemistry*, *37*, 239–314. http://dx.doi.org/10.1016/s0065-3160(02)37004-7.

Biarnés, X., Ardevol, A., Iglesias-Fernández, J., Planas, A., & Rovira, C. (2011). Catalytic itinerary in 1,3-1,4-β-glucanase unraveled by QM/MM metadynamics. Charge is not yet fully developed at the oxocarbenium ion-like transition state. *Journal of the American Chemical Society*, *133*(50), 20301–20309. http://dx.doi.org/10.1021/ja207113e.

Biarnés, X., Ardevol, A., Planas, A., Rovira, C., Laio, A., & Parrinello, M. (2007). The conformational free energy landscape of β-D-glucopyranose. Implications for substrate preactivation in β-glucoside hydrolases. *Journal of the American Chemical Society*, *129*(35), 10686–10693. http://dx.doi.org/10.1021/ja068411o.

Biarnés, X., Nieto, J., Planas, A., & Rovira, C. (2006). Substrate distortion in the Michaelis complex of *Bacillus* 1, 3–1,4-β-glucanase insight from first principles molecular dynamics simulations. *The Journal of Biological Chemistry*, *281*(3), 1432–1441. http://dx.doi.org/10.1074/jbc.M507643200.

Blake, C. C. F., Koenig, D. F., Mair, G. A., North, A. C. T., Phillips, D. C., & Sarma, V. R. (1965). Structure of hen egg-white lysozyme: A three-dimensional Fourier synthesis at 2 Å resolution. *Nature, 206*(4986), 757–761. http://dx.doi.org/10.1038/206757a0.

Bowman, A. L., Grant, I. M., & Mulholland, A. J. (2008). QM/MM simulations predict a covalent intermediate in the hen egg white lysozyme reaction with its natural substrate. *Chemical Communications, 37*(37), 4425–4427. http://dx.doi.org/10.1039/B810099C.

Breton, C., Fournel-Gigleux, S., & Palcic, M. M. (2012). Recent structures, evolution and mechanisms of glycosyltransferases. *Current Opinion in Structural Biology, 22*(5), 540–549. http://dx.doi.org/10.1016/j.sbi.2012.06.007.

Caines, M. E., Hancock, S. M., Tarling, C. A., Wrodnigg, T. M., Stick, R. V., Stütz, A. E., et al. (2007). The structural basis of glycosidase inhibition by five-membered iminocyclitols: The clan a glycoside hydrolase endoglycoceramidase as a model system. *Angewandte Chemie International Edition, 46*(24), 4474–4476. http://dx.doi.org/10.1002/anie.200700268.

Cekic, N., Heinonen, J. E., Stubbs, K. A., Roth, C., He, Y., Bennet, A. J., et al. (2016). Analysis of transition state mimicry by tight binding aminothiazoline inhibitors provides insight into catalysis by human O-GlcNAcase. *Chemical Science*, 7(6), 3742–3750. http://dx.doi.org/10.1039/C6SC00370B.

Çetinbaş, N., Macauley, M. S., Stubbs, K. A., Drapala, R., & Vocadlo, D. J. (2006). Identification of Asp174 and Asp175 as the key catalytic residues of human O-GlcNAcase by functional analysis of site-directed mutants. *Biochemistry, 45*(11), 3835–3844. http://dx.doi.org/10.1021/bi052370b.

Chen, F., Liu, H., Sun, H., Pan, P., Li, Y., Li, D., et al. (2016). Assessing the performance of the MM/PBSA and MM/GBSA methods. 6. Capability to predict protein-protein binding free energies and re-rank binding poses generated by protein-protein docking. *Physical Chemistry Chemical Physics, 18*(32), 22129–22139. http://dx.doi.org/10.1039/C6CP03670H.

Chipot, C., & Pohorille, A. (2007). *Free energy calculations: Theory and applications in chemistry and Biology*. Berlin, Germany: Springer.

Christensen, A. S., Kubař, T., Cui, Q., & Elstner, M. (2016). Semiempirical quantum mechanical methods for noncovalent interactions for chemical and biochemical applications. *Chemical Reviews, 116*(9), 5301–5337. http://dx.doi.org/10.1021/acs.chemrev.5b00584.

Cremer, D., & Pople, J. A. (1975). General definition of ring puckering coordinates. *Journal of the American Chemical Society, 97*(6), 1354–1358. http://dx.doi.org/10.1021/ja00839a011.

Davies, G., Ducros, V.-A., Varrot, A., & Zechel, D. (2003). Mapping the conformational itinerary of β-glycosidases by X-ray crystallography. *Biochemical Society Transactions, 31*(3), 523–527. http://dx.doi.org/10.1042/bst0310523.

Davies, G. J., Planas, A., & Rovira, C. (2012). Conformational analyses of the reaction coordinate of glycosidases. *Accounts of Chemical Research, 45*(2), 308–316. http://dx.doi.org/10.1021/ar2001765.

Deng, Y., & Roux, B. (2006). Calculation of standard binding free energies: Aromatic molecules in the T4 lysozyme L99A mutant. *Journal of Chemical Theory and Computation, 2*(5), 1255–1273. http://dx.doi.org/10.1021/ct060037v.

Dennis, R. J., Taylor, E. J., Macauley, M. S., Stubbs, K. A., Turkenburg, J. P., Hart, S. J., et al. (2006). Structure and mechanism of a bacterial bold β-glucosaminidase having O-GlcNAcase activity. *Nature Structural & Molecular Biology, 13*(4), 365–371. http://dx.doi.org/10.1038/nsmb1079.

Deshairs, C., Kessler, N., Aymard, M., & Quash, G. A. (1986). Catalytic properties of the A/H3N2 influenza neuraminidases: Influence of antigenic variations. *Journal of General Virology, 67*(3), 409–418. http://dx.doi.org/10.1099/0022-1317-67-3-409.

Dorfmueller, H. C., Borodkin, V. S., Schimpl, M., Shepherd, S. M., Shpiro, N. A., & van Aalten, D. M. F. (2006). GlcNAcstatin: A picomolar, selective O-GlcNAcase inhibitor that modulates intracellular O-GlcNAcylation levels. *Journal of the American Chemical Society, 128*(51), 16484–16485. http://dx.doi.org/10.1021/ja066743n.

Ducros, V. M. A., Zechel, D. L., Murshudov, G. N., Gilbert, H. J., Szabó, L., Stoll, D., et al. (2002). Substrate distortion by a β-mannanase: Snapshots of the Michaelis and covalent-intermediate complexes suggest a $B_{2,5}$ conformation for the transition state. *Angewandte Chemie International Edition, 41*(15), 2824–2827. http://dx.doi.org/10.1002/1521-3773(20020802)41:15<2824::AID-ANIE2824>3.0.CO;2-G.

Elstner, M., Porezag, D., Jungnickel, G., Elsner, J., Haugk, M., Frauenheim, T., et al. (1998). Self-consistent-charge density-functional tight-binding method for simulations of complex materials properties. *Physical Review B, 58*(11), 7260–7268. http://dx.doi.org/10.1103/PhysRevB.58.7260.

Errey, J. C., Lee, S. S., Gibson, R. P., Martinez Fleites, C., Barry, C. S., Jung, P. M. J., et al. (2010). Mechanistic insight into enzymatic glycosyl transfer with retention of configuration through analysis of glycomimetic inhibitors. *Angewandte Chemie International Edition, 49*(7), 1234–1237. http://dx.doi.org/10.1002/anie.200905096.

Fadda, E., & Woods, R. J. (2010). Molecular simulations of carbohydrates and protein–carbohydrate interactions: Motivation, issues and prospects. *Drug Discovery Today, 15*(15–16), 596–609. http://dx.doi.org/10.1016/j.drudis.2010.06.001.

Fushinobu, S., Mertz, B., Hill, A. D., Hidaka, M., Kitaoka, M., & Reilly, P. J. (2008). Computational analyses of the conformational itinerary along the reaction pathway of GH94 cellobiose phosphorylase. *Carbohydrate Research, 343*(6), 1023–1033. http://dx.doi.org/10.1016/j.carres.2008.02.026.

Garcia-Viloca, M., Gao, J., Karplus, M., & Truhlar, D. G. (2004). How enzymes work: Analysis by modern rate theory and computer simulations. *Science, 303*(5655), 186–195. http://dx.doi.org/10.1126/science.1088172.

Gaus, M., Cui, Q., & Elstner, M. (2011). DFTB3: Extension of the self-consistent-charge density-functional tight-binding method (SCC-DFTB). *Journal of Chemical Theory and Computation,* 7(4), 931–948. http://dx.doi.org/10.1021/ct100684s.

Genheden, S., & Ryde, U. (2015). The MM/PBSA and MM/GBSA methods to estimate ligand-binding affinities. *Expert Opinion on Drug Discovery, 10*(5), 449–461. http://dx.doi.org/10.1517/17460441.2015.1032936.

Gloster, T. M., & Vocadlo, D. J. (2012). Developing inhibitors of glycan processing enzymes as tools for enabling glycobiology. *Nature Chemical Biology, 8*(8), 683–694. http://dx.doi.org/10.1038/nchembio.1029.

Gómez, H., Polyak, I., Thiel, W., Lluch, J. M., & Masgrau, L. (2012). Retaining glycosyltransferase mechanism studied by QM/MM methods: Lipopolysaccharyl-α-1,4-galactosyltransferase C transfers α-galactose via an oxocarbenium ion-like transition state. *Journal of the American Chemical Society, 134*(10), 4743–4752. http://dx.doi.org/10.1021/ja210490f.

Govender, K., Gao, J., & Naidoo, K. J. (2014). AM1/d-CB1: A semiempirical model for QM/MM simulations of chemical glycobiology systems. *Journal of Chemical Theory and Computation, 10*(10), 4694–4707. http://dx.doi.org/10.1021/ct500372s.

Gubareva, L. V., Robinson, M. J., Bethell, R. C., & Webster, R. G. (1997). Catalytic and framework mutations in the neuraminidase active site of influenza viruses that are resistant to 4-guanidino-Neu5Ac2en. *Journal of Virology, 71*(5), 3385–3390. http://dx.doi.org/10.1006/viro.1997.8810.

Guérin, D. M., Lascombe, M.-B., Costabel, M., Souchon, H., Lamzin, V., Béguin, P., et al. (2002). Atomic (0.94 Å) resolution structure of an inverting glycosidase in complex with substrate. *Journal of Molecular Biology, 316*(5), 1061–1069. http://dx.doi.org/10.1006/jmbi.2001.5404.

Hadden, J. A., Tessier, M. B., Fadda, E., & Woods, R. J. (2015). Calculating binding free energies for protein–carbohydrate complexes. In T. Lütteke & M. Frank (Eds.), *Glycoinformatics* (pp. 431–465). New York, NY: Springer New York.

Hansen, N., & van Gunsteren, W. F. (2014). Practical aspects of free-energy calculations: A review. *Journal of Chemical Theory and Computation, 10*(7), 2632–2647. http://dx.doi.org/10.1021/ct500161f.

Harduin-Lepers, A., Vallejo-Ruiz, V., Krzewinski-Recchi, M. A., Samyn-Petit, B., Julien, S., & Delannoy, P. (2001). The human sialyltransferase family. *Biochimie, 83*(8), 727–737. http://dx.doi.org/10.1016/S0300-9084(01)01301-3.

Harris, T. K., & Turner, G. J. (2002). Structural basis of perturbed pKa values of catalytic groups in enzyme active sites. *IUBMB Life, 53*(2), 85–98. http://dx.doi.org/10.1080/15216540211468.

Homeyer, N., & Gohlke, H. (2012). Free energy calculations by the molecular mechanics Poisson−Boltzmann surface area method. *Molecular Informatics, 31*(2), 114–122. http://dx.doi.org/10.1002/minf.201100135.

Huber, T., Torda, A. E., & van Gunsteren, W. F. (1994). Local elevation: A method for improving the searching properties of molecular dynamics simulation. *Journal of Computer-Aided Molecular Design, 8*(6), 695–708. http://dx.doi.org/10.1007/BF00124016.

Iglesias-Fernández, J., Raich, L., Ardèvol, A., & Rovira, C. (2015). The complete conformational free energy landscape of β-xylose reveals a two-fold catalytic itinerary for β-xylanases. *Chemical Science, 6*(2), 1167–1177. http://dx.doi.org/10.1039/C4SC02240H.

Im, W., Bernèche, S., & Roux, B. (2001). Generalized solvent boundary potential for computer simulations. *The Journal of Chemical Physics, 114*(7), 2924–2937. http://dx.doi.org/10.1063/1.1336570.

IUPAC-IUB. (1980). Conformational nomenclature for five and six-membered ring forms of monosaccharides and their derivatives. *European Journal of Biochemistry, 111*(2), 295–298. http://dx.doi.org/10.1111/j.1432-1033.1980.tb04941.x.

Jancan, I., & Macnaughtan, M. A. (2012). Acid dissociation constants of uridine-5′-diphosphate compounds determined by ^{31}phosphorus nuclear magnetic resonance spectroscopy and internal pH referencing. *Analytica Chimica Acta, 749*(63–9), 63–69. http://dx.doi.org/10.1016/j.aca.2012.08.052.

Jiang, W., Hodoscek, M., & Roux, B. (2009). Computation of absolute hydration and binding free energy with free energy perturbation distributed replica-exchange molecular dynamics. *Journal of Chemical Theory and Computation, 5*(10), 2583–2588. http://dx.doi.org/10.1021/ct900223z.

Jin, T., Yu, H., & Huang, X.-F. (2016). Selective binding modes and allosteric inhibitory effects of lupane triterpenes on protein tyrosine phosphatase 1B. *Scientific Reports, 6*, 20766. http://dx.doi.org/10.1038/srep20766.

Jongkees, S. A. K., & Withers, S. G. (2014). Unusual enzymatic glycoside cleavage mechanisms. *Accounts of Chemical Research, 47*(1), 226–235. http://dx.doi.org/10.1021/ar4001313.

Joshi, M. D., Hedberg, A., & Mcintosh, L. P. (1997). Complete measurement of the pKa values of the carboxyl and imidazole groups in *Bacillus circulans* xylanase. *Protein Science, 6*(12), 2667–2670. http://dx.doi.org/10.1002/pro.5560061224.

Joshi, M. D., Sidhu, G., Nielsen, J. E., Brayer, G. D., Withers, S. G., & McIntosh, L. P. (2001). Dissecting the electrostatic interactions and pH-dependent activity of a family

11 glycosidase. *Biochemistry, 40*(34), 10115–10139. http://dx.doi.org/10.1021/bi0105429.

Joshi, M. D., Sidhu, G., Pot, I., Brayer, G. D., Withers, S. G., & McIntosh, L. P. (2000). Hydrogen bonding and catalysis: A novel explanation for how a single amino acid substitution can change the pH optimum of a glycosidase. *Journal of Molecular Biology, 299*(1), 255–279. http://dx.doi.org/10.1006/jmbi.2000.3722.

Kamerlin, S. C. L., Haranczyk, M., & Warshel, A. (2009). Progress in ab initio QM/MM free-energy simulations of electrostatic energies in proteins: Accelerated QM/MM studies of pKa, redox reactions and solvation free energies. *The Journal of Physical Chemistry B, 113*(5), 1253–1272. http://dx.doi.org/10.1021/jp8071712.

Klahn, M., Braun-Sand, S., Rosta, E., & Warshel, A. (2005). On possible pitfalls in ab initio quantum mechanics/molecular mechanics minimization approaches for studies of enzymatic reactions. *The Journal of Physical Chemistry B, 109*(32), 15645–15650. http://dx.doi.org/10.1021/jp0521757.

Kleinjung, J., & Fraternali, F. (2014). Design and application of implicit solvent models in biomolecular simulations. *Current Opinion in Structural Biology, 25*, 126–134. http://dx.doi.org/10.1016/j.sbi.2014.04.003.

Knott, B. C., Haddad Momeni, M., Crowley, M. F., Mackenzie, L. F., Götz, A. W., Sandgren, M., et al. (2013). The mechanism of cellulose hydrolysis by a two-step, retaining cellobiohydrolase elucidated by structural and transition path sampling studies. *Journal of the American Chemical Society, 136*(1), 321–329. http://dx.doi.org/10.1021/ja410291u.

Kollman, P. A., Massova, I., Reyes, C., Kuhn, B., Huo, S., Chong, L., et al. (2000). Calculating structures and free energies of complex molecules: Combining molecular mechanics and continuum models. *Accounts of Chemical Research, 33*(12), 889–897. http://dx.doi.org/10.1021/ar000033j.

Koshland, D. E. (1953). Stereochemistry and the mechanism of enzymatic reactions. *Biological Reviews, 28*(4), 416–436. http://dx.doi.org/10.1111/j.1469-185X.1953.tb01386.x.

Kuhn, B., Benz, J., Greif, M., Engel, A. M., Sobek, H., & Rudolph, M. G. (2013). The structure of human α-2,6-sialyltransferase reveals the binding mode of complex glycans. *Acta crystallographica Section D, Biological crystallography, 69*, 1826–1828. http://dx.doi.org/10.1107/S0907444913015412.

Kumari, M., Kozmon, S., Kulhanek, P., Stepan, J., Tvaroska, I., & Koca, J. (2015). Exploring reaction pathways for O-GlcNAc transferase catalysis. A string method study. *The Journal of Physical Chemistry B, 119*(12), 4371–4381. http://dx.doi.org/10.1021/jp511235f.

Lairson, L. L., Henrissat, B., Davies, G. J., & Withers, S. G. (2008). Glycosyltransferases: Structures, functions, and mechanisms. *Annual Review of Biochemistry*, 77(1), 521–555. http://dx.doi.org/10.1146/annurev.biochem.76.061005.092322.

Lameira, J., Alves, C. N., Tuñón, I., Martí, S., & Moliner, V. (2011). Enzyme molecular mechanism as a starting point to design new inhibitors: A theoretical study of O-GlcNAcase. *The Journal of Physical Chemistry B, 115*(20), 6764–6775. http://dx.doi.org/10.1021/jp202079e.

Lazarus, M. B., Jiang, J., Gloster, T. M., Zandberg, W. F., Whitworth, G. E., Vocadlo, D. J., et al. (2012). Structural snapshots of the reaction coordinate for O-GlcNAc transferase. *Nature Chemical Biology, 8*(12), 966–968. http://dx.doi.org/10.1038/nchembio.1109.

Lazarus, M. B., Nam, Y., Jiang, J., Sliz, P., & Walker, S. (2011). Structure of human O-GlcNAc transferase and its complex with a peptide substrate. *Nature, 469*(7331), 564–567. http://dx.doi.org/10.1038/nature09638.

Lee, J. K., Bain, A. D., & Berti, P. J. (2004). Probing the transition states of four glucoside hydrolyses with 13C kinetic isotope effects measured at natural abundance by NMR spectroscopy. *Journal of the American Chemical Society*, *126*(12), 3769–3776. http://dx.doi.org/10.1021/ja0394028.

Lee, S. S., Hong, S. Y., Errey, J. C., Izumi, A., Davies, G. J., & Davis, B. G. (2011). Mechanistic evidence for a front-side, S_Ni-type reaction in a retaining glycosyltransferase. *Nature Chemical Biology*, 7(9), 631–638. http://dx.doi.org/10.1038/nchembio.628.

Li, Y., & Chen, X. (2012). Sialic acid metabolism and sialyltransferases: Natural functions and applications. *Applied Microbiology and Biotechnology*, *94*(4), 887–905. http://dx.doi.org/10.1007/s00253-012-4040-1.

Li, G., & Cui, Q. (2003). pKa calculations with QM/MM free energy perturbations. *The Journal of Physical Chemistry B*, *107*(51), 14521–14528. http://dx.doi.org/10.1021/jp0356158.

Li, W., Niu, Y., Xiong, D.-C., Cao, X., & Ye, X.-S. (2015). Highly substituted cyclopentane–CMP conjugates as potent sialyltransferase inhibitors. *Journal of Medicinal Chemistry*, *58*(20), 7972–7990. http://dx.doi.org/10.1021/acs.jmedchem.5b01181.

Lira-Navarrete, E., Iglesias-Fernández, J., Zandberg, W. F., Compañón, I., Kong, Y., Corzana, F., et al. (2014). Substrate-guided front-face reaction revealed by combined structural snapshots and metadynamics for the polypeptide N-acetylgalactosaminyltransferase 2. *Angewandte Chemie International Edition*, *53*(31), 8206–8210. http://dx.doi.org/10.1002/anie.201402781.

Lombard, V., Golaconda Ramulu, H., Drula, E., Coutinho, P. M., & Henrissat, B. (2014). The carbohydrate-active enzymes database (CAZy) in 2013. *Nucleic Acids Research*, *42*(D1), D490–D495. http://dx.doi.org/10.1093/nar/gkt1178.

Ludwiczek, M. L., D'Angelo, I., Yalloway, G. N., Brockerman, J. A., Okon, M., Nielsen, J. E., et al. (2013). Strategies for modulating the pH-dependent activity of a family 11 glycoside hydrolase. *Biochemistry*, *52*(18), 3138–3156. http://dx.doi.org/10.1021/bi400034m.

Macauley, M. S., Whitworth, G. E., Debowski, A. W., Chin, D., & Vocadlo, D. J. (2005). O-GlcNAcase uses substrate-assisted catalysis: Kinetic analysis and development of highly selective mechanism-inspired inhibitors. *The Journal of Biological Chemistry*, *280*(27), 25313–25322. http://dx.doi.org/10.1074/jbc.M413819200.

Massova, I., & Kollman, P. A. (2000). Combined molecular mechanical and continuum solvent approach (MM-PBSA/GBSA) to predict ligand binding. *Perspectives in Drug Discovery and Design*, *18*(1), 113–135. http://dx.doi.org/10.1023/A:1008763014207.

Mayes, H. B., Broadbelt, L. J., & Beckham, G. T. (2014). How sugars pucker: Electronic structure calculations map the kinetic landscape of five biologically paramount monosaccharides and their implications for enzymatic catalysis. *Journal of the American Chemical Society*, *136*(3), 1008–1022. http://dx.doi.org/10.1021/Ja410264d.

McIntosh, L. P., Hand, G., Johnson, P. E., Joshi, M. D., Korner, M., Plesniak, L. A., et al. (1996). The pKa of the general acid/base carboxyl group of a glycosidase cycles during catalysis: A ^{13}C-NMR study of *Bacillus circulans* xylanase. *Biochemistry*, *35*(31), 9958–9966. http://dx.doi.org/10.1021/bi9613234.

McKimm-Breschkin, J. L., Williams, J., Barrett, S., Jachno, K., McDonald, M., Mohr, P. G., et al. (2013). Reduced susceptibility to all neuraminidase inhibitors of influenza H1N1 viruses with haemagglutinin mutations and mutations in non-conserved residues of the neuraminidase. *Journal of Antimicrobial Chemotherapy*, *68*(10), 2210–2221. http://dx.doi.org/10.1093/jac/dkt205.

McQuarrie, D. A. (2000). *Statistical mechanics*. Sausalito, California: University Science Books.

Miao, S., Ziser, L., Aebersold, R., & Withers, S. G. (1994). Identification of glutamic acid 78 as the active site nucleophile in *Bacillus subtilis* xylanase using electrospray tandem mass spectrometry. *Biochemistry*, *33*(23), 7027–7032. http://dx.doi.org/10.1021/bi00189a002.

Monegal, A., & Planas, A. (2006). Chemical rescue of α3-galactosyltransferase. Implications in the mechanism of retaining glycosyltransferases. *Journal of the American Chemical Society*, *128*(50), 16030–16031. http://dx.doi.org/10.1021/ja0659931.

Montgomery, A., Szabo, R., Skropeta, D., & Yu, H. (2016). Computational characterisation of the interactions between human ST6Gal I and transition-state analogue inhibitors: Insights for inhibitor design. *Journal of Molecular Recognition*, *29*(5), 210–222. http://dx.doi.org/10.1002/jmr.2520.

Offen, W. A., Zechel, D. L., Withers, S. G., Gilbert, H. J., & Davies, G. J. (2009). Structure of the Michaelis complex of β-mannosidase, Man2A, provides insight into the conformational itinerary of mannoside hydrolysis. *Chemical Communications*, *14*(18), 2484–2486. http://dx.doi.org/1039/b902240f.

Olsson, M. H. M., Søndergaard, C. R., Rostkowski, M., & Jensen, J. H. (2011). PROPKA3: Consistent treatment of internal and surface residues in empirical pKa predictions. *Journal of Chemical Theory and Computation*, *7*(2), 525–537. http://dx.doi.org/10.1021/ct100578z.

Paes, G., Berrin, J. G., & Beaugrand, J. (2012). GH11 xylanases: Structure/function/properties relationships and applications. *Biotechnology Advances*, *30*(3), 564–592. http://dx.doi.org/10.1016/j.biotechadv.2011.10.003.

Pan, A. C., Sezer, D., & Roux, B. (2008). Finding transition pathways using the string method with swarms of trajectories. *The Journal of Physical Chemistry B*, *112*(11), 3432–3440. http://dx.doi.org/10.1021/jp0777059.

Phillips, D. C. (1967). The hen egg-white lysozyme molecule. *Proceedings of the National Academy of Sciences of the United States of America*, *57*(3), 483–495.

Pokhrel, S., Joo, J. C., & Yoo, Y. J. (2013). Shifting the optimum pH of *Bacillus circulans* xylanase towards acidic side by introducing arginine. *Biotechnology and Bioprocess Engineering*, *18*(1), 35–42. http://dx.doi.org/10.1007/s12257-012-0455-x.

Pol-Fachin, L., Rusu, V. H., Verli, H., & Lins, R. D. (2012). GROMOS 53A6GLYC, an improved GROMOS force field for hexopyranose-based carbohydrates. *Journal of Chemical Theory and Computation*, *8*(11), 4681–4690. http://dx.doi.org/10.1021/ct300479h.

Pollet, A., Delcour, J. A., & Courtin, C. M. (2010). Structural determinants of the substrate specificities of xylanases from different glycoside hydrolase families. *Critical Reviews in Biotechnology*, *30*(3), 176–191. http://dx.doi.org/10.3109/07388551003645599.

Raich, L., Nin-Hill, A., Ardèvol, A., & Rovira, C. (2016). Chapter seven—Enzymatic cleavage of glycosidic bonds: Strategies on how to set up and control a QM/MM metadynamics simulation. *Methods in Enzymology*, *577*, 159–183. http://dx.doi.org/10.1016/bs.mie.2016.05.015.

Rao, F. V., Dorfmueller, H. C., Villa, F., Allwood, M., Eggleston, I. M., & van Aalten, D. M. F. (2006). Structural insights into the mechanism and inhibition of eukaryotic O-GlcNAc hydrolysis. *The EMBO Journal*, *25*(7), 1569–1578. http://dx.doi.org/10.1038/sj.emboj.7601026.

Rao, F. V., Rich, J. R., Raki, B., Buddai, S., Schwartz, M. F., Johnson, K., et al. (2009). Structural insight into mammalian sialyltransferases. *Nature Structural & Molecular Biology*, *16*(11), 1186–1188. http://dx.doi.org/10.1038/nsmb.1685.

Reif, M. M., & Oostenbrink, C. (2014). Net charge changes in the calculation of relative ligand-binding free energies via classical atomistic molecular dynamics simulation. *Journal of Computational Chemistry*, *35*(3), 227–243. http://dx.doi.org/10.1002/jcc.23490.

Riccardi, D., Schaefer, P., Yang, Y., Yu, H., Ghosh, N., Prat-Resina, X., et al. (2006). Development of effective quantum mechanical/molecular mechanical (QM/MM)

methods for complex biological processes. *The Journal of Physical Chemistry B, 110*(13), 6458–6469. http://dx.doi.org/10.1021/jp056361o.
Rocklin, G. J., Mobley, D. L., Dill, K. A., & Hünenberger, P. H. (2013). Calculating the binding free energies of charged species based on explicit-solvent simulations employing lattice-sum methods: An accurate correction scheme for electrostatic finite-size effects. *The Journal of Chemical Physics, 139*(18). 184103. http://dx.doi.org/10.1063/1.4826261
Rojas-Cervellera, V., Ardèvol, A., Boero, M., Planas, A., & Rovira, C. (2013). Formation of a covalent glycosyl–enzyme species in a retaining glycosyltransferase. *Chemistry—A European Journal, 19*(42), 14018–14023. http://dx.doi.org/10.1002/chem.201302898.
Roston, D., & Cui, Q. (2016). Chapter nine—QM/MM analysis of transition states and transition state analogues in metalloenzymes. In A. V. Gregory (Ed.), *Vol. 577. Methods in enzymology* (pp. 213–250): Academic Press. http://www.sciencedirect.com/science/article/pii/S0076687916300519.
Roux, B., & Yu, H. (2010). Assessing the accuracy of approximate treatments of ion hydration based on primitive quasichemical theory. *The Journal of Chemical Physics, 132*(23), 234101. http://dx.doi.org/10.1063/1.3436632.
Rye, C. S., & Withers, S. G. (2000). Glycosidase mechanisms. *Current Opinion in Chemical Biology, 4*(5), 573–580. http://dx.doi.org/10.1016/S1367-5931(00)00135-6.
Sabini, E., Sulzenbacher, G., Dauter, M., Dauter, Z., Jørgensen, P. L., Schülein, M., et al. (1999). Catalysis and specificity in enzymatic glycoside hydrolysis: A 2, 5B conformation for the glycosyl-enzyme intermediate revealed by the structure of the *Bacillus agaradhaerens* family 11 xylanase. *Chemistry & Biology, 6*(7), 483–492. http://dx.doi.org/10.1016/S1074-5521(99)80066-0.
Sattelle, B. M., & Almond, A. (2010). Less is more when simulating unsulfated glycosaminoglycan 3D-structure: Comparison of GLYCAM06/TIP3P, PM3-CARB1/TIP3P, and SCC-DFTB-D/TIP3P predictions with experiment. *Journal of Computational Chemistry, 31*(16), 2932–2947. http://dx.doi.org/10.1002/jcc.21589.
Schaefer, P., Riccardi, D., & Cui, Q. (2005). Reliable treatment of electrostatics in combined QM/MM simulation of macromolecules. *The Journal of Chemical Physics, 123*(1). 014905. http://dx.doi.org/10.1063/1.1940047.
Schimpl, M., Zheng, X., Borodkin, V. S., Blair, D. E., Ferenbach, A. T., Schuttelkopf, A. W., et al. (2012). O-GlcNAc transferase invokes nucleotide sugar pyrophosphate participation in catalysis. *Nature Chemical Biology, 8*(12), 969–974. http://dx.doi.org/10.1038/nchembio.1108.
Schramm, V. L. (2013). Transition states, analogues, and drug development. *ACS Chemical Biology, 8*, 71–81. http://dx.doi.org/10.1021/cb300631k.
Schwörer, R., & Schmidt, R. R. (2002). Efficient sialyltransferase inhibitors based on glycosides of N-acetylglucosamine. *Journal of the American Chemical Society, 124*(8), 1632–1637. http://dx.doi.org/10.1021/ja017370n.
Scudder, P., Doom, J., Chuenkova, M., Manger, I., & Pereira, M. (1993). Enzymatic characterization of β-D-galactoside α-2,3-trans-sialidase from Trypanosoma cruzi. *The Journal of Biological Chemistry, 268*(13), 9886–9891.
Senn, H. M., & Thiel, W. (2009). QM/MM methods for biomolecular systems. *Angewandte Chemie International Edition, 48*(7), 1198–1229. http://dx.doi.org/10.1002/anie.200802019.
Skropeta, D., Schwörer, R., Haag, T., & Schmidt, R. R. (2004). Asymmetric synthesis and affinity of potent Sialyltransferase inhibitors based on transition-state analogues. *Glycoconjugate Journal, 21*(5), 205–219. http://dx.doi.org/10.1023/B:GLYC.0000045093.96413.62.

Song, Y., Mao, J., & Gunner, M. R. (2009). MCCE2: Improving protein pKa calculations with extensive side chain rotamer sampling. *Journal of Computational Chemistry*, *30*(14), 2231–2247. http://dx.doi.org/10.1002/jcc.21222.

Soya, N., Fang, Y., Palcic, M. M., & Klassen, J. S. (2011). Trapping and characterization of covalent intermediates of mutant retaining glycosyltransferases. *Glycobiology*, *21*(5), 547–552. http://dx.doi.org/10.1093/glycob/cwq190.

Spiwok, V., Králová, B., & Tvaroška, I. (2010). Modelling of β-D-glucopyranose ring distortion in different force fields: A metadynamics study. *Carbohydrate Research*, *345*(4), 530–537. http://dx.doi.org/10.1016/j.carres.2009.12.011.

Stoddart, J. F. (1971). *Stereochemistry of carbohydrates*. New York: Wiley-Interscience.

Stubbs, J. M., & Marx, D. (2003). Glycosidic bond formation in aqueous solution: On the oxocarbenium intermediate. *Journal of the American Chemical Society*, *125*(36), 10960–10962. http://dx.doi.org/10.1021/ja035600n.

Szabo, R., & Skropeta, D. (2016). Advancement of sialyltransferase inhibitors: Therapeutic challenges and opportunities. *Medical Research Reviews*, *37*(2), 219–270. http://dx.doi.org/10.1002/med.21407.

Tailford, L. E., Money, V. A., Smith, N. L., Dumon, C., Davies, G. J., & Gilbert, H. J. (2007). Mannose foraging by *Bacteroides thetaiotaomicron* structure and specificity of the β-mannosidase, BtMan2A. *The Journal of Biological Chemistry*, *282*(15), 11291–11299. https://dx.doi.org/10.1074/jbc.M610964200.

Tailford, L. E., Offen, W. A., Smith, N. L., Dumon, C., Morland, C., Gratien, J., et al. (2008). Structural and biochemical evidence for a boat-like transition state in β-mannosidases. *Nature Chemical Biology*, *4*(5), 306–312. http://dx.doi.org/10.1038/nchembio.81.

Thiel, W. (2014). Semiempirical quantum–chemical methods. *Wiley Interdisciplinary Reviews: Computational Molecular Science*, *4*(2), 145–157. http://dx.doi.org/10.1002/wcms.1161.

Thompson, A. J., Dabin, J., Iglesias-Fernández, J., Ardèvol, A., Dinev, Z., Williams, S. J., et al. (2012). The reaction coordinate of a bacterial GH47 α-mannosidase: A combined quantum mechanical and structural approach. *Angewandte Chemie International Edition*, *51*(44), 10997–11001. http://dx.doi.org/10.1002/anie.201205338.

Thompson, A. J., Speciale, G., Iglesias-Fernández, J., Hakki, Z., Belz, T., Cartmell, A., et al. (2015). Evidence for a boat conformation at the transition state of GH76 α-1,6-mannanases—Key enzymes in bacterial and fungal mannoprotein metabolism. *Angewandte Chemie International Edition*, *54*(18), 5378–5382. http://dx.doi.org/10.1002/anie.201410502.

Torrie, G. M., & Valleau, J. P. (1977). Nonphysical sampling distributions in Monte Carlo free-energy estimation: Umbrella sampling. *Journal of Computational Physics*, *23*(2), 187–199. http://dx.doi.org/10.1016/0021-9991(77)90121-8.

van der Kamp, M. W., & Mulholland, A. J. (2013). Combined quantum mechanics/molecular mechanics (QM/MM) methods in computational enzymology. *Biochemistry*, *52*(16), 2708–2728. http://dx.doi.org/10.1021/bi400215w.

van Gunsteren, W. F., Bakowies, D., Baron, R., Chandrasekhar, I., Christen, M., Daura, X., et al. (2006). Biomolecular modeling: Goals, problems, perspectives. *Angewandte Chemie International Edition*, *45*(25), 4064–4092. http://dx.doi.org/10.1002/anie.200502655.

Varki, A., Cummings, R. D., Esko, J. D., Freeze, H. H., Stanley, P., Bertozzi, C. R., et al. (2009). *Essentials of glycobiology* (2nd ed). Cold Spring Harbor: Cold Spring Harbor Laboratory Press.

Vasella, A., Davies, G. J., & Böhm, M. (2002). Glycosidase mechanisms. *Current Opinion in Chemical Biology*, *6*(5), 619–629. http://dx.doi.org/10.1016/S1367-5931(02)00380-0.

Vipperla, B., Griffiths, T. M., Wang, X., & Yu, H. (2017). Theoretical pKa prediction of the α-phosphate moiety of uridine 5′-diphosphate-GlcNAc. *Chemical Physics Letters*, *667*, 220–225. http://dx.doi.org/10.1016/j.cplett.2016.12.001.
Vocadlo, D. J., & Davies, G. J. (2008). Mechanistic insights into glycosidase chemistry. *Current Opinion in Chemical Biology*, *12*(5), 539–555. http://dx.doi.org/10.1016/j.cbpa.2008.05.010.
Vocadlo, D. J., Davies, G. J., Laine, R., & Withers, S. G. (2001). Catalysis by hen egg-white lysozyme proceeds via a covalent intermediate. *Nature*, *412*(6849), 835–838. http://dx.doi.org/10.1038/35090602.
Vocadlo, D. J., Wicki, J., Rupitz, K., & Withers, S. G. (2002). Mechanism of *Thermoanaerobacterium saccharolyticum* β-xylosidase: Kinetic studies. *Biochemistry*, *41*(31), 9727–9735. http://dx.doi.org/10.1021/bi020077v.
Volkers, G., Worrall, L. J., Kwan, D. H., Yu, C.-C., Baumann, L., Lameignere, E., et al. (2015). Structure of human ST8SiaIII sialyltransferase provides insight into cell-surface polysialylation. *Nature Structural & Molecular Biology*, *22*(8), 627–635. http://dx.doi.org/10.1038/nsmb.3060.
von Itzstein, M. (2007). The war against influenza: Discovery and development of sialidase inhibitors. *Nature Reviews Drug discovery*, *6*, 967–974. http://dx.doi.org/10.1038/nrd2400.
Wang, J., Deng, Y., & Roux, B. (2006). Absolute binding free energy calculations using molecular dynamics simulations with restraining potentials. *Biophysical Journal*, *91*(8), 2798–2814. http://dx.doi.org/10.1529/biophysj.106.084301.
Wang, L., Wu, Y., Deng, Y., Kim, B., Pierce, L., Krilov, G., et al. (2015). Accurate and reliable prediction of relative ligand binding potency in prospective drug discovery by way of a modern free-energy calculation protocol and force field. *Journal of the American Chemical Society*, *137*(7), 2695–2703. http://dx.doi.org/10.1021/ja512751q.
Warshel, A. (1991). *Computer modeling of chemical reactions in enzymes and solution*. New York: Wiley.
Warshel, A., & Levitt, M. (1976). Theoretical studies of enzymic reactions: Dielectric, electrostatic and steric stabilization of the carbonium ion in the reaction of lysozyme. *Journal of Molecular Biology*, *103*(2), 227–249. http://dx.doi.org/10.1016/0022-2836(76)90311-9.
Williams, S. J., & Davies, G. J. (2001). Protein–carbohydrate interactions: Learning lessons from nature. *Trends in Biotechnology*, *19*(9), 356–362. http://dx.doi.org/10.1016/S0167-7799(01)01699-7.
Withers, S. G., & Davies, G. J. (2012). Glycobiology: The case of the missing base. *Nature Chemical Biology*, *8*(12), 952–953. http://dx.doi.org/10.1038/nchembio.1117.
Woods, C. J., Essex, J. W., & King, M. A. (2003). The development of replica-exchange-based free-energy methods. *The Journal of Physical Chemistry B*, *107*(49), 13703–13710. http://dx.doi.org/10.1021/jp0356620.
Woods, R. J., & Tessier, M. B. (2010). Computational glycoscience: Characterizing the spatial and temporal properties of glycans and glycan–protein complexes. *Current Opinion in Structural Biology*, *20*(5), 575–583. http://dx.doi.org/10.1016/j.sbi.2010.07.005.
Xiao, K., & Yu, H. (2016). Rationalising pKa shifts in *Bacillus circulans* xylanase with computational studies. *Physical Chemistry Chemical Physics*, *18*(44), 30305–30312. http://dx.doi.org/10.1039/C6CP02526A.
Yang, Y., Yu, H., & Cui, Q. (2008). Extensive conformational transitions are required to turn on ATP hydrolysis in myosin. *Journal of Molecular Biology*, *381*(5), 1407–1420. http://dx.doi.org/10.1016/j.jmb.2008.06.071.
Yang, Y., Yu, H., York, D., Cui, Q., & Elstner, M. (2007). Extension of the self-consistent-charge density-functional tight-binding method: Third-order expansion of the density

functional theory total energy and introduction of a modified effective Coulomb interaction. *The Journal of Physical Chemistry A*, *111*(42), 10861–10873. http://dx.doi.org/10.1021/jp074167r.
Yu, H., & Griffiths, T. M. (2014). pKa cycling of the general acid/base in glycoside hydrolase families 33 and 34. *Physical Chemistry Chemical Physics*, *16*(12), 5785–5792. http://dx.doi.org/10.1039/C4CP00351A.
Yu, H., & Roux, B. (2009). On the utilization of energy minimization to the study of ion selectivity. *Biophysical Journal*, *97*(8), L15–L17. http://dx.doi.org/10.1016/j.bpj.2009.08.005.
Yu, H., & van Gunsteren, W. F. (2005). Accounting for polarization in molecular simulation. *Computer Physics Communications*, *172*(2), 69–85. http://dx.doi.org/10.1016/j.cpc.2005.01.022.
Zechel, D. L., & Withers, S. G. (2000). Glycosidase mechanisms: Anatomy of a finely tuned catalyst. *Accounts of Chemical Research*, *33*(1), 11–18. http://dx.doi.org/10.1021/ar970172+.
Zhu, X., Yang, Y., & Yu, H. (2014). *In silico* enzyme modelling. *Australian Biochemist*, *45*, 12–15.
Zienau, J., & Cui, Q. (2012). Implementation of the solvent macromolecule boundary potential and application to model and realistic enzyme systems. *The Journal of Physical Chemistry B*, *116*(41), 12522–12534. http://dx.doi.org/10.1021/jp308218m.

CHAPTER THREE

Computational Biochemistry—Enzyme Mechanisms Explored

Martin Culka[2], Florian J. Gisdon[2], G. Matthias Ullmann[1]
Computational Biochemistry, University of Bayreuth, Bayreuth, Germany
[1]Corresponding author: e-mail address: Matthias.Ullmann@uni-bayreuth.de

Contents

1. Introduction 78
2. Structural Models 81
   2.1 Continuum-Electrostatics Models 82
   2.2 Quantum-Mechanical Models 83
   2.3 Empirical Molecular-Mechanical Models 84
   2.4 Hybrid QM/MM Models 86
   2.5 Pseudo-Atomistic Models 89
3. Calculating Enzymatic Mechanisms 91
   3.1 Thermodynamic Properties of Biological Systems 91
   3.2 Kinetic Properties of Biochemical Systems 93
   3.3 Mechanic and Energetic Properties of Molecular Reaction Paths 97
   3.4 Path Search Strategy 104
4. Going Beyond the Exploration of the Reaction Paths 104
References 105

Abstract

Understanding enzyme mechanisms is a major task to achieve in order to comprehend how living cells work. Recent advances in biomolecular research provide huge amount of data on enzyme kinetics and structure. The analysis of diverse experimental results and their combination into an overall picture is, however, often challenging. Microscopic details of the enzymatic processes are often anticipated based on several hints from macroscopic experimental data. Computational biochemistry aims at creation of a computational model of an enzyme in order to explain microscopic details of the catalytic process and reproduce or predict macroscopic experimental findings. Results of such computations are in part complementary to experimental data and provide an explanation of a biochemical process at the microscopic level. In order to evaluate the mechanism of an enzyme, a structural model is constructed which can be analyzed by several theoretical approaches. Several simulation methods can and should be combined to get a reliable picture of the process of interest. Furthermore, abstract models of

[2] These authors contributed equally.

Advances in Protein Chemistry and Structural Biology, Volume 109
ISSN 1876-1623
http://dx.doi.org/10.1016/bs.apcsb.2017.04.004

biological systems can be constructed combining computational and experimental data. In this review, we discuss structural computational models of enzymatic systems. We first discuss various models to simulate enzyme catalysis. Furthermore, we review various approaches how to characterize the enzyme mechanism both qualitatively and quantitatively using different modeling approaches.

1. INTRODUCTION

"Certainly no subject or field is making more progress on so many fronts at the present moment than biology, and if we were to name the most powerful assumption of all, which leads one on and on in an attempt to understand life, it is that all things are made of atoms, and that everything that living things do can be understood in terms of the jigglings and wigglings of atoms" (Feynman, 1964). These words were said by the well-known physicist Richard Feynman in his famous *Lectures on Physics* now more than 50 years ago. Until today, this sentence has not lost its validity and is the basis for much of the biomolecular research, maybe today even more than 50 years ago. Today, we are beginning to understand how the interplay of atoms and molecules lead to the complex processes that we find in living systems. Recent advances in genomics and systems biology help to gain more and more insights into the molecular organizations of the cell, which are nevertheless still too complex to allow a full overview of all the biochemical processes at atomic detail. New techniques in structural biology such as the free-electron laser allow to analyze structurally the kinetics of molecular processes at an atomic level (Nango et al., 2016; Pande et al., 2016). On the other hand, modern electron microscopy allows us to gain insights into large molecular assemblies that were not accessible to structural investigations a few years ago (Bartesaghi et al., 2015; Kühlbrandt, 2014). These new techniques complement traditional techniques of structural biology such as X-ray crystallography and NMR. In the past, a structural characterization of a protein was considered the final goal of an investigation. Nowadays to gain a deep understanding of an enzymatic mechanism, a structure is just the beginning. The structural information needs to be completed by other experimental information for instance from spectroscopy, kinetics, or electrochemistry. Often, all these different aspects are difficult to merge. Thus, it is important to approach the enzymatic mechanism also from a theoretical side. With the help of modern methods from computational chemistry, it is possible to gain insights into enzymatic mechanisms and to complete the picture.

In order to understand experimental data, we build models of reality and use these models in our theoretical considerations. Models are not only used for theoretical calculations, but also the interpretation of experimental data relies on models of the studied system. A model is a generalized hypothetical description used for analyzing or explaining a system. It is a simplified representation of a real system intended to enhance our ability to understand, predict, and control the behavior of the real system. When a model is made, there are always approximations required. Therefore a model is always an idealized representation of the real system. A model reproduces only certain aspects of the real system, only those that are relevant for the properties under study. Other aspects of the same real system may not be described equally well by the model, since it was originally made for another purpose. A model should be able to explain experimental data and make predictions about the outcome of new experiments. In order to be able to make predictions, a model has to be complicated enough to represent all important aspects of a real system at an appropriate level of description. However, the more complicated and complex a model is, the more difficult it becomes to interpret the results of the model. It is therefore required that the model is as complicated as necessary and not more complicated in order to give insights into the behavior of the real system. Most importantly, a model should promote our understanding of nature.

A model of a real system is generally constructed in several steps (see Fig. 1). The first and probably most important step is the construction of the conceptual model. In this step, the real system is translated into an idealized model system. Having in mind which properties of the system are interesting, the features of the system that are important to reproduce these properties are selected and it is decided how to use them to describe the desired aspects of the system. This first step requires a detailed inspection of the system that should be modeled in order to decide which details are required for the desired representation. When only a qualitative understanding is requested, the conceptual model is often enough to picture the system and thus it represents the final goal of the investigation. Instead, when a quantitative or semiquantitative understanding of the system is desired, the conceptual model needs to be translated into a mathematical model. A concrete physical theory is required to forge the mathematical model designed in the conceptual model. If this mathematical model can be solved analytically, the goal is reached and the behavior of the model system can be compared with the behavior of the real system. However, the mathematical models are often too complex to be solved analytically. In this case, the mathematical model needs to be translated into a discrete mathematical

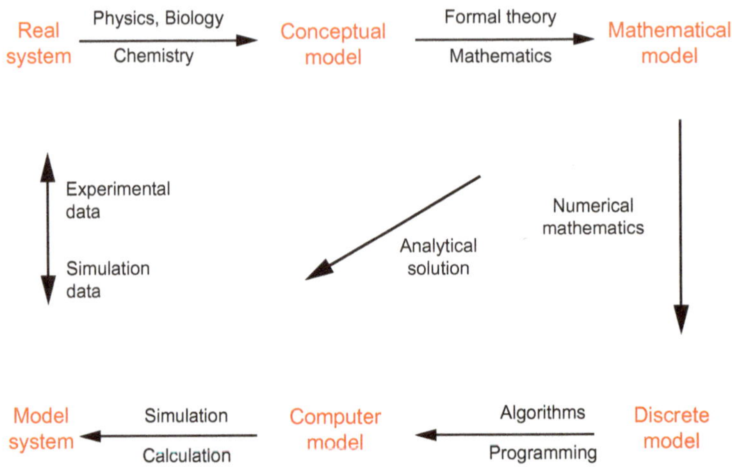

Fig. 1 Building theoretical models of real systems. This scheme depicts the stages of building theoretical models that can help to analyze real systems. These theoretical models are abstractions of the real systems that help to understand the behavior of the system. Mathematical modeling allows a quantitative comparison between the real system and the model.

model using methods from numerical mathematics. This discrete model can then be implemented in computer programs. The computer model can then be used to perform computer simulations and calculations using sets of parameters. The resulting simulation data can be compared with experimental data and so the quality of the model can be judged. The quantities that have been calculated from the model can lead to a better understanding of the real system. Moreover, the analysis of the effect of well-defined changes of the initial parameter can provide deeper insights into the behavior of the real system.

One of the most basic model that is widely used in the description of chemical systems is the Born–Oppenheimer approximation according to which one can separate the motion of nuclei from the motion of the electrons. This approximation allows us to describe molecular structures by defining the nuclear coordinates and to speak about different electronic states of a molecule. One fundamental concept that is used in molecular biophysics is the molecular energy landscape which describes the energy of a molecular system with N atoms in dependence of its $3N$ coordinates. This energy landscape contains minima which represent stable states of the systems, (first-order) saddle points which represent transition states of chemical reactions, and minimum energy paths which represent the trajectory along which chemical reactions occur. A large part of theoretical biochemistry is concerned with exploring this energy landscape and to extract different kind of data from it. In this article, we review some of the most important methods to explore this energy landscapes and to explain how they can be used to explore enzymatic mechanisms.

2. STRUCTURAL MODELS

The most computational enzymology techniques construct a structural model of the enzyme in question, usually using its experiment-derived 3D structure as a starting point. A structural model provides a way to calculate energy of the given structure, by using various approximations of the real picture to stay computationally feasible. Based on energy differences between different states within structural computer models of enzymes, thermodynamics, and kinetics of enzyme catalysis can be addressed. With the use of structural models (Fig. 2), researchers aim to track the events within the

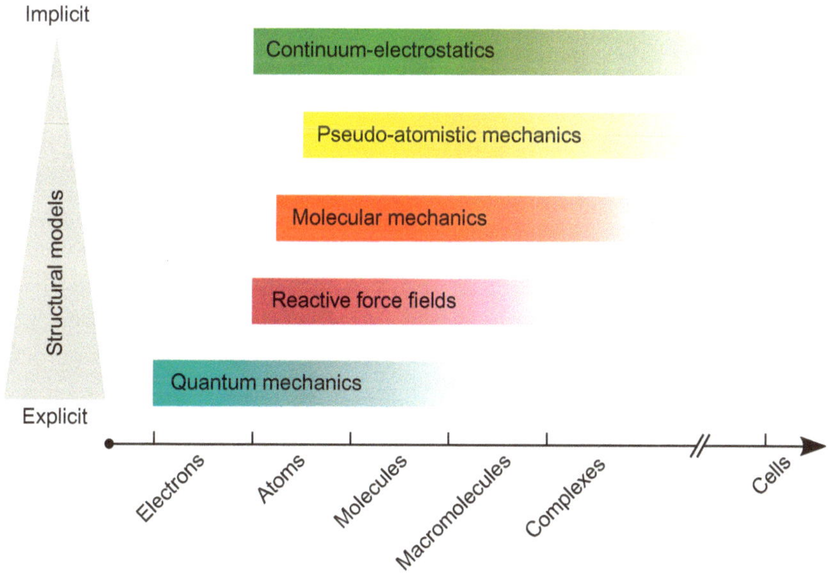

Fig. 2 Structural models for the simulation of catalytic events within enzymes. The *vertical axis* represents an ascending gradient from explicit to implicit models. The system description lose the detailed (explicit) character in favor of averaged (implicit) description of physical properties. The *horizontal axis* shows the information that can be obtained from the models, and together illustrates the size of the biological system, which is computationally feasible to treat. The *model bars* indicate possible applications. The *fading areas* show the potential of future developments enabled by improvements of algorithmic and computational power. All models can be combined into the so-called hybrid or multiscale models, which treat different parts of the biological system with the appropriate method, to enhance the effectiveness of the simulation and the quality of results.

enzyme that lead to catalysis. Models used to study enzyme mechanisms can be atomistic, i.e., they explicitly describe the behavior of electrons and nuclei, or pseudo-atomistic, i.e., they treat groups of atoms as one entity. In contrast to these particle-based models, continuum models describe a system in terms of continuous properties assigned to the space. Typical continuum models in enzymology are implicit solvent models. Their aim is to describe the average properties of a solvent environment, instead of discrete contributors of its parts. In this section, we briefly discuss various types of structural models and their combinations to simulate enzymatic activities.

2.1 Continuum-Electrostatics Models

Since most of the effects in biochemistry are dominated by electrostatics, a pure electrostatic model can be used to satisfactory describe many features of biomolecular systems while being computationally efficient. The most commonly used continuum-electrostatics model in biochemistry relies on the Poisson–Boltzmann equation. For a recent review the reader is referred to Ullmann and Bombarda (2014). The basic idea of this continuum-electrostatics model is to describe the protein as a low-dielectric region, which is embedded in (aqueous) solvent, described as a high-dielectric region. The charge distribution of the protein is described by a fixed charge distribution in the low-dielectric region, which is given by the molecular structure of the protein. The charge distribution of the protein is modeled by (fractional) point charges that are placed at the center of the atoms. The dissolved ions are represented by a Boltzmann-distributed charge density. The boundary of the low-dielectric region is defined by the solvent accessible surface of the protein.

The Poisson–Boltzmann equation is usually solved numerically (Honig & Nicholls, 1995; Warwicker & Watson, 1982). The solution of the Poisson–Boltzmann equation can be expressed as a potential $\phi(\mathbf{r})$, that is, composed of two contributions:

$$\phi(\mathbf{r}) = \sum_{i=1}^{M} \frac{q_i}{4\pi\varepsilon_p |\mathbf{r} - \mathbf{r}'_i|} + \phi_{\mathrm{rf}}(\mathbf{r}) \tag{1}$$

First, the Coulomb potential at the position $\mathbf{r}$ caused by M point charges q_i at positions $\mathbf{r}'_i$ in a medium with a permittivity ε_p, and second, the reaction field potential $\phi_{\mathrm{rf}}(\mathbf{r})$, arising from the M point charges q_i and the dielectric boundary between the protein and the solvent, as well as from the distribution of ions in the solution. Electrostatic energy can be obtained by integrating the potential distribution over the space.

Continuum-electrostatics models such as the above described Poisson–Boltzmann model have various applications. The most straightforward is calculation of solvation energies and visualization of electrostatic potentials (Baker, Sept, Joseph, Holst, & McCammon, 2001). More advanced applications allow to calculate energies of different protonation and oxidation microstates of the protein, as well as ligand binding energies and many other applications (Ullmann et al., 2008). Furthermore, the Poisson–Boltzmann model or other popular continuum models such as COSMO (Klamt & Schüürmann, 1993) are often combined with more detailed models to simulate solvent effects in a computationally affordable way (Chen, Noodleman, Case, & Bashford, 1994; Li, Nelson, Peng, Bashford, & Noodleman, 1998; Liu et al., 2004).

2.2 Quantum-Mechanical Models

The most detailed structural models that treat both electrons and nuclei explicitly are based on quantum mechanics (QM). In QM models, the energy of the structure is derived from solving an approximate Schrödinger equation. One way is to employ ab initio molecular orbital wave function approaches such as the single-determinant Hartree–Fock (HF) method (Cramer, 2004). The major drawback of the HF method is the mean-field treatment of electron repulsion referred to as the electron correlation problem. Møller–Plesset theory (Møller & Plesset, 1934) addresses the electron correlation effects by means of Rayleigh–Schrödinger perturbation theory. The most common second-order version is called MP2 (Head-Gordon, Pople, & Frisch, 1988). In coupled cluster (CC) methods (Čížek, 1966), multielectron wave-functions are constructed using the exponential cluster operator to account for the electron correlation. The latter two more accurate approaches are, however, still very costly for larger systems. For complex biological systems, fast semiempirical (SE) approximations of the Hartree–Fock theory have become very popular in the past and are still used and developed up to now (Stewart, 2013; Řezáč & Hobza, 2012; Yilmazer & Korth, 2015). The SE methods achieve considerable calculation speed-up by parameterizing various parts of the HF theory in order to reproduce experimental or high-level ab initio QM results.

Density functional theory (DFT) is an alternative ab initio approach that has become very popular for biological systems due to its favorable price/performance ratio (van Mourik, Bühl, & Gaigeot, 2014). Instead of searching for a multidimensional wave-function that describes the position of every electron, the whole problem is solved from the point of view of total electron distribution (density) in the space. The electron correlation is taken into account in DFT methods, albeit on approximate level. In addition, hybrid DFT–HF methods (Becke, 1993; Zhao & Truhlar, 2008) combine the best of the both worlds and have turned out to be very successful (Bryantsev, Diallo, Van Duin, & Goddard, 2009). Also DFT has a parameterized alternative called density-functional tight-binding (DFTB). DFTB is designed to reproduce DFT results rather than fit empirical data like SE methods (Elstner, 2006). QM methods play an inevitable role in computational studies of enzyme-catalyzed reactions, since some amount of quantum-mechanical treatment or QM-based parameterization is always needed in the active site.

Even with the enormous growth of computational power in last decades, biological systems such as enzymes can hardly be treated in the full extend by QM methods. Successful attempts have been made to simulate whole proteins purely by QM methods (Cole & Hine, 2016; Todorović, Bowler, Gillan, & Miyazaki, 2013). Nevertheless, in order to make the calculations feasible, researches often reduce the QM model of the enzyme to the active site residues only, by constructing a so-called cluster model. Continuum solvent is used to mimic the protein as a low-dielectric environment beyond the shell of the active site residues. Cluster models of enzyme active sites have been used to elucidate the mechanisms of various enzymes, including difficult cases such as radical enzymes (Feliks & Ullmann, 2012) or metalloenzymes (Li & Ryde, 2014; Manta, Raushel, & Himo, 2014). Although a rather simple approach, due to the absence of various sources of errors that more complex methods may introduce, the cluster models retain their role in computational enzymology (Georgieva & Himo, 2010). Considering the influence of the protein environment, the continuum solvent might be replaced by the protein residues simulated on an empirical level, see QM/MM approaches later.

2.3 Empirical Molecular-Mechanical Models

Due to the enormous computational cost of the QM approaches, alternative empirical methods that use principles of classical physics have been developed for treatment of complex systems. These methods are generally referred to as molecular mechanics (MM). In classical MM models, atoms are represented as spheres connected by springs. Empirical parameters have been derived to reproduce the expected behavior of biomolecules. These include spring constants for bond lengths, angles, and torsion angles as well as nonbonded interaction parameters such as van der Waals radii and partial atomic charges. The whole set of MM parameters is called force field (FF). The total potential energy of a system is calculated as a sum of all bonded and nonbonded contributions:

$$V(\vec{R}) = \sum_{\text{bonds}} K_b(b-b_0)^2 + \sum_{\text{angles}} K_\theta(\theta-\theta_0)^2 + \sum_{\text{torsions}} K_\varphi(1+\cos(n\varphi-\delta))$$
$$+ \sum_{\substack{\text{improper}\\\text{torsions}}} K_\omega(\omega-\omega_0)^2 + \sum_{\substack{\text{nonbonded}\\\text{pairs}}} \left\{ \underbrace{\varepsilon_{ij}^{\min}\left[\left(\frac{R_{ij}^{\min}}{r_{ij}}\right)^{12} - 2\left(\frac{R_{ij}^{\min}}{r_{ij}}\right)^{6}\right]}_{\substack{\text{Lennard—Jones}\\\text{potential}}} + \underbrace{\frac{q_i q_j}{4\pi\varepsilon_0\varepsilon r_{ij}}}_{\substack{\text{Coulomb}\\\text{law}}} \right\} \quad (2)$$

where b is the bond length, θ is the bond angle, ϕ the torsion angle, ω the improper torsion angle. The 0 indices represent the equilibrium values of given parameters and K the respective spring constants. Van der Waals interactions are represented by the Lennard–Jones potential where $R_{ij}^{\min}$ is the distance energy minimum position, r_{ij} is the distance between two atoms and $\varepsilon_{ij}^{\min}$ is the interaction constant. Electrostatic interactions are described by the Coulomb law where q_i and q_j are the partial charges of the interacting atoms, r_{ij} is their distance and $\varepsilon_0\varepsilon$ is the environment permittivity.

Several empirical MM force fields for biomolecules have been developed over the years, e.g., CHARMM (MacKerell et al., 1998), AMBER (Cornell et al., 1996), GROMOS (Oostenbrink, Villa, Mark, & Van Gunsteren, 2004), or OPLS (Jorgensen, Maxwell, & Tirado-Rives, 1996). MM models proved to be very useful especially for studying protein dynamics by molecular dynamics (MD) simulations. In context of enzyme catalysis, pure MM models can be used to study conformational changes or substrate binding dynamics (Costa, Batista, Bisch, & Perahia, 2015;

Gilson & Zhou, 2007). When one wants to model details of a catalytic mechanism in the enzyme active site, however, the conventional MM models are insufficient. They are, however, often used to model the non-reactive parts of the enzyme in hybrid QM/MM approaches, as described below.

Nevertheless, attempts have been made to allow for chemical reactivity within the empirical force fields without employing QM principles. These approaches are referred to as reactive force fields. One of the most widely used reactive force field, that has been also utilized for biological systems, is called ReaxFF (Senftle et al., 2016; van Duin, Dasgupta, Lorant, & Goddard III, 2001). In ReaxFF, the empirical energy equation (Eq. 2) is modified to follow a more complex bond order formalism rather than the balls-on-springs formalism. Although reactive force fields have been originally developed for material chemistry applications (Liang et al., 2013), first attempts have been made to model peptides, small proteins (Golkaram, Shin, & van Duin, 2014; Monti et al., 2013), and DNA (Verlackt et al., 2015) using ReaxFF.

Empirical force fields are popular because of their speed and their often realistic description of molecules, which is a consequence of extensive parameterization to reproduce experimentally derived structures and behaviors. It has been shown that MM methods are able to reproduce experimentally derived molecular structures, except for nonstandard regions (Kulik, Luehr, Ufimtsev, & Martínez, 2012). In such regions, ab initio QM methods consistently perform better. Despite this benefit, as already mentioned above, full ab initio studies of proteins are restricted to rather small systems (Kulik et al., 2012). Instead, hybrid approaches (referred to as QM/MM) have been developed to combine accurate ab initio models and fast empirical models.

2.4 Hybrid QM/MM Models

In QM/MM, the active site, where the reaction takes place, is treated on QM level, while the rest of the enzyme is modeled by an empirical MM force field. Since its development in 1976 (Warshel & Levitt, 1976), QM/MM has become the most popular approach to study enzymatic reactions (Senn & Thiel, 2007, 2009).

The key feature that QM/MM methods have to deal with is the interaction between QM and MM regions. For the description of a QM/MM system, two energy schemes evolved, a subtractive scheme and an additive scheme. The energy in the subtractive QM/MM scheme (Eq. 3) is obtained by an MM calculation of the entire system (*S*) with the inner part (*I*) cut out and replaced by a QM calculation.

$$E^{sub}_{\mathrm{QM/MM}} = E_{\mathrm{MM}}(S) + E_{\mathrm{QM}}(I) - E_{\mathrm{MM}}(I) \qquad (3)$$

The QM energy of the quantum mechanically treated inner part ($E_{\mathrm{QM}}(I)$) is added to the MM energy of the entire system ($E_{\mathrm{MM}}(S)$), but the MM energy of the inner part ($E_{\mathrm{MM}}(I)$) has to be subtracted to avoid double counting. The most widely used approach that employs the subtractive scheme is ONIOM (Maseras & Morokuma, 1995; Svensson et al., 1996), which is capable of combining *n* layers of any implemented QM or MM approach. The potential drawback of the subtractive scheme is that a proper MM description of the active site region is required and this is often difficult to achieve for substrate or enzyme cofactors. In the additive QM/MM scheme (Field, 2007), the MM energy is only calculated on the outer system (*O*) instead of *S* (Eq. 4). The QM energy is added and additionally, a coupling term ($E_{\mathrm{QM/MM}}(I, O)$) is introduced, which treats the interaction of the QM and the MM part.

$$E^{add}_{\mathrm{QM/MM}} = E_{\mathrm{MM}}(O) + E_{\mathrm{QM}}(I) + E_{\mathrm{QM/MM}}(I, O) \qquad (4)$$

The $E_{\mathrm{QM/MM}}(I, O)$ term itself is composed of van der Waals, electrostatic, and bonded contributions. Van der Waals interactions are fully described on the MM level. The common description with the Lennard–Jones potential is also applied to QM atoms. Therefore it is necessary to have suitable parameters for all QM atoms. Since the main impact of van der Waals interactions occurs in a short range, atoms near the boundary should be well parametrized. Thus, changes of QM atom properties during catalysis have no meaningful influence on $E_{\mathrm{QM/MM}}(I, O)$, as long as they appear not close to the boundary. In proteins, important catalytic contributions are mediated by electrostatic interactions. All charges and partial charges cause electrostatic forces. Within one model, the interactions of charges are well defined, but the coupling between two models needs to be adapted.

The electrostatic term within $E_{\mathrm{QM/MM}}(I, O)$ describes the coupling between the QM charge density and the MM charge model. One efficient but simple method to treat electrostatic interaction is mechanical embedding (Bakowies & Thiel, 1991), where the MM model is directly applied to the QM region. Since the charge density of the QM region is then represented within the MM model, the MM region can react to it. A major drawback is, however, that the QM calculation is performed in the absence of the electrostatic MM environment, thus the atoms in the QM part cannot react to their full environment. But especially this electrostatic environment is crucial for enzymatic catalysis (Zhang, 2013). The electrostatic embedding approach (Bakowies & Thiel, 1991) treats this important interaction by providing the electrostatic environment for the QM calculation. The electrostatic environment appears as a one-electron term in the QM Hamiltonian. Thus the charge distribution in the QM region is polarized according to the MM charges. Despite the higher accuracy in the calculation, charge leakage effects can occur at the boundary of the QM region, where the QM charge density is polarized in immediate proximity by MM charges. In both embedding schemes the MM charges are rigid and do not react to the QM charge density. In a polarized embedding scheme the polarization happens in both, the QM region direction and the MM region direction. For the polarization of the MM region, a polarizable force field has to be applied on the MM level (Thompson & Schenter, 1995). This approach of polarized embedding is in general the most accurate, though computational demanding one.

In the majority of QM/MM simulations of enzymes, one has to deal with covalent bonds on QM/MM boundary (e.g., bond between an amino acid side chain and the protein backbone) in addition to the nonbonded interactions. The boundary QM atom valency needs to be saturated to allow for proper electronic structure calculation. The simplest approach to treat the QM/MM boundary is the link atom approach (Field, Bash, & Karplus, 1990; Singh & Kollman, 1986), where the QM atom is capped by an auxiliary atom (usually hydrogen) that is constraint in the direction of the QM/MM bond. This generates a problem of QM density overpolarization near the boundary that has to be treated. Alternative approaches are based on frozen hybrid orbitals (Amara, Field, Alhambra, & Gao, 2000; Théry, Rinaldi, Rivail, Maigret, & Ferenczy, 1994). A frontier atom is chosen in the QM/MM boundary and a set of suitably oriented localized orbitals is placed on it. This treatment allows to converge on more proper electronic structure in the boundary region. On the other hand frozen orbital approaches are more technically demanding and require calibration for the specific bond and QM method.

A conceptually different QM/MM approach that has been used to study enzyme catalysis is the empirical valence bond (EVB) (Kamerlin & Warshel, 2010, 2011; Warshel, 2003). EVB uses the valence-bond (VB) approach for quantum description of the enzyme active site. The QM methods discussed so far are based on molecular orbital theory that combines the atomic orbitals into a molecular orbital wave-function. The VB methods instead describe the system as a linear combination of all possible states where electrons occupy localized orbitals. In EVB, the stationary points along the reaction path are described by an MM force field, while the transitions between them are treated by an SE valence-bond QM approach (Shurki, Derat, Barrozo, & Kamerlin, 2015). Thus for every reaction step simulated by EVB, a set of empirical parameters has to be derived. This is usually done on a model reaction in solution or in gas phase, where the parameters are fitted to reproduce the experimental data or ab initio QM results (Åqvist & Warshel, 1993). Once well calibrated, extensive conformational sampling along the reaction pathway can be achieved in reasonable time with EVB in order to get a proper free energy landscape (see Section 3.3.2). On the other hand, the major disadvantage of EVB in comparison to common QM and QM/MM methods is the need of specific calibration for every reaction step. In fact, prior knowledge of the reaction mechanism is required to perform an EVB simulation, while unknown mechanism alternatives can be discovered within conventional ab initio QM and QM/MM models.

2.5 Pseudo-Atomistic Models

Although the computational and algorithmic power is increasing, modeling of larger protein systems remains difficult. Bridging between molecular behavior and biological system function requires different levels of abstraction to manage the huge amount of data in a reasonable time. Pseudo-atomistic models provide one level of abstraction by introducing pseudo atoms, which comprise groups of atoms, several amino acids or even whole molecules. These pseudo atoms are represented as one entity with all atoms within such an entity considered to be frozen. They are modeled to simulate the essential or averaged behavior of such entities and their interactions. Pure pseudo-atomistic models contain merely homogeneous or heterogeneous pseudo atom species. Besides these pure models, multiscale models are developed, which combine atomistic resolution of some molecular parts with pseudo-atomistic simplification. Because of the simplification, to reduce multiple individual atoms to single entities, comparable to coarser grains, pseudo-atomistic models are referred to as coarse-grained (CG) models. The advantage of a CG model is, that the degrees of freedom are decreased. Therefore larger systems can be studied and longer time scales can be reached than using classical atomistic models. In the first application of such a pseudo-atomistic model on proteins (Levitt & Warshel, 1975), it is pointed out that this simplification brings additional benefit in the interpretation of results. With the reduction in degrees of freedom, the energy landscape for the system is smoothened. That is, less important movements or details are averaged, and essential features can be focused. Since CG models allow the treatment of bigger biological systems, and can simplify the analysis, their development was increasing in the last years, as some recent reviews show (Ingólfsson et al., 2014; Kar & Feig, 2014; Kmiecik et al., 2016; Meier et al., 2013; Noid, 2013; Riniker, Allison, & van Gunsteren, 2012; Saunders & Voth, 2013).

The concept of a coarse-grained model is a reduction of degrees of freedom. This reduction is achieved by replacing several individual atoms by pseudo atoms. For an optimal simulation it is important that the coarse-grained system keeps the overall character of the all-atom system. Therefore it is necessary to provide rules for the coarse-grained particles to behave. Analogue to the all-atom MM force fields described above, the behavior is evaluated by an empirical potential energy function. Therefore a protein is represented as an elastic network of coarse-grained beads connected to each other by elastic springs. In a simple case, each bead represents one amino acid. Such a representation is useful to investigate domain motions in large systems, where longer time scales have to be achieved (Rücker, Wieninger, Ullmann, & Sticht, 2012). A harmonic bond potential is applied to each pair of beads, to allow them to move with respect to their surrounding. The potential energy function (Eq. 5) is a power series expansion near a minimum structure $\mathbf{r}_0$, represented as a $3N$ dimensional Cartesian coordinate vector.

$$V(\mathbf{r}) = V(0) + \nabla V(\mathbf{r}) + \frac{1}{2}\mathbf{r}^T\mathbf{H}\mathbf{r} \tag{5}$$

All movements of the system can be evaluated relative to that minimum. The constant term $V(0)$ describes the energy at the minimum position and can be set to zero. The first derivative of the potential $V(\mathbf{r})$ is the gradient, which is zero at a minimum. The elastic network potential simplifies to the second-order term, which is a sum of pairwise potentials, with the second derivative matrix $\mathbf{H}$ providing the force constants.

In recent years more and more complex models have been developed, which better and better represent the nature of biological systems (Tozzini, 2005). A further development in the treatment of big biological systems is the combination with more accurate methods, such as MM or

QM/MM. Such so-called multiscale or multiresolution models are suitable for the description of catalytic effects within a biological relevant surrounding. In a recent study (Sokkar, Boulanger, Thiel, & Sanchez-Garcia, 2015) chorismate mutase and p-hydroxybenzoate hydroxylase were investigated with a QM/MM/CG approach, by modeling the catalytically relevant part of the enzyme in QM, the remaining amino acids in MM and the surrounding solvent in CG. In such multiscale models it is important to define efficient data exchange between the different potentials. A proper exchange between the different resolutions becomes even more important in the adaptive resolution multiscale models (Heyden & Truhlar, 2008; Shen & Hu, 2014; Zavadlav, Melo, Marrink, & Praprotnik, 2015). Here, specific regions are defined, similarly to the QM/MM/CG approach, but with connecting buffer zones, which enable the entering particles to change their resolution. Such approaches incorporate all benefits of the multiscale models and the flexibility to adapt to major changes within the modeled biological system. A detailed list of CG models and available programs with an extensive description of the current state of the art can be found in a recent review (Kmiecik et al., 2016).

3. CALCULATING ENZYMATIC MECHANISMS

As discussed in the introduction, in computational enzymology one is interested in explaining macroscopic thermodynamic and kinetic data derived from experiment by microscopic models. In Section 2, we reviewed types of structural model environments to study the enzyme catalysis. In this section, we review how to derive thermodynamic and kinetic parameters within these models.

3.1 Thermodynamic Properties of Biological Systems

Each simulation of biological systems requires the analysis of its thermodynamic equilibrium states. At equilibrium, no driving forces act on the system. Thus, these states are thermodynamically stable. The relative stability of an initial state and a final state, for instance, allows to predict, if a reaction is endergonic or exergonic.

The direction in which a system changes can be evaluated by changes in free energy. Dependent on the simulation, the Gibbs free energy ΔG (Eq. 6) at constant temperature T and pressure P, or the Helmholtz free energy ΔA (Eq. 7) at constant temperature and volume V can be calculated. ΔA is comprised of the internal energy ΔU of the system, and temperature scaled entropy ΔS, a measure for thermally accessible configurations. For ΔG the system energy is the enthalpy ΔH, which is comprised of internal energy ΔU of a system and the work for its volume expansion PV at adjusted pressure. Since experiments are mainly performed at constant pressure, we will hereafter refer to free energy as the Gibbs free energy ΔG.

$$\Delta G = \Delta H - T\Delta S = \Delta U + PV - T\Delta S \tag{6}$$

$$\Delta A = \Delta U - T\Delta S \tag{7}$$

The computational prediction of thermodynamic properties is based on the analysis of ensembles. An ensemble is a large number of virtual copies of a system with identical macroscopic properties. To obtain reasonable ensembles, sampling methods such as MD approaches or Monte Carlo (MC) approaches are applied (Paquet & Viktor, 2015). While MD approaches are usually based on Newtonian mechanics, MC methods are based on repeated random sampling using certain thermodynamics principles. MD simulations are used to investigate the time evolution of biological systems. Applying Newton's second law of motion (Eq. 8) the movements of atoms in a system are described with the classical equation of motion.

$$F_i = -\nabla_i E = m_i a_i \tag{8}$$

The force F_i acting on atom i with mass m_i determines its acceleration a_i, while the force is defined by the negative gradient of the potential energy function $\nabla_i E$ with respect to the coordinates of atom i. By integrating the equations of motion for all atoms over small time steps, it is possible to obtain their time dependent positions, just providing the initial positions and the initial velocities. The initial positions are directly given by the coordinates of the atoms, and the initial velocities are usually distributed randomly with a certain probability distribution. Only the acceleration is needed, which can be obtained by the gradient of the potential energy function and the atomic mass (Eq. 8). The velocities of the atoms define the temperature, which is an important thermodynamic property in MD simulations. An equilibration simulation is performed until the system reaches an equilibrium state, which is a global minimum on the energy surface. During this process it is important that the system has enough time and energy (i.e., temperature) to escape local minima. The system can in some cases anyway remain kinetically trapped in a local minimum. Annealing simulations start at higher temperature to overcome such energy barriers and gradually decrease it to the desired temperature, allowing the system to reach an equilibrium state.

In contrast to MD simulations, statistical approaches, such as Monte Carlo sampling, are independent of force evaluation. The sampling is based on random movements of the system from a set of possible movements, while new configurations are accepted in the case of lower energy, $\Delta E < 0$. Configurations with higher energy are accepted according to the Metropolis criterion, which ensures importance sampling of the microstates by a Boltzmann factor $\exp\left(^{-\Delta E}/_{k_B T}\right)$. Monte Carlo simulations create ensembles by energetically evaluating random movements of a system with the possibility to permit less favorable movements. In contrast to MD, Monte Carlo moves are not bound to small motions that can happen during the time integration step. Thus in Monte Carlo, it is often easier to overcome energy barriers, which prevents the system to be kinetically trapped. On the other hand in standard Metropolis Monte Carlo, dynamic properties of the system are not accessible, since the time dimension is not considered in Monte Carlo. If only sampling is considered, the same level of convergence is achieved faster by Monte Carlo simulations than with MD, provided efficient Monte Carlo moves are chosen (Jorgensen & Tirado-Rives, 1996). Monte Carlo sampling can also predict thermodynamic properties such as preferred protonation states or redox states under specific environmental conditions such as pH, solution redox potential, or membrane potential (Bombarda, Becker, & Ullmann, 2006; Calimet & Ullmann, 2004; Ullmann, 2000; Ullmann & Ullmann, 2012). These are equally important properties to define a state, which can be used for further kinetic analysis of biological systems.

3.2 Kinetic Properties of Biochemical Systems

Once thermodynamics of the stable conformations and protonation or oxidation states of the enzyme are identified, the interest of a computational enzymologist turns to kinetic properties. Enzyme kinetics aims at tracking the rates of chemical or physical processes in the enzyme. In computational enzymology, discrete reaction steps are first addressed within a structural model before the rate of overall reactions can be related to experimental measurements.

Reaction rate is characterized by a temperature-dependent rate constant, which can be viewed as a probability factor for overcoming a free energy

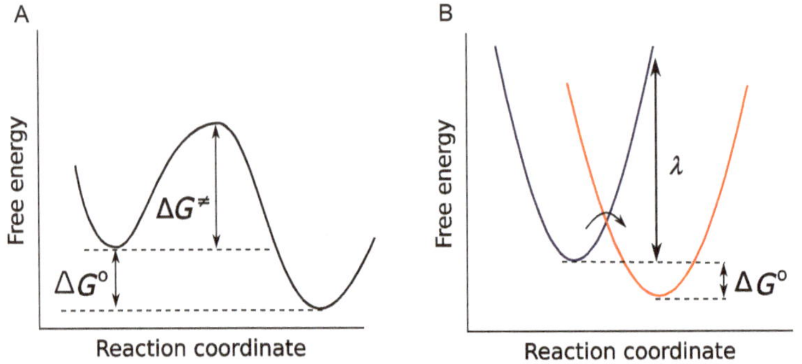

Fig. 3 (A) Reaction coordinate for interconversion between two stable states is thermodynamically characterized by reaction free energy ΔG° and kinetically by the free activation energy $\Delta G^{\neq}$. (B) Marcus model of two parabolas representing the two stable electronic configurations of the system. Reaction free energy ΔG° and the reorganization energy λ used for calculation of the free activation energy $\Delta G^{\neq}$ of the electron transfer are highlighted.

barrier between two stable states of the system (Fig. 3A). The reaction barrier can be related to the reaction rate by the Arrhenius law:

$$k(T) = A \ \ exp\left(-\beta \Delta G^{\neq}\right) \tag{9}$$

where $\Delta G^{\neq}$ is the free activation energy, β is $^{1}/_{k_B T}$ (k_B is the Boltzmann constant, T is the absolute temperature), and A is a preexponential factor. Eq. (9) is further adopted to theoretically describe rates of different processes involved in the enzyme function.

The first category are long-range electron transfer reactions, which can be regarded as nonadiabatic processes. A Marcus model of two harmonic potentials representing an initial and a final electronic configuration (Fig. 3B) can be used to describe the reaction coordinate. The free activation energy can be calculated from reaction free energy ΔG° and reorganization energy λ, which is the energy needed to perform the structural changes within the system:

$$\Delta G^{\neq} = \frac{(\Delta G^{\circ} + \lambda)^2}{4\lambda} \tag{10}$$

The reaction rate constant (9) can then be adopted for electron transfers using Marcus theory for the activation energy and Fermi's golden rule (Marcus & Sutin, 1985) for the preexponential factor:

$$k_{ET}(T) = \frac{2\pi}{\hbar} H_{DA}^2 \frac{1}{\sqrt{4\pi\beta^{-1}\lambda}} \exp\left(-\beta \frac{(\Delta G^{\circ} + \lambda)^2}{4\lambda}\right) \tag{11}$$

where $\hbar$ is the reduced Planck constant and H_{DA}^2 is the electronic coupling between the reactant state and the product state, which, to a good approximation, decreases exponentially with the distance between donor and acceptor (Gamow, 1928; Gray & Winkler, 2005). Marcus theory for instance in the form of the Moser–Dutton ruler (Moser, Keske, Warncke, Farid, & Dutton, 1992; Page, Moser, Chen, & Dutton, 1999) can be combined with a microstate model used for calculating microscopic redox potentials (Ullmann, 2000), and an electrostatic model for calculating reorganization energies (Sharp, 1998) in order to calculate charge transfer energies in complex systems. This approach was successfully applied within continuum-electrostatics models to describe electron transfer reactions in complex systems such as photosynthetic center (Becker, Ullmann, & Ullmann, 2007; Bombarda & Ullmann, 2011) which was recently reviewed (Ullmann, Mueller, & Bombarda, 2016).

The second category, which will be discussed more extensively here, is chemical reactions occurring in the enzyme active site that can be described by transition state theory (TST). TST assumes that the free activation energy $\Delta G^{\neq}$ can be derived from quasi-equilibrium between the reactant and transition states. Thus, intermediate structures along the reaction coordinate (Fig. 3A) including the transition state structure are explicitly analyzed. The reaction rate constant (Eq. 9) can be expressed by Eyring–Polanyi equation:

$$k(T) = \zeta\beta^{-1}h^{-1}exp\left(-\beta\Delta G^{\neq}\right) \tag{12}$$

where h is the Planck constant and ζ in the transmission coefficient which corresponds to the probability of being reactive once the transition state is reached. Even though the transmission coefficient has to be considered in general (i.e., for quantum tunneling effects), the lowering of $\Delta G^{\neq}$ is a dominant effect of enzymes in catalysis (Gao et al., 2006) and majority of computational enzymology studies focus on it. To get the rate-limiting energy barrier of a chemical process, one has to investigate the reaction path between substrate and product state. On the basis of the models described in the previous section, potential energy of any conformation in a molecular structure (here enzyme–substrate complex) can be calculated. When searching for the reaction path, the 3N-dimensional energy function (with N the number of atoms of the system) is portrayed as a potential energy surface (PES). A PES can be viewed as a landscape with valleys and mountain passes, which correspond to stable states and easiest transition pathways between them. A reaction coordinate of an elementary step follows the path

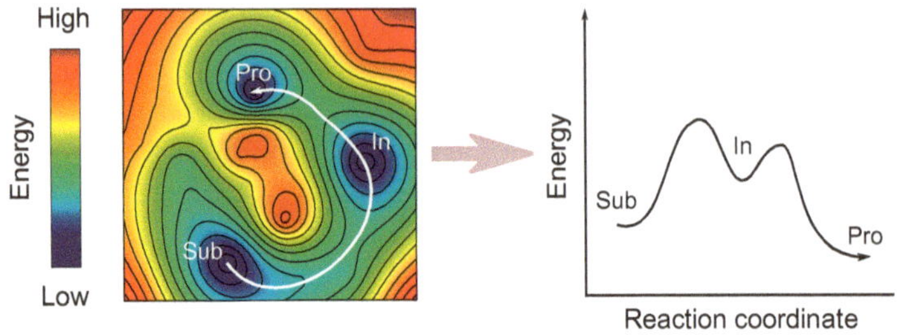

Fig. 4 Reaction path can be imagined as a trail through an energy landscape where basins represent the stable states and the passes represent the transition states. The path from substrate (*Sub*) to product (*Pro*) usually involves several intermediate (*In*) states. The energy change plotted against the position on the path is referred to as reaction profile.

from a minimum along the shallowest ascent toward the transition state and from there along the deepest descent toward the product state. Most of the chemical reactions are composed of several elementary steps forming together the minimum energy path (MEP) from reactant to product (Fig. 4).

Many methods have been developed to identify reaction paths and to distinguish among mechanism variants based on corresponding energy profiles. In general, one has to first identify the mechanistic options and then get corresponding accurate energy profiles. At the beginning of a computational kinetic study, optimization techniques are usually used to find minima and saddle points on the PES, which together define MEPs. The obtained energy profile corresponds to the enthalpic part of the free energy (Eq. 6). If the enthalpic differences among mechanism variants are big, one can already make qualitative conclusions about the preferred mechanism. The entropic part of Eq. (6) has to be considered to obtain proper reaction free energy profile at finite temperature in general, although the effect on the reaction barrier height is not always that profound (Kazemi, Himo, & Åqvist, 2016). The entropy-corrected PES is referred to as free energy surface at certain temperature. To get a quantitative free energy surface, one needs to perform extensive sampling along the reaction path, which can be extremely costly or even prohibitive in complex structural models. Therefore, the sampling is usually performed only when the most probable mechanism is identified on PES.

In the next part, we first discuss methods to identify a reaction path on PES. We then turn to the sampling methods that aim to get full free energy reaction profiles. In the end, we return to the preexponential transmission coefficient of Eq. (12) and discuss the role of nuclear quantum effects.

3.3 Mechanic and Energetic Properties of Molecular Reaction Paths

Computational investigation of an enzyme mechanism begins with a structural mechanistic model of the enzyme. As described earlier, stable experimentally determined structures corresponding to minima on the PES are commonly used as starting states. Approaches for searching reaction paths can be categorized as single ended or double ended. Double-ended approaches require the knowledge of two states, usually the initial state and the final state, to search for mechanistic possibilities in between. In most cases, however, one first explores the PES with a single-ended method and then loads the first path estimate into a double-ended method for further refinement.

3.3.1 Methods to Explore and Investigate Mechanistic Possibilities

3.3.1.1 Single-Ended Reaction Path Search Methods

The simplest single-ended approach is the adiabatic mapping of the PES. Based on chemical intuition and experimental data, a set of movements that lead toward a desired final state is applied. A function characterizing a desired movement is called collective variable (CV). For instance, in a cysteine protease, the catalytic cysteine has to be deprotonated by a neighboring histidine before it can attack the peptide bond of the substrate. Therefore, a first collective variable to be investigated is the distance between the thiol hydrogen of the cysteine and the closest nitrogen of the histidine. During adiabatic mapping along such a CV, the distance is shrunk in discrete steps. The enzyme structure is optimized at each step, while the CV is biased at given position by a harmonic potential. The mapping finishes by reaching the desired product state and the energies of the discrete scanning steps represent a rough potential energy reaction profile. The energy maxima and minima along this reaction profile should be further optimized without bias. Minima can be optimized by standard gradient-based energy minimization routines (Leach, 2001). Maxima can be refined to first-order saddle points using the eigenvector following method (Cerjan, 1981) that requires calculation of the Hessian matrix of second derivatives. The largest negative eigenvector of the Hessian points in direction of the reaction path and thus can be followed uphill to reach the saddle point. Since the saddle point is hard to reach in one step, the Hessian has to be updated. Although various approximate techniques exist (Munro & Wales, 1999; Wales & Walsh, 1998), the usage of purely single-ended eigenvector following for big systems with many degrees of freedom may be both unreliable and computationally unfeasible.

Alternatively, the single-ended growing string method (Zimmerman & Bowman, 2016) can be used to locate the nearest saddle point and next stable intermediate state. This method, like double-ended methods discussed further below, generates a chain of structures to sample the reaction coordinate. Unlike its double-ended sister, which uses both initial and final state structures for the chain generation, the single-ended growing string method uses a set of collective variables to indicate the initial direction from the initial state. Once, a saddle point estimate is located, the eigenvector following method is used to refine it. Interestingly, the single-ended growing string method can suggest the CVs automatically and thus explore various reaction possibilities from a given initial state.

In contrast to the above described PES mapping approaches, the reaction mechanism can be also explored at finite temperature without knowledge of the final state structure. In principle, sufficiently long MD simulations of the initial state at physiological temperature should lead to a reaction. Current computer power is, however, not sufficient to perform such a long simulation, especially for models with quantum essence that are required in most cases. Fortunately, MD simulations can be driven along predefined CVs in a similar way as in adiabatic PES mapping. One popular technique is called umbrella sampling (US) (Torrie & Valleau, 1977). A harmonic (umbrella) bias potential is applied in order to restrain the MD simulation around a certain value of the CV. The value of the CV can be changed in discrete steps, while a sufficiently long biased MD simulation is performed for each value till the desired final state is reached. Another elegant method how to sample reaction possibilities at finite temperature in accessible time is called metadynamics (Laio & Parrinello, 2002). In this method, Gaussian bias potential "hills" are added to the PES in certain time intervals along the values of a CV that has been already visited in the MD trajectory. In this fashion, the MD is discouraged to revisit already sampled CV values, and thus the simulation proceeds toward the desired final state. The bias introduced by metadynamics brings bigger flexibility compared to the US bias on one hand, but potentially slower convergence on the other.

Although the single-ended finite temperature methods provide reasonable conformational freedom and entropic aspects in one simulation, their use with higher QM models can be computationally unfeasible. Furthermore, while the choice of collective variables in stiff adiabatic PES mapping is usually trivial, in more flexible bias MD methods it can be rather elusive. Therefore, double-ended methods reviewed in the next paragraphs are often preferred. The entropic aspects can also be accessed a posteriori once the PES path is established as we shall see below.

3.3.1.2 Double-Ended Reaction Path Search Methods

Once the final state structure is discovered by a single-ended method or when it is known from an experimental structure or guessed by chemical intuition, so-called double-ended methods can be employed to construct a path between them. Alternatively, an estimated path derived, e.g., from adiabatic PES mapping described above can be loaded in and refined. The major double-ended method category that will be described here comprises of chain-of-states (COS) methods. A COS method discretizes the reaction path between the initial and final state into a set of intermediate structures and optimizes them in a connected manner. Once this procedure achieves convergence, the resulting path should represent the minimal energy pathway (MEP) between the initial and final state.

One popular COS approach is nudged elastic band (NEB) (Jonsson, Mills, & Jacobsen, 1998). NEB approximates the reaction path by a set of structures that are connected by springs into a chain. The parallel component of the spring force keeps the images distributed along the whole path, while the perpendicular component helps to push images to the MEP valley. Because NEB does not directly seek for the saddle point structures, a climbing image variant of NEB (CI-NEB) (Henkelman, Uberuaga, & Jónsson, 2000) has been developed. The converged NEB path is usually loaded into CI-NEB. The parallel component of the NEB spring potential acting on the highest image of the path is inversed to drive it uphill, while the perpendicular component is kept unchanged in order to keep the image on the path. Once the path is converged again, the highest image should represent the saddle point structure.

The string method (SM) (E, Ren, & Vanden-Eijnden, 2002) uses a concept on equal image distribution to cover the path instead of introducing spring constants between the images. The images are equally redistributed along a spline fit of the reaction path every optimization cycle. The path is converged once the redistribution does not significantly change the image position. An SM variant called growing string method (Peters, Heyden, Bell, & Chakraborty, 2004) construct the chain of states gradually from the initial state toward the final state and thus reduces the bias potentially introduced by initial path guess (e.g., linear interpolation). Another SM variant optimizes the COS at finite temperature, allowing for a better conformational relaxation of the MEP (Vanden-Eijnden & Venturoli, 2009;

E, Ren, & Vanden-Eijnden, 2005). Note that the single-ended string method described above is a variant of the growing string method with even less initial bias.

The above described COS methods do not a priori search for the transition state structures. Although CI-NEB makes a step in this direction, it is not actually guaranteed that it will reach a first-order saddle point. One can, of course, employ the above described eigenvector following approaches to refine the maxima of a COS path to saddle points. However, these methods are both computationally costly and also unreliable once the input structure is not close to the actual saddle point. An elegant solution is the conjugate peak refinement (CPR) (Fischer & Karplus, 1992; Gisdon, Culka, & Ullmann, 2016) method that gradually constructs chain of states between the initial and final states while it aims at locating the first-order saddle points. In contrast to NEB or SM, the number of states along the path is not fixed and thus the sampling in the saddle point region can be increased to facilitate its proper location. The CPR method is based on the fact that in the vicinity of a saddle point, there is one direction, which points to an energetic maximum, while all others lead to a minimum. The CPR algorithm picks the highest energy structure along the discretized path and performs a line maximization along the corresponding tangential path vector. This corrected maximum is then minimized in conjugate space similar to the conjugated gradient minimization method. By staying conjugate to the original path, falling to the neighboring minimum is prevented. The optimized structure is added into the chain of states. Gradually the saddle point is approached by providing more sampling in the transition region and occasionally the conjugate optimization procedure can converge to locate the first-order saddle point. A successful run ends, when all maxima along the reaction path between the initial and final state are identified and optimized to first-order saddle points.

3.3.1.3 Path-Based Reaction Path Search Methods

Once a reaction path is found, it is sometimes necessary to correct it for conformational flexibility, since a biological system has many possible degrees of freedom. Path-based methods include the conformational flexibility to evaluate alternative reaction paths or a set of possible reactive coordinates. These methods usually require an initial path, which does not necessarily have to be properly refined. Any path generated by double-ended or single-ended methods can be used. The concept of metadynamics, introduced as a single-ended search method, can also be applied for path-based approaches. Instead of using a trivial CV, such as bond distances or dihedral angles, a function that describes the whole reaction path is used (Branduardi, Gervasio, & Parrinello, 2007; Bernardi, Melo, & Schulten, 2015). The application of the path CV allows to dynamically sample the conformational space based on the input path estimate. This provides information on the conformational flexibility of the input intermediates and transition states, which is often crucial for reaction paths obtained from static approaches, such as adiabatic mapping, NEB, or CPR. If the distance from the initial path is taken as a second CV to be biased, even alternative paths can be identified and energetically evaluated.

A conceptually different approach is followed in transition path sampling (TPS) (Bolhuis, Dellago, & Chandler, 1998; Dellago, Bolhuis, Csajka, & Chandler, 1998). The aim of TPS is to connect two stable states by a collection of all likely transition pathways, which represent the transition path ensemble. As mentioned above, an input trajectory does not necessarily have to be properly refined, thus it may have a low weight in the transition path ensemble. Therefore, for a TPS simulation the initial trajectory has to be evaluated and equilibrated toward a representative transition path trajectory. Unlike US ensembles or metadynamics ensembles, a TPS simulation creates unbiased trajectories, since it does not enhance the sampling of rare events by bias potentials (Swenson & Bolhuis, 2014). TPS, however, is also dependent on a collective variable, yet it is not used to drive the simulation, but rather to discriminate the stable states of the system, and to monitor the progress along the trajectory. This makes the CV a crucial quantity in a TPS simulation. The enhancement of sampling rare events in TPS is achieved by importance sampling of trajectory space. All dynamic movements are preformed by a Monte Carlo approach, while a new trajectory is created from an existing one in the ensemble. For that, specific procedures are available to drive movements in trajectory space (Bolhuis, Chandler, Dellago, & Geissler, 2002; Rowley & Woo, 2009). TPS has been successfully applied to simulate biological systems (Dellago & Bolhuis, 2007) and is also recently used to study enzyme catalysis, such as the hydride transfer in a dihydrofolate reductase (Wang, Antoniou, Schwartz, & Schramm, 2016).

3.3.2 Methods to Obtain Free Energy

So far, we have been concentrating on the qualitative aspects of the reaction mechanism. Now we turn to the methods to get the free energy profile estimates. As noted earlier (Eq. 6), the free energy is composed of the potential energy (enthalpic) part that is calculated directly by the model, and the entropic part, for which extensive sampling at finite temperature along the reaction path is needed. In case of US and metadynamics the sampling is already included in the path search procedure, so entropic influence is explicitly included and thus free energy profile can be directly estimated. For US, weighted histogram analysis method (WHAM) (Kumar, Rosenberg, Bouzida, Swendsen, & Kollman, 1992) can be used to remove the umbrella bias and to integrate the simulation windows into a free energy profile. In case of metadynamics, the sum of added Gaussian hills plotted against the collective variable directly represents the free energy profile. TPS collected pathways are dynamic trajectories, thus kinetic information, such as rate constants, can be extracted (Dellago, 2007). But since only trajectories are considered that connect certain regions in configuration space, the configurations are not distributed according to the equilibrium distribution of the system. To determine free energy profiles, one has to obtain equilibrium-distributed configurations. One possibility is to apply biasing procedures, such as US variants, to rarely visited states, and divide the CV into overlapping windows (Dellago, 2007).

In case of static methods such as adiabatic PES mapping or chain-of-states methods, the entropic aspect has to be added by sampling method (MD or Monte Carlo) to account for entropic part of the free energy profile. One option is to use these static paths as an input to above described path-based approaches. A possible obstacle can be that achieving extensive sampling when high-level QM methods are used to treat the active site may be computationally prohibitive. In the same time, usage of, e.g., SE QM models can introduce additional bias into the results. A way out of this dilemma offers the free energy perturbation (FEP) methods. For instance, one can find the reaction path by a chain-of-states method within QM/MM model (Kästner, Senn, Thiel, Otte, & Thiel, 2006). Subsequently, an MD simulation is performed for every state of the PES reaction path with QM region kept frozen. Perturbation energy is calculated as energy for moving one step forward in the PES QM reaction path while staying in the same MM conformational ensemble:

$$\Delta E_{pert} = E_{\mathrm{QM/MM}}\left(\mathbf{r}^{i+1}_{\mathrm{QM}}, \mathbf{r}^{i}_{\mathrm{MM}}\right) - E_{\mathrm{QM/MM}}\left(\mathbf{r}^{i}_{\mathrm{QM}}, \mathbf{r}^{i}_{\mathrm{MM}}\right) \tag{13}$$

where i and $i + 1$ are the indices of adjacent path steps and $\mathbf{r}$ are the coordinate vectors. Free energy for the $i \rightarrow i+1$ step is then calculated by averaging over the MM ensemble at step i using Zwanzig equation (Zwanzig, 1954):

$$\Delta G^{i \to i+1} = -\frac{1}{\beta} \ln \langle \exp(-\beta \Delta E_{pert}) \rangle_i \tag{14}$$

The reaction free energy profile is calculated by adding QM energy corrected for zero-temperature vibrations to the perturbation-derived free energy. Another example where FEP concept is exploited are the FEP/US studies in EVB model (Warshel, 1991).

3.3.3 Methods to Simulate Nuclear Quantum Effects

The majority of methods commonly used to locate and sample the enzyme reaction rely on atomistic models that treat the nuclei as classical objects. Even in the common QM methods, just the electrons are treated as quantum objects. This approximation is necessary to make the calculations on bigger systems computationally feasible and in many cases also justified, since the heavy nuclei hardly show any quantum behavior. The quantum nature becomes in practice significant in case of proton transfer processes, which are often part of enzyme-catalyzed reaction coordinates. If a proton transfer is the rate-limiting step, the nuclear quantum-mechanical effects (NQM) play important role in the overall reaction rate. NQM effect are most pronounced when comparing protium and deuterium variants and calculating kinetic isotope effects.

Although small number of protons can be treated as quantum particles using nuclear-electronic orbital (Webb, Iordanov, & Hammes-Schiffer, 2002) methods, most of the approaches that deal with NQM in enzymes can be classified as correction methods that are applied, e.g., on US path ensembles. One direction is the ensemble-averaged variational transition state theory with multidimensional tunneling (EA-VTST/MT) (Garcia-Viloca, Alhambra, Truhlar, & Gao, 2001; Truhlar et al., 2002) that corrects the free energy barrier $\Delta G^{\neq}$ for quantum-mechanical vibrations. EA-VTST/MT in addition also calculates the transmission coefficient ζ (see Eq. 12) that corrects mainly for quantum-mechanical tunneling through the free energy barrier. Another family of approaches to NQM is based on Feynman path integrals. Quantum nature of the nuclei is approximated by transforming the classical spheres into rings of quasiparticles connected by springs. Path integral MD simulation techniques include quantized classical path (QCP) (Hwang & Warshel, 1993), centroid molecular dynamics (CMD) (Cao & Voth, 1994a, 1994b), or ring-polymer molecular dynamics (RPMD) (Braams & Manolopoulos, 2006; Craig & Manolopoulos, 2004).

3.4 Path Search Strategy

Taken together, enormous amount of methods to locate reaction path and calculate its energy profile has been developed. It is not easy to objectively find the one and only correct strategy, although different groups certainly have their preferred approaches. The choice of the method is also influenced by the primary question that the researcher is asking. If the task is to find the correct catalytic mechanism, many potential variants of the reaction mechanism need to be tested in a reasonable time. If the individual steps of the mechanism are already known, the task might be to get proper rate constant to relate the model to experimental parameters. In many cases, big energy barriers on PES can rule the unfeasible mechanism variants out and find the most promising set of reaction steps in fraction of time in more accurate QM models compared to direct usage of sampling approaches in approximate models. The PES path can be further optimized and corrected by much more demanding sampling methods, or a reaction-specific SE potential (e.g., EVB) can be constructed based on the previous PES investigation in high-level QM models.

The kinetic and thermodynamic parameters determined by the afore-described methods should be combined in order to get a more reliable picture. The different methods should not be viewed as competitive approaches but rather as complementary and one should search for synergy among different methods with limitations of the models in mind. The limitations and synergy should also be considered when comparing computational results with experimental data.

4. GOING BEYOND THE EXPLORATION OF THE REACTION PATHS

In order to understand the biological systems, it is not enough to understand the mechanism of an enzyme. It is required to analyze the enzyme in its physiological context and how the rate of catalysis is influenced by parameters such as pH or metabolite concentration. Today's systems biology is using kinetic models with stretched exponential or noninteger stoichiometry in order to describe complex metabolic networks. Even if such models are of certain practical use to solve some research problems, they are not satisfactory from a theoretical point of view, since mass and energy conservation is not guaranteed. Other models are taking kinetic parameters from databases. However, such parameters were usually determined under specific circumstances and cannot account for all possible effects. Considering the constantly increasing number of complete genomes and partially reconstructed metabolisms, it comes more and more important to get a more realistic view of the metabolic reaction in its context. The challenge for computational biochemistry today and in the future is to derive enzymatic parameters from structure models by using methods that we reviewed in this article. However, it will be required to go beyond such information. The kinetic parameters can be combined in master equation approaches (Becker et al., 2007; Bombarda & Ullmann, 2011) or kinetic Monte Carlo simulations (Till, Becker, Essigke, & Ullmann, 2008) in order to simulate complete catalytic cycles that are influenced by environmental parameters such as pH or membrane potential. To carry the approach further, it might be possible to model the whole cellular context in reaction–diffusion equations, which may allow in the future to model complex biochemical reactions. Combining structural biology and systems biology may thus be a promising direction, especially considering the pace in which both fields make progress in recent years. With the help of computer models that rely on a solid experimental basis, we may more and more understand how the jigglings and wigglings of atoms leads to the complex phenomenon we call *life*.

REFERENCES

Amara, P., Field, M. J., Alhambra, C., & Gao, J. (2000). The generalized hybrid orbital method for combined quantum mechanical/molecular mechanical calculations: Formulation and tests of the analytical derivatives. *Theoretical Chemistry Accounts*, *104*, 336–343.

Åqvist, J., & Warshel, A. (1993). Simulation of enzyme reactions using valence bond force fields and other hybrid quantum/classical approaches. *Chemical Reviews*, *93*, 2523–2544.

Baker, N. A., Sept, D., Joseph, S., Holst, M. J., & McCammon, J. A. (2001). Electrostatics of nanosystems: Application to microtubules and the ribosome. *Proceedings of the National Academy of Sciences*, *98*, 10037–10041.

Bakowies, D., & Thiel, W. (1991). MNDO study of large carbon clusters. *Journal of the American Chemical Society*, *113*, 3704–3714.

Bartesaghi, A., Merk, A., Banerjee, S., Matthies, D., Wu, X., Milne, J. L. S., & Subramaniam, S. (2015). 2.2 Å resolution cryo-EM structure of Î^{2}-galactosidase in complex with a cell-permeant inhibitor. *Science*, *348*, 1147–1151.

Becke, A. (1993). Density functional thermochemistry. III. The role of exact exchange. *The Journal of Chemical Physics*, *98*, 5648–5652.

Becker, T., Ullmann, R. T., & Ullmann, G. M. (2007). Simulation of the electron transfer between the tetraheme subunit and the special pair of the photosynthetic reaction center using a microstate description. *The Journal of Physical Chemistry. B*, *111*, 2957–2968.

Bernardi, R. C., Melo, M. C. R., & Schulten, K. (2015). Enhanced sampling techniques in molecular dynamics simulations of biological systems. *Biochimica et Biophysica Acta*, *1850*, 872–877.

Bolhuis, P. G., Chandler, D., Dellago, C., & Geissler, P. L. (2002). Transition path sampling: Throwing ropes over rough mountain passes, in the dark. *Annual Review of Physical Chemistry*, *53*, 291–318.

Bolhuis, P. G., Dellago, C., & Chandler, D. (1998). Sampling ensembles of deterministic transition pathways. *Faraday Discussions*, *110*, 421–436.

Bombarda, E., Becker, T., & Ullmann, G. M. (2006). The influence of the membrane potential on the protonation of bacteriorhodopsin: Insights from electrostatic calculations into the regulation of proton pumping. *Journal of the American Chemical Society*, *128*, 12129–12139.

Bombarda, E., & Ullmann, G. M. (2011). Continuum electrostatic investigations of charge transfer processes in biological molecules using a microstate description. *Faraday Discussions*, *148*, 173–193.

Braams, B. J., & Manolopoulos, D. E. (2006). On the short-time limit of ring polymer molecular dynamics. *The Journal of Chemical Physics*, *125*, 124105.

Branduardi, D., Gervasio, F. L., & Parrinello, M. (2007). From A to B in free energy space. *The Journal of Chemical Physics*, *126*, 054103. http://dx.doi.org/10.1063/1.2432340.

Bryantsev, V. S., Diallo, M. S., Van Duin, A. C. T., & Goddard, W. A. (2009). Evaluation of B3LYP, X3LYP, and M06-class density functionals for predicting the binding energies of neutral, protonated, and deprotonated water clusters. *Journal of Chemical Theory and Computation*, *5*, 1016–1026.

Calimet, N., & Ullmann, G. M. (2004). The influence of a transmembrane pH gradient on protonation probabilities of bacteriorhodopsin: The structural basis of the back-pressure effect. *Journal of Molecular Biology*, *339*, 571–589.

Cao, J., & Voth, G. A. (1994a). The formulation of quantum statistical mechanics based on the Feynman path centroid density. I. Equilibrium properties. *The Journal of Chemical Physics*, *100*, 5093–5105.

Cao, J., & Voth, G. A. (1994b). The formulation of quantum statistical mechanics based on the Feynman path centroid density. II. Dynamical properties. *The Journal of Chemical Physics*, *100*, 5106–5117.

Cerjan, C. J. (1981). On finding transition states. *The Journal of Chemical Physics*, *75*, 2800.

Chen, J. L., Noodleman, L., Case, D., & Bashford, D. (1994). Incorporating solvation effects into density functional electronic structure calculations. *The Journal of Physical Chemistry*, *98*, 11059–11068.

Čížek, J. (1966). On the correlation problem in atomic and molecular systems. Calculation of wavefunction components in Ursell-type expansion using quantum-field theoretical methods. *The Journal of Chemical Physics*, *45*, 4256–4266.

Cole, D. J., & Hine, N. D. M. (2016). Applications of large-scale density functional theory in biology. *Journal of Physics. Condensed Matter*, *28*, 393001.

Cornell, W. D., Cieplak, P., Bayly, C. I., Gould, I. R., Merz, K. M., Ferguson, D. M., … Kollman, P. A. (1996). A second generation force field for the simulation of proteins, nucleic acids, and organic molecules. *Journal of the American Chemical Society*, *118*, 2309–2309.

Costa, M. G. S., Batista, P. R., Bisch, P. M., & Perahia, D. (2015). Exploring free energy landscapes of large conformational changes: Molecular dynamics with excited normal modes. *Journal of Chemical Theory and Computation*, *11*, 2755–2767.

Craig, I. R., & Manolopoulos, D. E. (2004). Quantum statistics and classical mechanics: Real time correlation functions from ring polymer molecular dynamics. *The Journal of Chemical Physics*, *121*, 3368–3373.

Cramer, C. J. (2004). *Essentials of computational chemistry: Theories and models* (2nd ed.). Chichester: Wiley.

Dellago, C. (2007). Transition path sampling and the calculation of free energies. In C. Chipot & A. Pohorille (Eds.), *Free energy calculations* (pp. 249–276). Berlin, Heidelberg: Springer.

Dellago, C., & Bolhuis, P. G. (2007). Transition path sampling simulations of biological systems. In M. Reiher (Ed.), *Atomistic approaches in modern biology* (Vol. 268, pp. 291–317). Berlin, Heidelberg: Springer.
Dellago, C., Bolhuis, P. G., Csajka, F. S., & Chandler, D. (1998). Transition path sampling and the calculation of rate constants. *The Journal of Chemical Physics*, *108*, 1964.
E, W., Ren, W., & Vanden-Eijnden, E. (2002). String method for the study of rare events. *Physical Review B*, *66*, 052301.
E, W., Ren, W., & Vanden-Eijnden, E. (2005). Finite temperature string method for the study of rare events. *The Journal of Physical Chemistry B*, *109*, 6688–6693.
Elstner, M. (2006). The SCC-DFTB method and its application to biological systems. *Theoretical Chemistry Accounts*, *116*, 316–325.
Feliks, M., & Ullmann, G. M. (2012). Glycerol dehydratation by the B12-independent enzyme may not involve the migration of a hydroxyl group: A computational study. *The Journal of Physical Chemistry. B*, *116*, 7076–7087.
Feynman, R. (1964). *The Feynman lectures on physics*. New York, USA: Basic Books. Retrieved from http://www.feynmanlectures.caltech.edu/.
Field, M. J. (2007). *A practical introduction to the simulation of molecular systems* (2nd ed.). Cambridge, UK: Cambridge University Press.
Field, M. J., Bash, P. A., & Karplus, M. (1990). A combined quantum mechanical and molecular mechanical potential for molecular dynamics simulations. *Journal of Computational Chemistry*, *11*, 700–733.
Fischer, S., & Karplus, M. (1992). Conjugate peak refinement: An algorithm for finding reaction paths and accurate transition states in systems with many degrees of freedom. *Chemical Physics Letters*, *194*, 252–261.
Gamow, G. (1928). Zur quantentheorie des atomkernes. *Zeitschrift fr Physiotherapie*, *51*, 204–212.
Gao, J., Ma, S., Major, D. T., Nam, K., Pu, J., & Truhlar, D. G. (2006). Mechanisms and free energies of enzymatic reactions. *Chemical Reviews*, *106*, 3188–3209.
Garcia-Viloca, M., Alhambra, C., Truhlar, D. G., & Gao, J. (2001). Inclusion of quantum-mechanical vibrational energy in reactive potentials of mean force. *The Journal of Chemical Physics*, *114*, 9953–9958.
Georgieva, P., & Himo, F. (2010). Quantum chemical modeling of enzymatic reactions: The case of histone lysine methyltransferase. *Journal of Computational Chemistry*, *31*, 1707–1714.
Gilson, M. K., & Zhou, H.-X. (2007). Calculation of protein-ligand binding affinities. *Annual Review of Biophysics and Biomolecular Structure*, *36*, 21–42.
Gisdon, F. J., Culka, M., & Ullmann, G. M. (2016). PyCPR—A python-based implementation of the Conjugate Peak Refinement (CPR) algorithm for finding transition state structures. *Journal of Molecular Modeling*, *22*, 242.
Golkaram, M., Shin, Y. K., & van Duin, A. C. T. (2014). Reactive molecular dynamics study of the pH-dependent dynamic structure of α-helix. *The Journal of Physical Chemistry. B*, *118*, 13498–13504.
Gray, H. B., & Winkler, J. R. (2005). Long-range electron transfer. *Proceedings of the National Academy of Sciences*, *102*, 3534–3539.
Head-Gordon, M., Pople, J. A., & Frisch, M. J. (1988). MP2 energy evaluation by direct methods. *Chemical Physics Letters*, *153*, 503–506.
Henkelman, G., Uberuaga, B. P., & Jónsson, H. (2000). A climbing image nudged elastic band method for finding saddle points and minimum energy paths. *The Journal of Chemical Physics*, *113*, 9901.
Heyden, A., & Truhlar, D. G. (2008). Conservative algorithm for an adaptive change of resolution in mixed atomistic/coarse-grained multiscale simulations. *Journal of Chemical Theory and Computation*, *4*, 217–221.

Honig, B., & Nicholls, A. (1995). Classical electrostatics in biology and chemistry. *Science*, *268*, 1144–1149.

Hwang, J. K., & Warshel, A. (1993). A quantized classical path approach for calculations of quantum mechanical rate constants. *The Journal of Physical Chemistry*, *97*, 10053–10058.

Ingólfsson, H. I., Lopez, C. A., Uusitalo, J. J., de Jong, D. H., Gopal, S. M., Periole, X., & Marrink, S. J. (2014). The power of coarse graining in biomolecular simulations. *Wiley Interdisciplinary Reviews: Computational Molecular Science*, *4*, 225–248.

Jonsson, H., Mills, G., & Jacobsen, K. W. (1998). Nudged elastic band method for finding minimum energy paths of transitions. In B. J. Berne, G. Ciccotti, & D. F. Coker (Eds.), *Classical and quantum dynamics in condensed phase simulations* (pp. 385–404). Singapore: World Scientific.

Jorgensen, W. L., Maxwell, D. S., & Tirado-Rives, J. (1996). Development and testing of the OPLS all-atom force field on conformational energetics and properties of organic liquids. *Journal of the American Chemical Society*, *118*, 11225–11236.

Jorgensen, W. L., & Tirado-Rives, J. (1996). Monte Carlo vs molecular dynamics for conformational sampling. *The Journal of Physical Chemistry*, *100*, 14508–14513.

Kamerlin, S. C. L., & Warshel, A. (2010). The EVB as a quantitative tool for formulating simulations and analyzing biological and chemical reactions. *Faraday Discussions*, *145*, 71–106.

Kamerlin, S. C. L., & Warshel, A. (2011). The empirical valence bond model: Theory and applications. *Wiley Interdisciplinary Reviews: Computational Molecular Science*, *1*, 30–45.

Kar, P., & Feig, M. (2014). Recent advances in transferable coarse-grained modeling of proteins. *Advances in Protein Chemistry and Structural Biology*, *96*, 143–180.

Kästner, J., Senn, H. M., Thiel, S., Otte, N., & Thiel, W. (2006). QM/MM free-energy perturbation compared to thermodynamic integration and umbrella sampling: Application to an enzymatic reaction. *Journal of Chemical Theory and Computation*, *2*, 452–461.

Kazemi, M., Himo, F., & Åqvist, J. (2016). Enzyme catalysis by entropy without Circe effect. *Proceedings of the National Academy of Sciences of the United States of America*, *113*, 2406–2411.

Klamt, A., & Schüürmann, G. (1993). COSMO: A new approach to dielectric screening in solvents with explicit expressions for the screening energy and its gradient. *Journal of the Chemical Society, Perkin Transactions 2* (799–805).

Kmiecik, S., Gront, D., Kolinski, M., Wieteska, L., Dawid, A. E., & Kolinski, A. (2016). Coarse-grained protein models and their applications. *Chemical Reviews*, *116*, 7898–7936.

Kühlbrandt, W. (2014). Cryo-EM enters a new era. *eLife*, *3*, e03678.

Kulik, H. J., Luehr, N., Ufimtsev, I. S., & Martínez, T. J. (2012). Ab initio quantum chemistry for protein structures. *The Journal of Physical Chemistry. B*, *116*, 12501–12509.

Kumar, S., Rosenberg, J. M., Bouzida, D., Swendsen, R. H., & Kollman, P. A. (1992). The weighted histogram analysis method for free-energy calculations on biomolecules. I. The method. *Journal of Computational Chemistry*, *13*, 1011–1021.

Laio, A., & Parrinello, M. (2002). Escaping free-energy minima. *Proceedings of the National Academy of Sciences of the United States of America*, *99*, 12562–12566.

Leach, A. R. (2001). *Molecular modelling: Principles and applications* (2nd ed.). New Jersey: Prentice Hall.

Levitt, M., & Warshel, A. (1975). Computer simulation of protein folding. *Nature*, *253*, 694–698.

Li, J., Nelson, M. R., Peng, C. Y., Bashford, D., & Noodleman, L. (1998). Incorporating protein environments in density functional theory: A self-consistent reaction field calculation of redox potentials of [2Fe2S] clusters in ferredoxin and phthalate dioxygenase reductase. *The Journal of Physical Chemistry. B*, *102*, 6311–6324.

Li, J., & Ryde, U. (2014). Comparison of the active-site design of molybdenum oxo-transfer enzymes by quantum mechanical calculations. *Inorganic Chemistry, 53*, 11913–11924.

Liang, T., Shin, Y. K., Cheng, Y.-T., Yilmaz, D. E., Vishnu, K. G., Verners, O., … van Duin, A. C. (2013). Reactive potentials for advanced atomistic simulations. *Annual Review of Materials Research, 43*, 109–129.

Liu, T., Han, W.-G., Himo, F., Ullmann, G. M., Bashford, D., Toutchkine, A., … Noodleman, L. (2004). Density functional vertical self-consistent reaction field theory for solvatochromism studies of solvent-sensitive dyes. *The Journal of Physical Chemistry. B, 108*, 11157–11169.

MacKerell, A. D., Bashford, D., Bellott, M., Dunbrack, R. L., Evanseck, J. D., Field, M. J., … Karplus, M. (1998). All-atom empirical potential for molecular modeling and dynamics studies of proteins. *The Journal of Physical Chemistry. B, 102*, 3586–3616.

Manta, B., Raushel, F. M., & Himo, F. (2014). Reaction mechanism of zinc-dependent cytosine deaminase from *Escherichia coli*: A quantum-chemical study. *The Journal of Physical Chemistry. B, 118*, 5644–5652.

Marcus, R., & Sutin, N. (1985). Electron transfers in chemistry and biology. *Biochimica et Biophysica Acta, Reviews on Bioenergetics, 811*, 265–322.

Maseras, F., & Morokuma, K. (1995). IMOMM: A new integrated ab initio + molecular mechanics geometry optimization scheme of equilibrium structures and transition states. *Journal of Computational Chemistry, 16*, 1170–1179.

Meier, K., Choutko, A., Dolenc, J., Eichenberger, A. P., Riniker, S., & Van Gunsteren, W. F. (2013). Multi-resolution simulation of biomolecular systems: A review of methodological issues. *Angewandte Chemie, International Edition, 52*, 2820–2834.

Møller, C., & Plesset, M. S. (1934). Note on an approximation treatment for many-electron systems. *Physics Review, 46*, 618–622.

Monti, S., Corozzi, A., Fristrup, P., Joshi, K. L., Shin, Y. K., Oelschlaeger, P., … Bourne, P. E. (2013). Exploring the conformational and reactive dynamics of biomolecules in solution using an extended version of the glycine reactive force field. *Physical Chemistry Chemical Physics, 15*, 15062.

Moser, C. C., Keske, J. M., Warncke, K., Farid, R. S., & Dutton, P. L. (1992). Nature of biological electron transfer. *Nature, 355*, 796–802.

Munro, L. J., & Wales, D. J. (1999). Defect migration in crystalline silicon. *Physical Review B, 59*, 3969–3980.

Nango, E., Royant, A., Kubo, M., Nakane, T., Wickstrand, C., Kimura, T., … Iwata, S. (2016). A three-dimensional movie of structural changes in bacteriorhodopsin. *Science, 354*, 1552–1557.

Noid, W. G. (2013). Perspective: Coarse-grained models for biomolecular systems. *The Journal of Chemical Physics, 139*, 090901.

Oostenbrink, C., Villa, A., Mark, A. E., & Van Gunsteren, W. F. (2004). A biomolecular force field based on the free enthalpy of hydration and solvation: The GROMOS force-field parameter sets 53A5 and 53A6. *Journal of Computational Chemistry, 25*, 1656–1676.

Page, C. C., Moser, C. C., Chen, X., & Dutton, P. L. (1999). Natural engineering principles of electron tunneling in biological oxidation-reduction. *Nature, 402*, 47–52.

Pande, K., Hutchison, C. D. M., Groenhof, G., Aquila, A., Robinson, J. S., Tenboer, J., … Schmidt, M. (2016). Femtosecond structural dynamics drives the trans/cis isomerization in photoactive yellow protein. *Science, 352*, 725–729.

Paquet, E., & Viktor, H. L. (2015). Molecular dynamics, Monte Carlo simulations, and Langevin dynamics: A computational review. *BioMed Research International, 2015*, 183918.

Peters, B., Heyden, A., Bell, A. T., & Chakraborty, A. (2004). A growing string method for determining transition states: Comparison to the nudged elastic band and string methods. *The Journal of Chemical Physics*, *120*, 7877–7886.

Riniker, S., Allison, J. R., & van Gunsteren, W. F. (2012). On developing coarse-grained models for biomolecular simulation: A review. *Physical Chemistry Chemical Physics*, *14*, 12423.

Rowley, C. N., & Woo, T. K. (2009). New shooting algorithms for transition path sampling: Centering moves and varied-perturbation sizes for improved sampling. *The Journal of Chemical Physics*, *131*, 234102.

Rücker, P., Wieninger, S. A., Ullmann, G. M., & Sticht, H. (2012). pH-dependent molecular dynamics of vesicular stomatitis virus glycoprotein G. *Proteins*, *80*, 2601–2613.

Saunders, M. G., & Voth, G. A. (2013). Coarse-graining methods for computational biology. *Annual Review of Biophysics*, *42*, 73–93.

Senftle, T. P., Hong, S., Islam, M. M., Kylasa, S. B., Zheng, Y., Shin, Y. K., … van Duin, A. C. T. (2016). The ReaxFF reactive force-field: Development, applications and future directions. *Computational Materials*, *2*, 15011.

Senn, H. M., & Thiel, W. (2007). QM/MM methods for biological systems. In M. Reiher (Ed.), *Atomistic approaches in modern biology* (Vol. 268, pp. 173–290). Berlin, Heidelberg: Springer.

Senn, H. M., & Thiel, W. (2009). QM/MM methods for biomolecular systems. *Angewandte Chemie, International Edition*, *48*, 1198–1229.

Sharp, K. E. (1998). Calculation of electron transfer reorganization energies using the Finite Difference Poisson-Boltzmann model. *Biophysical Journal*, *73*, 1241–1250.

Shen, L., & Hu, H. (2014). Resolution-adapted all-atomic and coarse-grained model for biomolecular simulations. *Journal of Chemical Theory and Computation*, *10*, 2528–2536.

Shurki, A., Derat, E., Barrozo, A., & Kamerlin, S. C. L. (2015). How valence bond theory can help you understand your (bio) chemical reaction. *Chemical Society Reviews*, *44*, 1037–1052.

Singh, U. C., & Kollman, P. A. (1986). A combined ab initio quantum mechanical and molecular mechanical method for carrying out simulations on complex molecular systems: Applications to the CH3Cl + Cl exchange reaction and gas phase protonation of polyethers. *Journal of Computational Chemistry*, *7*, 718–730.

Sokkar, P., Boulanger, E., Thiel, W., & Sanchez-Garcia, E. (2015). Hybrid quantum mechanics/molecular mechanics/coarse grained modeling: A triple-resolution approach for biomolecular systems. *Journal of Chemical Theory and Computation*, *11*, 1809–1818.

Stewart, J. J. P. (2013). Optimization of parameters for semiempirical methods VI: More modifications to the NDDO approximations and re-optimization of parameters. *Journal of Molecular Modeling*, *19*, 1–32.

Svensson, M., Humbel, S., Froese, R. D. J., Matsubara, T., Sieber, S., & Morokuma, K. (1996). ONIOM: A multilayered integrated MO + MM method for geometry optimizations and single point energy predictions. A test for Diels–Alder reactions and $Pt(P(t\text{-}Bu)_3)_2 + H_2$ oxidative addition. *The Journal of Physical Chemistry*, *100*, 19357–19363.

Swenson, D. W. H., & Bolhuis, P. G. (2014). A replica exchange transition interface sampling method with multiple interface sets for investigating networks of rare events. *The Journal of Chemical Physics*, *141*, 044101.

Théry, V., Rinaldi, D., Rivail, J.-L., Maigret, B., & Ferenczy, G. G. (1994). Quantum mechanical computations on very large molecular systems: The local self-consistent field method. *Journal of Computational Chemistry*, *15*, 269–282.

Thompson, M. A., & Schenter, G. K. (1995). Excited ates of the bacteriochlorophyll b dimer of Rhodopseudomonas viridis: A QM/MM study of the photosynthetic reaction center that includes MM polarization. *The Journal of Physical Chemistry*, *99*, 6374–6386.

Till, M. S., Becker, T., Essigke, T., & Ullmann, G. M. (2008). Simulating the proton transfer in Gramicidin A by a sequential dynamical Monte Carlo method. *The Journal of Physical Chemistry. B*, *112*, 13401–13410.

Todorović, M., Bowler, D. R., Gillan, M. J., & Miyazaki, T. (2013). Density-functional theory study of gramicidin A ion channel geometry and electronic properties. *Journal of The Royal Society Interface*, *10*, 20130547.

Torrie, G., & Valleau, J. (1977). Nonphysical sampling distributions in Monte Carlo free-energy estimation: Umbrella sampling. *Journal of Computational Chemistry*, *23*, 187–199.

Tozzini, V. (2005). Coarse-grained models for proteins. *Current Opinion in Structural Biology*, *15*, 144–150.

Truhlar, D. G., Gao, J., Alhambra, C., Garcia-Viloca, M., Corchado, J., Sánchez, M. L., & Villà, J. (2002). The incorporation of quantum effects in enzyme kinetics modeling. *Accounts of Chemical Research*, *35*, 341–349.

Ullmann, G. M. (2000). The coupling of protonation and reduction in proteins with multiple redox centers: Theory, computational method, and application to cytochrome c_3. *The Journal of Physical Chemistry. B*, *104*, 6293–6301.

Ullmann, G. M., & Bombarda, E. (2014). Continuum electrostatic analysis of proteins. In G. Naray-Szabo (Ed.), *Protein modelling* (pp. 135–163). Cham, Switzerland: Springer International Publishing.

Ullmann, G. M., Kloppmann, E., Essigke, T., Krammer, E.-M., Klingen, A. R., Becker, T., & Bombarda, E. (2008). Investigating the mechanisms of photosynthetic proteins using continuum electrostatics. *Photosynthesis Research*, *97*, 33–53.

Ullmann, G. M., Mueller, L., & Bombarda, E. (2016). Theoretical analysis of electron transfer in proteins: From simple proteins to complex machineries. In W. Cramer (Ed.), *Cytochrome complexes: Evolution, structures, energy transduction, and signaling* (Vol. 37, pp. 99–127). Dordrecht, Netherlands: Springer International Publishing.

Ullmann, R. T., & Ullmann, G. M. (2012). GMCT: A Monte Carlo simulation package for macromolecular receptors. *Journal of Computational Chemistry*, *33*, 887–900.

Vanden-Eijnden, E., & Venturoli, M. (2009). Revisiting the finite temperature string method for the calculation of reaction tubes and free energies. *The Journal of Chemical Physics*, *130*, 194103.

van Duin, A. C. T., Dasgupta, S., Lorant, F., & Goddard, W. A., III. (2001). ReaxFF: A reactive force field for hydrocarbons. *The Journal of Physical Chemistry. B*, *105*, 9396–9409.

van Mourik, T., Bühl, M., & Gaigeot, M.-P. (2014). Density functional theory across chemistry, physics and biology. *Philosophical Transactions. Series A, Mathematical, Physical, and Engineering Sciences*, *372*, 20120488.

Verlackt, C. C. W., Neyts, E. C., Jacob, T., Fantauzzi, D., Golkaram, M., Shin, Y.-K., … Bogaerts, A. (2015). Atomic-scale insight into the interactions between hydroxyl radicals and DNA in solution using the ReaxFF reactive force field. *New Journal of Physics*, *17*, 103005.

Řezáč, J., & Hobza, P. (2012). Advanced corrections of hydrogen bonding and dispersion for semiempirical quantum mechanical methods. *Journal of Chemical Theory and Computation*, *8*, 141–151.

Wales, D. J., & Walsh, T. R. (1998). Theoretical study of the water pentamer. *American Institute of Physics*, *105*, 6957–6971.

Wang, Z., Antoniou, D., Schwartz, S. D., & Schramm, V. L. (2016). Hydride transfer in DHFR by transition path sampling, kinetic isotope effects, and heavy enzyme studies. *Biochemistry*, *55*, 157–166.

Warshel, A. (1991). *Computer modeling of chemical reactions in enzymes and solutions*. New York: Wiley & Sons.

Warshel, A. (2003). Computer simulations of enzyme catalysis: Methods, progress, and insights. *Annual Review of Biophysics and Biomolecular Structure*, *32*, 425–443.
Warshel, A., & Levitt, M. (1976). Theoretical studies of enzymic reactions: Dielectric, electrostatic and steric stabilization of the carbonium ion in the reaction of lysozyme. *Journal of Molecular Biology*, *103*, 227–249.
Warwicker, J., & Watson, H. C. (1982). Calculation of the electrostatic potential in the active site cleft due the α-Helix dipols. *Journal of Molecular Biology*, *186*, 671–679.
Webb, S. P., Iordanov, T., & Hammes-Schiffer, S. (2002). Multiconfigurational nuclear-electronic orbital approach: Incorporation of nuclear quantum effects in electronic structure calculations. *The Journal of Chemical Physics*, *117*, 4106–4118.
Yilmazer, N. D., & Korth, M. (2015). Enhanced semiempirical QM methods for biomolecular interactions. *Computational and Structural Biotechnology Journal*, *13*, 169–175.
Zavadlav, J., Melo, M. N., Marrink, S. J., & Praprotnik, M. (2015). Adaptive resolution simulation of polarizable supramolecular coarse-grained water models. *The Journal of Chemical Physics*, *142*, 244118.
Zhang, Y. (2013). Electrostatic interaction of the electrostatic-embedding and mechanical-embedding schemes for QM/MM calculations. *Communications in Computational Chemistry*, *1*, 109–117.
Zhao, Y., & Truhlar, D. G. (2008). The M06 suite of density functionals for main group thermochemistry, thermochemical kinetics, noncovalent interactions, excited states, and transition elements: Two new functionals and systematic testing of four M06-class functionals and 12 other function. *Theoretical Chemistry Accounts*, *120*, 215–241.
Zimmerman, M., & Bowman, G. (2016). Chapter nine—How to run FAST simulations. *Methods in Enzymology*, *578*, 213–225.
Zwanzig, R. W. (1954). High-temperature equation of state by a perturbation method. I. Nonpolar gases. *The Journal of Chemical Physics*, *22*, 1420–1426.

CHAPTER FOUR

A Paradigm for C—H Bond Cleavage: Structural and Functional Aspects of Transition State Stabilization by Mandelate Racemase

Stephen L. Bearne*[,1], Martin St. Maurice†[,1]
*Dalhousie University, Halifax, NS, Canada
†Marquette University, Milwaukee, WI, United States
[1]Corresponding authors: e-mail address: sbearne@dal.ca; martin.stmaurice@marquette.edu

Contents

1. Introduction 115
   1.1 Brief Historical Perspective 115
   1.2 MR as a "Model" Enzyme 115
2. Structure and Function Within the MR Subgroup 117
   2.1 Overview of the ENS 117
   2.2 Structural Features of the MR Subgroup 118
   2.3 Mechanistic Features of the MR Subgroup 124
3. MR Catalysis: An Uphill Battle 125
4. Meeting the Challenges: Solutions Afforded by MR 126
   4.1 The Thermodynamic Challenge 126
   4.2 Transition State Stabilization 128
   4.3 The Kinetic Challenge 129
5. Mandelate Racemase Structure: Ligand Binding Within a Catalytic Cage 133
   5.1 Ligand Binding Modes 134
   5.2 The Hydrophobic Cavity and the Active Site Flap 141
   5.3 Substrate and Enzyme Motion During Catalysis 143
6. Evolutionary Relationships: A Mechanistic Link to Aldolases? 148
   6.1 Inactivation of MR by 3-Hydroxypyruvate 148
   6.2 Mechanistic Similarities 150
7. Concluding Remarks 151
Acknowledgments 152
References 152

Advances in Protein Chemistry and Structural Biology, Volume 109
ISSN 1876-1623
http://dx.doi.org/10.1016/bs.apcsb.2017.04.007

Abstract

Mandelate racemase (MR) from *Pseudomonas putida* catalyzes the Mg^{2+}-dependent, 1,1-proton transfer reaction that racemizes (*R*)- and (*S*)-mandelate. MR shares a partial reaction (i.e., the metal ion-assisted, Brønsted base-catalyzed proton abstraction of the α-proton of carboxylic acid substrates) and structural features ($(\beta/\alpha)_7\beta$-barrel and N-terminal $\alpha + \beta$ capping domains) with a vast group of homologous, yet functionally diverse, enzymes in the enolase superfamily. Mechanistic and structural studies have developed this enzyme into a paradigm for understanding how enzymes such as those of the enolase superfamily overcome kinetic and thermodynamic barriers to catalyze the abstraction of an α-proton from a carbon acid substrate with a relatively high pK_a value. Structural studies on MR bound to intermediate/transition state analogues have delineated those structural features that MR uses to stabilize transition states and enhance reaction rates of proton abstraction. Kinetic, site-directed mutagenesis, and structural studies have also revealed that the phenyl ring of the substrate migrates through the hydrophobic cavity within the active site during catalysis and that the Brønsted acid–base catalysts (Lys 166 and His 297) may be utilized as binding determinants for inhibitor recognition. In addition, structural studies on the adduct formed from the irreversible inhibition of MR by 3-hydroxypyruvate revealed that MR can form and deprotonate a Schiff-base with 3-hydroxypyruvate to yield an enol(ate)-aldehyde adduct, suggesting a possible evolutionary link between MR and the Schiff-base forming aldolases. As the archetype of the enolase superfamily, mechanistic and structural studies on MR will continue to enhance our understanding of enzyme catalysis and furnish insights into the evolution of enzyme function.

ABBREVIATIONS

3-HP 3-hydroxypyruvate
BzH benzohydroxamate
CHCHA cyclohexanecarbohydroxamate
ENS enolase superfamily
FucD L-fuconate dehydratase
GalrD3 galactarate dehydratase III
GCI D-galactaro-1,4-lactone cycloisomerase
ITC isothermal titration calorimetry
LyxD L-lyxonate dehydratase
MLE muconate lactonizing enzyme
MR mandelate racemase
NAL *N*-acetylneuraminate lyase
RhamD L-rhamnonate dehydratase
TarD D-tartrate dehydratase
TFHTP 3,3,3-trifluoro-2-hydroxy-2-(trifluoromethyl)propanoic acid
TGD L-talarate/galactarate dehydratase
TS transition state

1. INTRODUCTION

1.1 Brief Historical Perspective

Mandelate racemase (MR) was initially identified by Stanier and coworkers (Gunsalus, Stanier, & Gunsalus, 1953; Gunsalus, Gunsalus, & Stanier, 1953) following the observation that *Pseudomonas putida* strain A.3.12 was uniquely capable of growing on media containing either of the two naturally occurring enantiomers of mandelic acid as a sole carbon source. While a few other organisms have since been identified that are capable of utilizing both the (*S*)- and (*R*)-enantiomers of mandelic acid (Bhat & Vaidyanathan, 1976; Durham, 1984; Jamaluddin, Rao, & Vaidyanathan, 1970), *P. putida* remains the only confirmed organism to express MR in nature.

Early investigations into catalysis by MR revealed that the enzyme employs a cofactor-independent, two-base mechanism that promotes C—H bond cleavage through the stabilization of a carbanionic intermediate (Hegeman, Rosenberg, & Kenyon, 1970; Kenyon, Gerlt, Petsko, & Kozarich, 1995; Kenyon & Hegeman, 1970, 1977, 1979; Lin et al., 1988; Powers, Koo, Kenyon, Gerlt, & Kozarich, 1991). Throughout the 1990s, Kenyon, Gerlt, Petsko, Kozarich, and colleagues thoroughly characterized the mechanism and structure of the enzyme. These studies culminated in the discovery that MR was structurally and mechanistically related to a much broader group of homologous enzymes that constitute the enolase superfamily (ENS) (Babbitt & Gerlt, 1997, 2000; Babbitt et al., 1996, 1995; Gerlt & Babbitt, 1998, 2001; Gerlt, Babbitt, Jacobson, & Almo, 2012; Gerlt, Babbitt, & Rayment, 2005; Gerlt, Kenyon, Kozarich, Neidhart, & Petsko, 1992; Hasson et al., 1998). By virtue of its being the first and most thoroughly studied enzyme of its type, MR became the archetype for the MR subgroup, which is now the most functionally heterogeneous subgroup in the ENS (Gerlt et al., 2012).

1.2 MR as a "Model" Enzyme

MR has served as an excellent "model" enzyme in two main capacities. First, the enzyme catalyzes the heterolytic cleavage of a C—H bond, a particularly difficult reaction considering that the pK_a of the substrate mandelate is ~29 (Chiang, Kresge, Pruszynski, Schepp, & Wirz, 1990; Gerlt,

Kozarich, Kenyon, & Gassman, 1991; Renaud & Fox, 1988). Studies focused on the catalytic mechanism underlying this reaction have done much to enhance our understanding of enzymes that catalyze rapid proton abstractions from carbon acid substrates with high pK_a values (Babbitt & Gerlt, 1997; Babbitt et al., 1996; Gerlt, 1998; Gerlt & Gassman, 1992, 1993a, 1993b; Gerlt et al., 1992, 1991). This is particularly significant considering the highly pervasive occurrence of such reactions in enzyme mechanisms (Richard & Amyes, 2001). Second, MR is the archetype for the ENS (Neidhart, Kenyon, Gerlt, & Petsko, 1990), and, outside of enolase, it is possibly the most thoroughly characterized enzyme in the ENS. Since it shares both mechanistic and structural features with the approximately 49,000 enzymes currently annotated as ENS enzymes (Akiva et al., 2014), MR has served as a paradigm and a benchmark for understanding substrate specificity and catalysis within this superfamily.

In addition to its use as a model enzyme for understanding fundamental aspects of enzyme catalysis and evolution, MR has received attention for its potential utility as a biocatalyst in dynamic kinetic resolutions (El Gihani & Williams, 1999) because of its ability to catalyze the racemization of mandelate derivatives and a variety of α-hydroxy-β,γ-unsaturated carboxylic acids (Fan et al., 2015; Felfer et al., 2005; Felfer, Strauss, Kroutil, Fabian, & Faber, 2001; Gadler et al., 2006; Goriup, Strauss, Felfer, Kroutil, & Faber, 2001; Kaftzik, Kroutil, Faber, & Kragl, 2004; Larissegger-Schnell, Kroutil, & Faber, 2005; Li et al., 2013; Resch, Fabian, & Kroutil, 2010; Stecher, Felfer, & Faber, 1997; Strauss & Faber, 1999). MR has been immobilized for application in fixed-bed reactors and organic solvents (Bauer, Boy, Faber, Felfer, & Voss, 2001; Strauss, Kandelbauer, & Faber, 2000; Wrzosek, García Rivera, Battenbrock, & Seidel-Morgenstern, 2016), and subjected to directed evolution and computational studies with the goal of enhancing its activity with nonnatural substrates (Gu et al., 2014; Gu & Yu, 2012; Yang et al., 2017; Zhuohang, Ji, & Hongwei, 2011). A detailed understanding of MR catalysis will inform and facilitate development of its use in biotechnology applications.

In this review, we focus our discussion primarily on the results obtained from recent structural studies of MR in complex with a variety of ground state and intermediate/transition state (TS) analogues. These findings augment our understanding of how MR stabilizes the altered substrate in the TS to effect catalysis and help define the origin of substrate specificity within the MR subgroup of the ENS.

2. STRUCTURE AND FUNCTION WITHIN THE MR SUBGROUP

2.1 Overview of the ENS

In 1990, the remarkable similarity of the secondary, tertiary, and quaternary structure of MR and muconate lactonizing enzyme (MLE) was recognized, suggesting that these two enzymes likely evolved from a common ancestor (Neidhart et al., 1990). Although these two enzymes catalyze different overall reactions, it was also recognized that they share a partial reaction: the divalent cation-assisted, Brønsted base-catalyzed abstraction of a proton from a carbon acid substrate to form an enolic intermediate (Babbitt et al., 1995). Subsequently, sequence and structural analyses led to the identification of a number of genes encoding homologous enzymes that possessed highly conserved amino acid residues that bind the essential metal ion and act as the Brønsted base; this group of homologous proteins was designated as the "enolase superfamily" (ENS), bearing the name of the family member that plays a central role in glycolysis (Babbitt et al., 1996). Over the past two decades, with the growing abundance of genomic information, the ENS has grown to include ~49,000 members as listed by the Structure Function Linkage Database (http://sfld.rbvi.ucsf.edu/) (Akiva et al., 2014). These enzymes catalyze a variety of reactions, including 1,1-proton transfers (racemization and epimerization), β-elimination (i.e., dehydration or elimination of ammonia), or intramolecular addition/elimination (i.e., cycloisomerization) (Babbitt et al., 1996; Gerlt & Babbitt, 1998, 2001; Gerlt et al., 2012, 2005) but still share the same initial partial reaction (Scheme 1A). At present, the ENS is divided into seven subgroups based on the identities of the Brønsted acid–base catalysts and the ligands of the metal ion: enolase, MR, MLE, β-methylaspartate ammonia lyase, D-glucarate dehydratase, D-mannonate dehydratase, and galactarate dehydratase 2 (Akiva et al., 2014; Rakus et al., 2007). Of these, the enolase subgroup is the largest with ~14,000 members, but it is a homofunctional subgroup composed entirely of enolases. The next two largest subgroups are the MR and MLE subgroups with ~12,000 and ~13,000 members, respectively (Akiva et al., 2014). Although the reactions catalyzed by MR subgroup members are diverse, such as 1,1-proton transfers (e.g., MR) and cycloisomerizations (e.g., D-galactaro-1,4-lactone cycloisomerase (GCI) (Vetting, Bouvier, Gerlt, & Almo, 2016)), most of the members appear to catalyze dehydration reactions

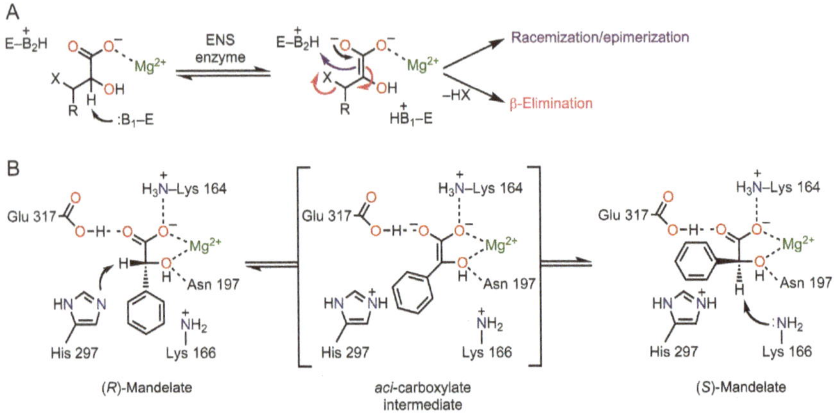

Scheme 1 Reaction mechanisms. (A) General reactions catalyzed by the ENS enzymes, where B_1 and B_2 are the Brønsted acid–base catalysts. (B) Reaction catalyzed by mandelate racemase.

(e.g., D-tartrate dehydratase (TarD) (Yew, Fedorov, Fedorov, Wood, et al., 2006), L-fuconate dehydratase (FucD) (Wichelecki et al., 2014; Yew, Fedorov, Fedorov, Rakus, et al., 2006), L-talarate/galactarate dehydratase (TGD) (Yew, Fedorov, Fedorov, Almo, & Gerlt, 2007), L-rhamnonate dehydratase (RhamD) (Rakus et al., 2008), L-lyxonate dehydratase (LyxD) (Ghasempur et al., 2014), galactarate dehydratase III (GalrD3) (Groninger-Poe et al., 2014), among others) (Table 1). Within the MR subgroup, MR has received the most attention.

2.2 Structural Features of the MR Subgroup

The assignment of members of the ENS into subgroups is based on sequence and structural considerations rather than on mechanism or substrate specificity. Thus, MR includes several structural features that are shared with all other members of the subgroup. The structurally characterized members of the MR subgroup are tightly packed homooctamers with a quaternary structure that is best described as a tetramer of dimers (Fig. 1A). Cross-linking studies with dimethyl suberimidate further suggest that MR is octameric in solution (Fee, Hegeman, & Kenyon, 1974; Tsou, Ransom, Gerlt, Powers, & Kenyon, 1989), although more recent gel filtration experiments indicate that it might also be tetrameric under certain conditions (E. Sagy & S.L. Bearne, unpublished). In the crystal structures, each of the four homodimers exhibits a tight dimerization interface (~1400 Å^2 of buried surface area) with an interdigitating loop that extends into the active

Table 1 Structurally Characterized Enzymes in the MR Subgroup

Enzyme	Representative PDB (Organism)	Mg^{2+} Binding Ligands (β-Strands 3, 4, and 5)	First Acid–Base Catalytic Dyad (β-Strands 6 and 7)	Second Acid–Base Catalyst (β-Strand 2)	References
Mandelate racemase (MR)	1MDR (*Pseudomonas putida*)	D195 (β3) E221 (β4) E222-H_2O (β4) E247 (β5)	D247 (β6) H297 (β7)	K166 (β2)	Landro et al. (1994)
L-Fuconate dehydratase (FucD)	2HXT (*Xanthomonas campestris*)	D248 (β3) E274 (β4) E275-H_2O (β4) E301 (β5)	D324 (β6) H351 (β7)	K220 (β2)	Yew, Fedorov, Fedorov, Rakus, et al. (2006)
D-Tartrate dehydratase (TGD)	2DW7 (*Bradyrhizobium japonicum*)	D213 (β3) E239 (β4) E240-H_2O (β4) E265 (β5)	D292 (β6) H322 (β7)	K184 (β2)	Yew, Fedorov, Fedorov, Wood, et al. (2006)
L-Rhamnonate dehydratase (RhamD)	3CXO (*Salmonella typhimurium*)	D226 (β3) E252 (β4) E253-H_2O (β4) E280 (β5)	D302 (β6) H329 (β7)	None verified experimentally	Rakus et al. (2008)
L-Talarate/galactarate dehydratase (TGD)	2PP1 (*Salmonella typhimurium*)	D226 (β3) E252 (β4) E253-H_2O (β4) E278 (β5)	D301 (β6) H328 (β7)	K197 (β2)	Yew et al. (2007)

Continued

Table 1 Structurally Characterized Enzymes in the MR Subgroup—cont'd

Enzyme	Representative PDB (Organism)	Mg^{2+} Binding Ligands (β-Strands 3, 4, and 5)	First Acid–Base Catalytic Dyad (β-Strands 6 and 7)	Second Acid–Base Catalyst (β-Strand 2)	References
L-Lyxonate dehydratase (LyxD)	3STP (*Pseudomonas aeruginosa*)	E214 (β3) E240 (β4) E241-H_2O (β4) E266 (β5)	D289 (β6) H316 (β7)	R179 (β2)	Ghasempur et al. (2014)
Galactarate dehydratase III (GalrD3)	4JN7 (*Agrobacterium tumefaciens*)	D189 (β3) E216 (β4) E217-H_2O (β4) E 242 (β5)	D265 (β6) H292 (β7)	H191 (β3) (not verified experimentally)	Groninger-Poe et al. (2014)
D-Galactaro-1,4-lactone cycloisomerase (GCI)	4HPN (*Agrobacterium tumefaciens*)	D194 (β3) E220 (β4) E221-H_2O (β4) E246 (β5)	D269 (β6) H296 (β7)	K166 (β2)	Vetting et al. (2016)

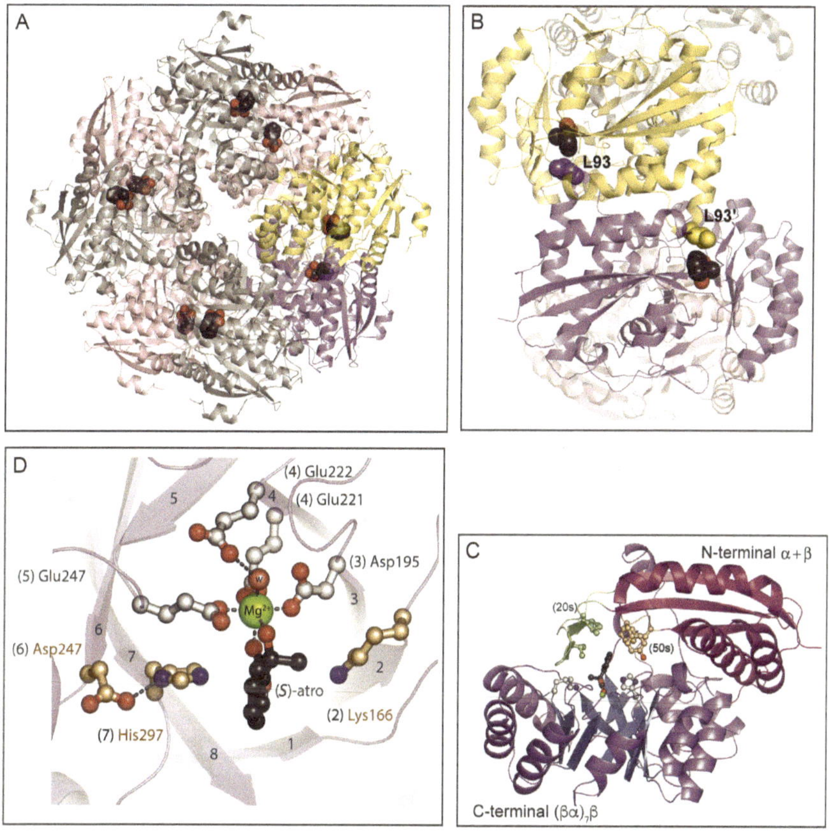

Fig. 1 Structural features of MR. (A) Most ENS enzymes are homooctamers that can best be described as a tetramer of dimers. The overall octameric arrangement in the MR subgroup is displayed for *P. putida* MR (PDB ID 1MDR), with three protomers colored in *pink*, three protomers colored in *gray*, and one homodimeric unit colored in *blue* and *gold*. The eight active sites of the homooctamer are denoted by the presence of (*S*)-atrolactate, illustrated as CPK-colored *black spheres*. (B) Enzymes in the MR subgroup are relatively unique in having an interdigitating loop that connects the dimer pairs. This loop contributes specificity determinants and/or structural reinforcement to the active site of the neighboring protomer. In the example of *P. putida* MR, Leu 93 (*blue* and *gold spheres*) contributes to forming the hydrophobic pocket at the active site of the neighboring protomer. The location of the active sites are denoted by the presence of (*S*)-atrolactate, illustrated as CPK-colored *black spheres* (PDB ID 1MDR) (Landro et al., 1994). (C) The bidomain structure of ENS enzymes. The N-terminal α + β capping domain, highlighted in *purple*, contributes the specificity determinants. In the specific case of *P. putida* MR, the *ball* and *sticks* projecting into the active site represent the side chains of Leu 18, Val 22, and Val 29 from the 20s loop (*green*) and Phe 52 and Tyr 54 from the 50s loop (*gold*). The location of the active site is denoted by the substrate analogue, (*S*)-atrolactate, illustrated in CPK-colored *black ball* and *sticks*. The C-terminal $(\beta/\alpha)_7\beta$-modified TIM barrel domain is colored *blue*. The two Brønsted acid–base catalysts Lys 166 and His 297

(Continued)

site of the neighboring protomer (Fig. 1B) (Landro et al., 1994; Lietzan et al., 2012; Mitra et al., 1995; Nagar, Lietzan, St. Maurice, & Bearne, 2014; Neidhart et al., 1990; Schafer et al., 1996). The interdigitating loop appears to be present in all members of the MR subgroup and is a relatively unique structural feature within the ENS; the ManD subgroup is the only other subgroup to include a similar feature (Bearne, 2017). The interdigitating loop contributes specificity determinants and/or structural features to the active site in the neighboring protomer. For example, the interdigitating loop of MR (92-CLAGY-96) emanates from the $\alpha + \beta$ capping domain of each symmetry-related protomer in the homodimer, contributing a binding determinant (Leu 93) to the hydrophobic cavity that surrounds the phenyl group of the substrate (Fig. 1B). Similarly, in the MR subgroup enzymes TarD and FucD, the side chains of Lys 102 and Trp 101, respectively, penetrate deeply into the adjacent protomer, serving as a binding determinant for either the distal carboxylate of D-tartrate or the methyl group of L-fuconate (Yew, Fedorov, Fedorov, Rakus, et al., 2006; Yew, Fedorov, Fedorov, Wood, et al., 2006).

The overall domain organization in the MR subgroup is consistent with all other members of the ENS: an N-terminal $\alpha + \beta$ capping domain precedes a C-terminal $(\beta/\alpha)_7\beta$-modified TIM barrel domain, with the active site positioned at the interface of these two domains. The $(\beta/\alpha)_7\beta$-barrel includes several highly conserved elements that contribute the essential Mg^{2+}-binding and catalytic residues to the active site, while the N-terminal capping domain contributes side chains that interact with the distal portion of the substrate, thereby influencing substrate binding and selectivity (Fig. 1C). The N-terminal $\alpha + \beta$ capping domain is composed of three β-strands followed by three α-helices. Two functionally important

Fig. 1—Cont'd are illustrated in CPK-colored *white ball* and *sticks* (PDB ID 1MDR) (Landro et al., 1994). (D) The $(\beta/\alpha)_7\beta$-barrel domain in MR is representative of the MR subgroup. The Brønsted acid–base catalysts (*orange ball* and *sticks*) and metal binding residues (*white ball* and *sticks*) in *P. putida* MR are labeled along with the strand number from which they originate. The catalytic and metal binding residues are located at the C-terminal ends of β-strands 2 through 7 and are highly conserved across the subgroup. An ordered water molecule (labeled "W") occupies the sixth coordination site of the active site Mg^{2+} in nearly all X-ray crystal structures of MR subgroup members. The second Glu on strand 4 (Glu 222 in MR) is present in nearly all members of the subgroup and serves to orient and position this water molecule (PDB ID 1MDR) (Landro et al., 1994).

loops connect these regions of defined secondary structure: these so-called 20s and 50s loops provide critical side chain binding determinants to the active site. The 20s loop, which spans residues 17–30 in MR and connects the first and second β-strands, is relatively flexible, permitting substrate access while also sequestering the active site from solvent. The 50s loop spans residues 51–55 in MR and connects the third β-strand to the first α-helix of the N-terminal capping domain. The 20s and 50s loops vary widely in length and sequence throughout the MR subgroup. For example, the 20s loop is extended to include more than 50 residues in RhamD and contributes all of the specificity binding determinants to the active site, while the 50s loop is significantly shortened (Rakus et al., 2008).

Octahedral coordination of the hexacoordinate Mg^{2+} ion at the active site of all MR subgroup enzymes is achieved through interactions with the substrate, conserved amino acid side chains, and an ordered water molecule in the $(\beta/\alpha)_7\beta$-barrel domain. An Asp residue at the C-terminal end of β-strand 3, and two Glu residues at the C-terminal ends of β-strands 4 and 5 arrange their carboxylate oxygens in a meridional coordination geometry around the Mg^{2+} cation (Neidhart et al., 1990). These amino acids are conserved throughout the MR subgroup and serve as the primary "fingerprint" for subgroup family assignment. An ordered water molecule and the carboxylate and α-hydroxyl oxygens of the substrate complete the Mg^{2+} coordination sphere (Fig. 1D). All MR subgroup enzyme structures reveal a second Glu on β-strand 4 that serves to position the ordered water molecule coordinating the Mg^{2+} ion. While this second Glu is found in some enzymes of other subgroups, it is not conserved across the ENS. For example, a conserved Asp residue serves this same function in the enolase subgroup.

Most enzymes in the MR subgroup use two Brønsted acid–base catalysts to effect reactions. A His residue at the C-terminal end of β-strand 7 and an Asp residue at the C-terminal end of β-strand 6 are fully conserved across all members of the MR subgroup and function as a catalytic dyad, with His acting as the Brønsted acid–base catalyst. A second Brønsted acid–base catalyst (e.g., Lys, Arg, His, Tyr) is also often present at the C-terminal end of the second β-strand. Most enzymes in the subgroup, including MR, have a Lys at this position though this amino acid is not conserved throughout the subgroup. Examples of Mg^{2+}-coordinating residues and Brønsted acid–base catalysts for well-characterized MR subgroup enzymes are listed in Table 1. Specifically, for MR, Lys 166 acts as the *S*-specific Brønsted base to abstract the α-proton from (*S*)-mandelate, while His 297 acts as a Brønsted acid to protonate the *aci*-carboxylate intermediate. When (*R*)-mandelate is the substrate, His 297 acts as the *R*-specific Brønsted base to abstract the α-proton, while the conjugate acid of Lys 166 acts as the Brønsted acid to protonate the intermediate (Kallarakal et al., 1995; Landro et al., 1994; Powers et al., 1991).

2.3 Mechanistic Features of the MR Subgroup

The MR subgroup is among the largest of subgroups within the ENS, currently comprising over 12,000 protein sequences as catalogued by the Structure Function Linkage Database (Akiva et al., 2014). Like all enzymes of the ENS, members of the MR subgroup catalyze the same first partial reaction: the Mg^{2+}-assisted abstraction of an α-proton from a carboxylic acid substrate to generate a stabilized enol(ate) intermediate (Babbitt & Gerlt, 1997; Babbitt et al., 1996; Gerlt & Babbitt, 2001). Once formed, the fate of the enol(ate) intermediate differs throughout the superfamily and within the MR subgroup, contributing to functional heterogeneity among enzymes of the subgroup (Scheme 1A). While MR is the archetype for the subgroup, in some ways the enzyme is an oddity since the vast majority of enzymes in the MR subgroup are acid sugar dehydratases that catalyze β-elimination rather than racemization. For example, other enzymes of the MR subgroup for which the catalyzed reactions have been confirmed and for which structural information is available are listed in Table 1. Although these enzymes all catalyze the initial abstraction of the α-proton, only TGD appears to catalyze a 1,1-proton transfer reaction that effects the epimerization of its substrates in a similar fashion to MR. When the dehydration of L-talarate is catalyzed by TGD, competing epimerization to galactarate occurs, although when galactartate is the substrate, minimal epimerization to L-talarate is observed (Yew et al., 2007). Interestingly, most of the dehydratases characterized to date catalyze the formation of the keto tautomer of their product. However, the vinylogous β-elimination catalyzed by TarD yields the enol of oxaloacetate, prompting Gerlt and coworkers to refer to TarD as "a link between the 1,1-proton transfer reaction catalyzed by MR and the dehydration reactions catalyzed by other acid sugar dehydratases" (Yew, Fedorov, Fedorov, Wood, et al., 2006).

Recently, a new cycloisomerization reaction catalyzed by GCI was identified within the MR subgroup (Vetting et al., 2016). Although the reaction is the related β-elimination of a carboxylate rather than water as for the dehydratases, this finding suggests that there could be yet unidentified activities within the subgroup.

3. MR CATALYSIS: AN UPHILL BATTLE

The heterolytic abstraction of a proton adjacent to a carboxylic acid group (α-proton of a carbon acid substrate as shown in Scheme 1) is a difficult reaction. This is illustrated by the observation that, in the absence of MR, the rate constant (k_{non}) for the racemization of (*R*)-mandelate is $3 \times 10^{-13}\ s^{-1}$ (Bearne & Wolfenden, 1997). To furnish the observed rate enhancement (k_{cat}/k_{non}) of approximately 15 orders of magnitude, MR must overcome two problems. The first problem is thermodynamic in nature and arises from the formation of the unstable enol or enolate intermediate (Amyes & Richard, 1996, 2007; Chiang, Kresge, Schepp, & Xie, 2000; Richard, Williams, O'Donoghue, & Amyes, 2002; Thibblin & Jencks, 1979). The second problem is kinetic in nature. The rate of nonenzymatic abstraction of the α-proton from a carbon acid is slower than the rate of abstraction of a proton from a heteroatom or normal acid (HX, where X = N, O, or S) of equal acidity (Albery, 1982; Bernasconi, 1992; Chiang, Kresge, Santaballa, & Wirz, 1988; Eigen, 1964). This difference is thought to arise from changes in the orientations of solvent dipoles as the negative charge develops on the carbonyl oxygen during α-proton abstraction (Bernasconi, 1992) and changes in the geometry of the carbon atom (Guthrie, 1998). The relative magnitude of these two problems may be estimated using Marcus formalism (Albery, 1980; Cohen & Marcus, 1968; Gerlt, 1998; Kresge & Silverman, 1999; Marcus, 1969; Silverman, 2000), wherein the free energy of activation ($\Delta G^{\ddagger}$) for a unimolecular proton transfer reaction is partitioned into two parts: (1) the thermodynamic barrier (ΔG°), and (2) the intrinsic kinetic barrier ($\Delta G^{\ddagger}_{int}$) as shown in Eq. 1 (where $|\Delta G^{\ddagger}| \leq 4\Delta G^{\ddagger}_{int}$) (Maskill, 1985). $\Delta G^{\ddagger}_{int}$ is the hypothetical activation barrier in the absence of a thermodynamic barrier (i.e., when $\Delta G^{\circ} = 0$). For example, the values of ΔG° and $\Delta G^{\ddagger}_{int}$ for the nonenzymatic imidazole-catalyzed racemization of mandelate at 25°C have been estimated to be approximately 30 and 9 kcal/mol, respectively, giving $\Delta G^{\ddagger} = 30.3$ kcal/mol (Bearne & Spiteri, 2005).

$$\Delta G^{\ddagger} = \Delta G^{\ddagger}_{int} + \frac{\Delta G^{\circ}}{2} + \frac{(\Delta G^{\circ})^2}{16\Delta G^{\ddagger}_{int}} \qquad (1)$$

As can be seen from the estimated relative magnitudes of ΔG° and $\Delta G^{\ddagger}_{int}$, there is much more catalytic advantage to be gained by overcoming the thermodynamic barrier by stabilization of the intermediate (i.e., decreased ΔG°) through electrostatic stabilization and H-bonding interactions (Guthrie & Kluger, 1993; Richard et al., 2002). With regard to the kinetic problem, Gerlt and Gassman (1993a, 1993b) pointed out that the solvent reorganization accompanying the nonenzymatic reaction need not occur within an enzyme active site and that the optimal placement of an electrophilic catalyst adjacent to the carbonyl oxygen can reduce the intrinsic kinetic barrier (i.e., $\Delta G^{\ddagger}_{int}$) by stabilizing the developing negative charge on the carbonyl oxygen. The intrinsic kinetic barrier may also be reduced through concerted general acid catalysis (Gerlt & Gassman, 1992, 1993a, 1993b; Gerlt, Kreevoy, Cleland, & Frey, 1997). We will now review how MR overcomes these two problems.

4. MEETING THE CHALLENGES: SOLUTIONS AFFORDED BY MR

The first crystal structure of MR was reported by Petsko and coworkers in 1990 (Neidhart et al., 1990, 1988). Numerous structures of both the wild-type and mutant variants of MR have been solved since that time, providing valuable insights into how MR overcomes the thermodynamic and kinetic problems that accompany proton abstraction from a carbon acid substrate.

4.1 The Thermodynamic Challenge

MR is an extremely proficient enzyme and can discriminate between the substrate in the ground state and the altered substrate in the TS, binding the latter species with an association constant equal to $5 \times 10^{18}\ M^{-1}$, corresponding to a free energy of TS stabilization equal to 26 kcal/mol (Bearne & Wolfenden, 1997; St. Maurice & Bearne, 2002). X-ray crystal structures of MR with bound substrate (Kallarakal et al., 1995), analogues of the substrate (Landro et al., 1994; Mitra et al., 1995; Neidhart et al., 1991; Schafer et al., 1996), and intermediate analogues (Lietzan et al., 2012) all reveal multiple polar interactions that "solvate" the ligands, contributing to the enthalpic stabilization of the intermediate and the altered substrate in the TS, and facilitating catalysis by reducing the ΔG° component of $\Delta G^{\ddagger}$. Specifically, both Glu 317 (Mitra et al., 1995) and Lys 164 (Neidhart et al., 1991) interact with the carboxyl group of mandelate,

Asn 197 interacts with the α-hydroxyl group of mandelate (St. Maurice & Bearne, 2000), and the active site divalent metal ion is believed to stabilize the additional negative charge present in the intermediate and the TS (Scheme 1B) (Fee et al., 1974; Guthrie & Kluger, 1993). In addition to these conspicuous electrostatic interactions, van der Waals interactions contribute to the binding, particularly for the phenyl ring, which is bound within a predominately hydrophobic cavity lined by residues from the 20s and 50s loops. For example, the crystal structure of wild-type MR with bound intermediate/TS analogue benzohydroxamate (BzH, PDB 3UXK; Fig. 2) shows that the phenyl group binds in the hydrophobic pocket lined by residues from the N-terminal domain (Leu 18, Val 22, Val 29, Phe 52, and Tyr 54), the barrel domain (Leu 298, Leu 319, and Leu 321), and Leu 93 from the adjacent protomer (Lietzan et al., 2012). Comparison of the binding constants of (*R*)- and (*S*)-mandelate with that of either glycolate, (*R*)-lactate, or (*S*)-lactate revealed that interactions between the hydrophobic pocket

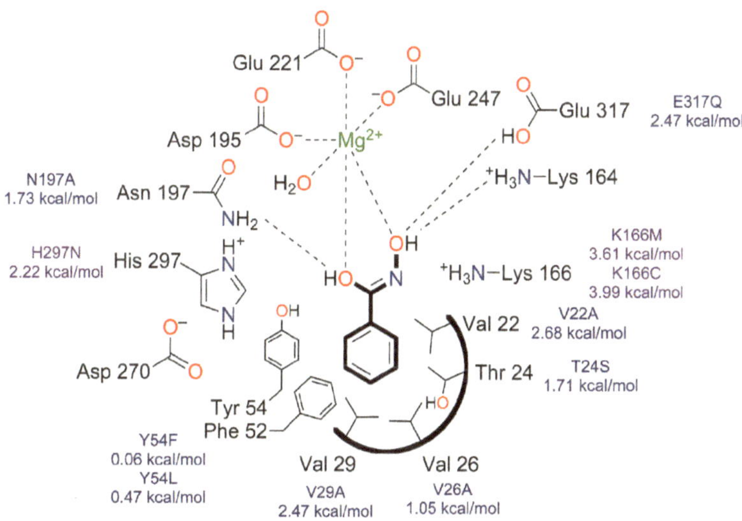

Fig. 2 Schematic representation of the active site of MR. The schematic is based on the X-ray crystal structure of MR with the intermediate/TS analogue inhibitor benzohydroxamate (BzH) bound at the active site (PDB 3UXK) (Lietzan et al., 2012). The Brønsted acid–base catalysts (Lys 166 and His 297) are shown along with residues from the 20s and 50s loops that form the hydrophobic pocket where the phenyl ring is bound, polar residues that interact with the hydroxamate moiety (Glu 317, Lys 164, Asn 197), and those residues that coordinate the magnesium ion (Asp 195, Glu 221, Glu 247). The observed changes in the binding free energies are shown for those residues that have been replaced using site-directed mutagenesis and for which the changes in binding affinity for BzH have been measured.

and the phenyl group contribute ~2 kcal/mol of free energy to ground state binding (St. Maurice & Bearne, 2004). Indeed, isothermal titration calorimetry (ITC) experiments revealed that BzH binding is accompanied by a negative ΔC_p value of -358 ± 3 cal mol K^{-1}, consistent with release of water or formation of an increased system of hydrophobic interactions (Guan et al., 2011; Myslinski, DeLorbe, Clements, & Martin, 2011; Nagar & Bearne, 2015; Parker et al., 1999).

In order to estimate the potential contribution of cation-π/NH-π interactions between the two Brønsted acid–base catalysts and the phenyl ring of either the substrate or the intermediate, ITC experiments were conducted to examine the ability of the MR mutants H297N MR and K166M MR to bind the ground state analogues (*S*)- and (*R*)-atrolactate and the intermediate/TS analogue inhibitors BzH and cyclohexanecarbohydroxamate (CHCHA), which lacks an aromatic π system. The contribution of the individual cation-π/NH-π interactions between the phenyl ring of BzH and His 297 and Lys 166 was estimated by taking the difference of the $\Delta\Delta G$ values for the binding of H297N MR with BzH and with CHCHA (i.e., 1.17–0.26 = 0.91 kcal/mol), and of K166M MR binding with BzH and with CHCHA (i.e., 3.48–1.74 = 1.74 kcal/mol), respectively (Nagar & Bearne, 2015). Hence, the interactions between His 297 and Lys 166 with the phenyl ring of BzH contribute a total of 2.65 kcal/mol to the binding free energy.

4.2 Transition State Stabilization

TS analogue inhibitors can serve as excellent probes of molecular interactions that stabilize the altered substrate in the TS (Radzicka & Wolfenden, 1995; Schramm, 1998, 2011; Wolfenden, 1972, 1976; Wolfenden & Frick, 1987). We explored the ability of MR to bind a variety of ground state and intermediate/TS analogues (Bourque, Burley, & Bearne, 2007; Burley & Bearne, 2005; St. Maurice & Bearne, 2000, 2004; St. Maurice, Bearne, Lu, & Taylor, 2003) and found that the *aci*-carboxylate analogues afforded the best inhibition with BzH ($K_\mathrm{i} = 9.3$ μM) (St. Maurice & Bearne, 2000), cupferron ($K_\mathrm{i} = 2.7$ μM) (Bourque et al., 2007), and *N*-hydroxyformanilide ($K_\mathrm{i} = 2.8$ μM) (Bourque et al., 2007) being the most potent competitive inhibitors of MR identified to-date. Much of our subsequent studies have utilized BzH because it is readily commercially available. Of course, tight-binding alone is not sufficient evidence to conclude that an enzyme active site is recognizing a ligand as an analogue of the TS, even though both BzH and cupferron are bound in a manner that is anticipated for the actual intermediate/TS, i.e., both chelate the active site divalent metal ion and are bound in a conformation with the phenyl ring coplanar with the hydroxamate and diazeniumdiolate moieties, respectively (Lietzan et al., 2012). Moreover, BzH appears to be a reasonable mimic of the intermediate/TS because its binding affinity for 21 MR variants correlates well with changes in the free energy of TS stabilization afforded by these variants (Bartlett & Marlowe, 1983; Lietzan et al., 2012; Phillips, Kaplan, Rutter, & Bartlett, 1992). Some of those binding interactions are shown in Fig. 2. Summing the changes in free energies accompanying the various indicated mutations yields a value of ~18 kcal/mol, which greatly exceeds the free energy of binding (6.8 kcal/mol) by 11.2 kcal/mol. This result is not surprising and indicates that there is a high degree of binding cooperativity within the active site of MR upon binding the altered substrate in the TS (Kraut, Carroll, & Herschlag, 2003).

4.3 The Kinetic Challenge

Gerlt and Gassman (1993a, 1993b) pointed out that the solvent reorganization that retards nonenzymatic carbon acid deprotonation need not occur within an enzyme active site and that an electrophilic catalyst placed adjacent to the carbonyl oxygen can stabilize the developing negative charge on the carbonyl oxygen thereby reducing the intrinsic kinetic barrier (i.e., $\Delta G^{\ddagger}_{int}$). Indeed, Glu 317 is ideally positioned within the active site to fulfill this role through concerted general acid catalysis (Mitra et al., 1995). Replacement of Glu 317 by Gln reduced the free energy of TS stabilization by 6.1 kcal/mol (Mitra et al., 1995). Bearne and Wolfenden estimated that the effective molarity of the side chain of Glu 317 was $\geq 3 \times 10^5$ M within the active site (Bearne & Wolfenden, 1997). While this proposed role for Glu 317 affords one mechanism for reducing the intrinsic kinetic barrier, it does not provide a means for stabilizing the double bond character of the enol(ate) intermediate.

Anslyn and coworkers (Houk, Anslyn, & Stanton, 2006; Houk, Monzingo, & Anslyn, 2008) proposed that enzymes could use cation-π/NH-π interactions to activate their carbon acid substrate (i.e., reduce the pK_a of the α-proton) through electrophilic coordination directed toward the π-bond of the carbonyl group. This coordination destabilizes the α-C–H bond via backside coordination, while enhancing the stability of the emerging delocalized π-electron density on the enolate (Houk et al., 2006). Interestingly, examination of the structure of MR with bound BzH revealed that the Brønsted acid–base catalysts Lys 166 and His 297 are positioned so that they form an obtuse angle of 135° with the "α-carbon", which places them adjacent to the hydroxamate/hydroximate bond of BzH (Fig. 3) (Nagar & Bearne, 2015). The distances from the ζN of Lys 166 and the ε^2N of the imidazole ring of His 297 to the center of the potential C=N bond of the hydroximate moiety are 3.0 and 3.1 Å, respectively, and are well within the reported range (3.4–6.0 Å) for cation-π or NH-π interactions (Burley & Petsko, 1986; Mitchell, Nandi, McDonald, Thornton, & Price, 1994; Scrutton & Raine, 1996; Zacharias & Dougherty, 2002). Consequently, one or both of these residues could fulfill the role of stabilizing the developing π-bond of the enol(ate) intermediate/TS.

To test this hypothesis, site-directed mutagenesis and ITC studies were conducted similar to the ones described above. Comparison of the binding affinities revealed that cation-π/NH-π interactions between His 297 and the hydroxamate/hydroximate moiety contribute approximately 0.26 kcal/mol to binding, while interactions with Lys 166 contribute approximately 1.74 kcal/mol (Nagar & Bearne, 2015). Thus, the conjugate acids of Lys 166 and His 297 may serve to lower the intrinsic free energy of activation ($\Delta G^{\ddagger}_{int}$) for enol(ate) formation relative to the greater value of $\Delta G^{\ddagger}_{int}$ that arises from the changes in solvent reorganization and hybridization that accompany the slow, nonenzymatic proton dissociation from carbon acids (Gerlt & Gassman, 1993b).

Is this structural feature unique to MR? Examples of other enzymes from the MR subgroup of the ENS for which structural information is available (including bound substrate or intermediate/TS analogues) suggest that it is not a unique feature as illustrated by Fig. 3. In MR (Lietzan et al., 2012), TarD (Yew, Fedorov, Fedorov, Wood, et al., 2006), FucD (Yew, Fedorov, Fedorov, Rakus, et al., 2006), TGD (Yew et al., 2007), and GalrD3 (Groninger-Poe et al., 2014), an obtuse angle ranging between 118 and 164° is formed between the Brønsted acid–base catalysts and the "α-carbon" of the ligands, and the distances from the ζN of Lys and the ε^2N of the imidazole ring of His to the center of where the π-bond forms ranges between ~3.3 and 3.5 Å. Thus, it would appear that the conjugate acids of the Brønsted acid–base catalysts could play a role in reducing $\Delta G^{\ddagger}_{int}$ for other enzymes of the MR subgroup. Indeed, this observation may be extended to other enzymes of the ENS, including D-glucarate dehydratase (Gulick et al., 2000), MLE (Sakai et al., 2009), and enolase (Larsen et al., 1996) as shown in Fig. 3. However, stabilization of the enol(ate) by this mechanism does not appear to be universal, as highlighted by the example of D-mannonate dehydratase (Rakus et al., 2007).

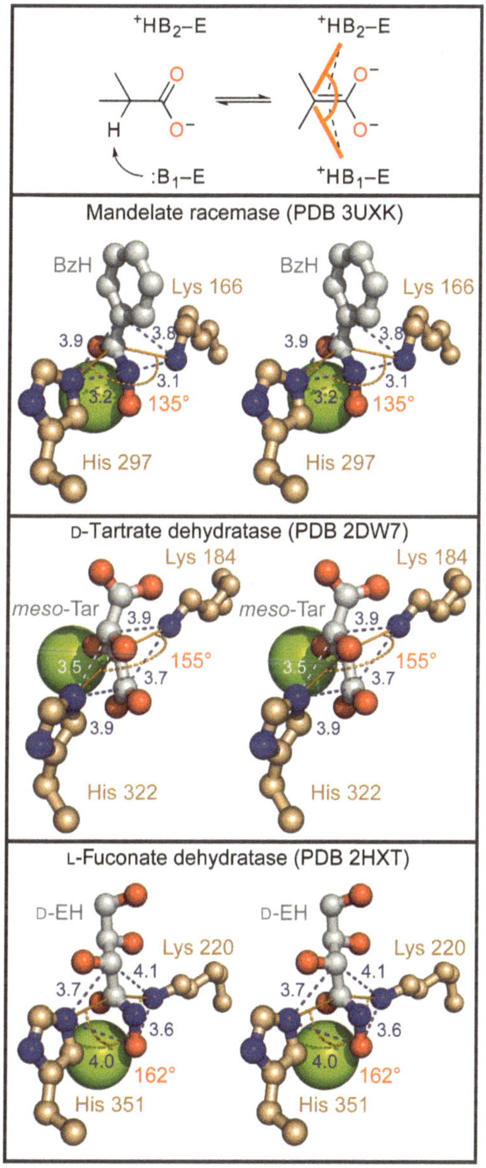

Fig. 3 Positioning of the Brønsted acid–base catalysts. The active sites of enzymes from the MR subgroup (MR (Lietzan et al., 2012), TarD (Yew, Fedorov, Fedorov, Wood, et al., 2006), FucD (Yew, Fedorov, Fedorov, Rakus, et al., 2006), TGD (Yew et al., 2007), and GalrD3 (Groninger-Poe et al., 2014)), as well as other members from other subgroups of the ENS (D-glucarate dehydratase (Gulick, Hubbard, Gerlt, & Rayment, 2000), MLE (Sakai et al., 2009), D-mannonate dehydratase (Rakus et al., 2007), and enolase

(Continued)

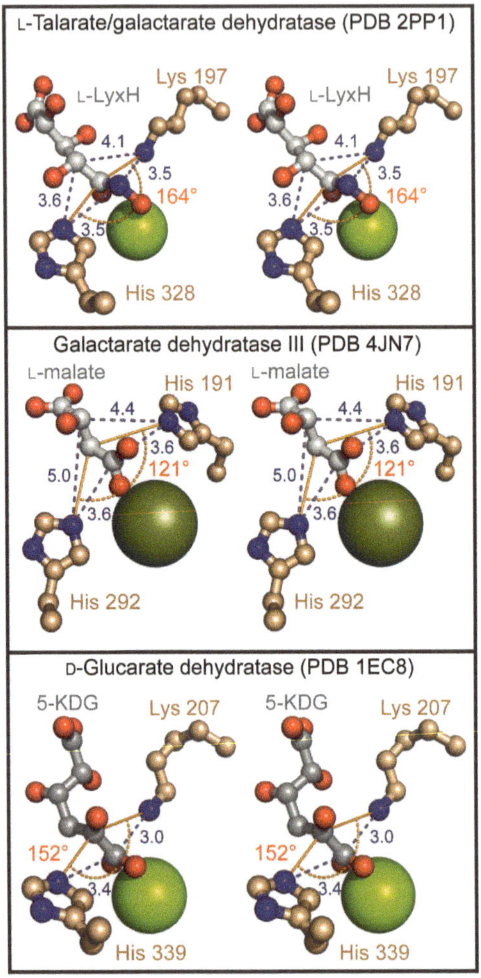

Fig 3—Cont'd (Larsen, Wedekind, Rayment, & Reed, 1996)) are shown for comparison of the orientation of the Brønsted acid–base catalysts with respect to the C—C bond which develops π-bond character upon forming the enol(ate) during the reactions catalyzed by these enzymes. The distances (Å) between the heteroatoms of the Brønsted acid–base catalysts and the β-carbon and the atom that corresponds to the carboxylate carbon of substrates are indicated where applicable. Also shown are the angles formed between the Brønsted acid–base catalysts and the "α-carbon." In most active sites, the Brønsted acid–base catalysts are positioned so that they may interact with the developing π-character of the enol(ate) intermediate, although this does not appear to occur universally (e.g., L-mannonate dehydratase). *Ligand abbreviations*: *2-PG*, 2-phosphoglycerate; *5-KDG*, 5-keto-4-deoxy-D-glucarate; *BzH*, benzohydroxamate; *D-EH*, D-eythronohydroxamate; *D-man*, D-mannonate; *meso-Tar*, (2*R*,3*S*)-tartrate; *L-LyxH*, L-lyxarohydroxamate.

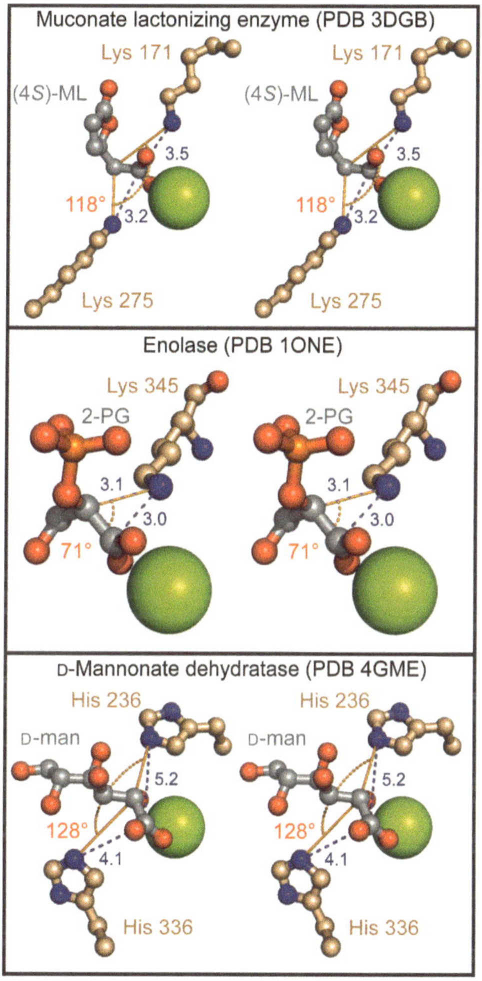

Fig 3—Cont'd

5. MANDELATE RACEMASE STRUCTURE: LIGAND BINDING WITHIN A CATALYTIC CAGE

To control the local environment of bound ligands, enzymes often sequester their ligands in water-occluded active sites, referred to by Richard as *catalytic cages* (Richard, Amyes, Goryanova, & Zhai, 2014). Such protein cages furnish the opportunity for a larger number of interactions between the enzyme and the ligand, as well as permit control of the dielectric of the microenvironment at the active site, which can greatly strengthen polar interactions between the enzyme and the ligand. Given the amenability of MR to crystallization, a variety of X-ray crystal structures for MR have been determined over the past 25 years (Table 2) that have highlighted how the enzyme functions as a catalytic cage.

Initially, recombinant MR was purified from either *Pseudomonas aeruginosa* (Kallarakal et al., 1995; Mitra et al., 1995; Neidhart et al., 1991, 1990, 1988; Schafer et al., 1996; Tsou et al., 1989) or *Escherichia coli* (Landro et al., 1994) transformed with the pSCR1 plasmid containing the gene encoding the enzyme from *P. putida*. When recombinant *P. putida* MR was expressed from *E. coli* as a fusion protein bearing an N-terminal hexahistidine affinity tag (Bearne, St Maurice, & Vaughan, 1999), it was not possible to concentrate the protein to levels suitable for crystallization experiments because the protein was prone to precipitation—a problem that is not unique to MR within the ENS (Vetting et al., 2016; Yew, Fedorov, Fedorov, Wood, et al., 2006). This problem was overcome by switching the affinity tag to an N-terminal StrepII-tag (Narmandakh & Bearne, 2010), which provided ready access to the protein concentrations required for crystallization trials. Subsequently, a number of additional MR structures were solved with a variety of bound ground state and intermediate/TS analogues. These structures have revealed the role(s) of various binding determinants in ground state binding, catalytic turnover, and TS stabilization, and inspired efforts such as those described in Section 4 to quantify the individual side chain contributions to binding and catalysis.

5.1 Ligand Binding Modes

The structures of both wild-type and various mutated forms of MR have been determined using X-ray crystallography. To-date, however, there is no structure of MR without a ligand bound at its active site. The closest description of "unbound MR" comes from the structure of MR with a sulfate anion bound at its active site (PDB ID 2MNR) (Neidhart et al., 1991). In this structure, however, the sulfate anion occupies a similar binding mode to all other substrate analogues with respect to Mg^{2+} coordination and the bridging of the two Brønsted acid–base catalysts Lys 166 and His 297. In the absence of a structure for the true unbound state, it is not presently possible to determine whether ligand binding is accompanied by a conformational change in MR. In addition to the sulfate-bound MR, many structures of MR have been reported with a variety of bound substrate and

Table 2 Structures of MR With Their Bound Ligands and Corresponding Binding Affinities

Ligand	K_i (μM)[a]	$\Delta G_{binding}$ (kcal/mol)	MR Variant	PDB ID
Ligand-free				
Sulfate anion	–	–	Wild-type	2MNR[a]
Ground state ligands				
(*R*)-Mandelate[b]	810[c]	4.2	K166R	1MDL[d,e]
(*S*)-Mandelate	620[c]	4.4	K166R[g]	1MDL[d,e]
(*S*)-Atrolactate	150[f,g] 140[g,h]	5.2 5.3	Wild-type	1MNS[g]
			E317Q	1DTN[i]
			D270N	1MRA[j]
Tartronate	1800[h,k]	3.7	Wild-type	4M6U[k]
Benzilate	670[f,l]	4.3	C92S/K166C/ C264S	4HNC[k]
3,3,3-Trifluoro-2-hydroxy-2-(trifluoromethyl) propanoate	27[h,k]	6.2	Wild-type	4FP1[k]
Intermediate/transition state analogues				

Continued

Table 2 Structures of MR With Their Bound Ligands and Corresponding Binding Affinities—cont'd

Ligand	K_i (μM)	$\Delta G_{binding}$ (kcal/mol)	MR Variant	PDB ID
Benzohydroxamate	9.3[h,m]	6.9	Wild-type	3UXK[n]
Cupferron	2.7[h,o]	7.6	Wild-type	3UXL[n]
Inactivators				
3-Hydroxypyruvate[h]	$k_{inact}/K_i = 83\ M^{-1}\ s^{-1}$	–	Wild-type	4X2P[p]
(*R*)-α-Phenylglycidate[h]	$k_{inact} = 0.011\ s^{-1}$, $K_i = 67\ \mu M$[g]	5.7	Wild-type	1MDR[g]

[a]Neidhart et al. (1991).
[b]Binding affinities are for binding to the wild-type enzyme. For (*R*)- and (*S*)-mandelate, the K_m values are reported, where $K_m = K_S$ (St. Maurice & Bearne, 2002).
[c]St. Maurice and Bearne (2000).
[d]This structure has both (*S*)-mandelate bound (chelating the Mg^{2+}) and (*R*)-atrolactate bound (sandwiched between the 20s loop and the phenyl ring of (*S*)-mandelate).
[e]Kallarakal et al. (1995).
[f]Assayed in the $S \rightarrow R$ reaction direction.
[g]Landro et al. (1994).
[h]Assayed in the $R \rightarrow S$ reaction direction.
[i]Mitra et al. (1995).
[j]Schafer et al. (1996).
[k]Nagar et al. (2014).
[l]Siddiqi et al. (2005).
[m]St. Maurice and Bearne (2000).
[n]Lietzan et al. (2012).
[o]Bourque et al. (2007).
[p]Nagar, Wyatt, St. Maurice, and Bearne (2015).

intermediate/TS analogues (Table 2). Superposition of all of these structures reveals, with one notable exception, no gross conformational changes between MR bound with sulfate, with (*S*)-mandelate, with substrate analogues, or with intermediate/TS analogues (Fig. 4A). The lone exception is the structure of K166R MR with both (*S*)- and (*R*)-mandelate bound (PDB ID 1MDL) (Kallarakal et al., 1995). (*R*)-Mandelate was cocrystallized with K166R MR with the hope of obtaining a structure with the (*R*)-enantiomer bound in the active site; however, the mutant enzyme slowly turned over (*R*)-mandelate to (*S*)-mandelate, resulting in (*S*)-mandelate binding at the active site and (*R*)-mandelate occupying a nonproductive conformation in the hydrophobic cavity. As a consequence of (*R*)-mandelate occupying much of the hydrophobic cavity, the 20s loop is occluded from its normal conformation and occupies an artificially "open" conformation (Fig. 4A; blue trace). It is not clear whether this conformational outlier is relevant to the function of the enzyme; the potential role of conformational changes in the 20s and 50s loops during catalytic turnover will be addressed in more detail in Sections 5.2 and 5.3.

5.1.1 Structures of MR With Bound Ground-State Analogues

(*S*)-Mandelate and the (*S*)-mandelate analogues (*R*)-α-phenylglycidate and (*S*)-atrolactate are bound at the active site in identical conformations. As is the case for all structurally characterized MR subgroup enzymes, the substrate contributes a carboxylate oxygen and an α-OH to complete the octahedral coordination sphere of the active site Mg^{2+} ion (Figs. 1D and 2). This chelating interaction with Mg^{2+} serves to anchor and orient the substrate at the active site and is presumably maintained irrespective of which enantiomer is bound at the active site. A small constellation of active site residues (Glu 317, Lys 164, and Asn 197) provides additional, direct electrostatic and H-bonding interactions to the polar end of the substrate molecule, further anchoring it in place (Fig. 2). The active site is further defined by *R*-specific and *S*-specific pockets within the hydrophobic cavity for binding the phenyl group of (*R*)- and (*S*)-mandelate, respectively (vide infra) (Siddiqi et al., 2005; St. Maurice & Bearne, 2004). In all structures with bound (*S*)-mandelate and (*S*)-mandelate analogues, the α-H or α-H-equivalent group projects toward Lys 166, the *S*-specific base. The *R*-specific base, His 297, is positioned on the opposite side of the active site, poised to abstract the α-proton from (*R*)-mandelate (Fig. 4B–D).

As described in Section 4, in addition to serving as the *R*- and *S*-specific Brønsted acid–base catalysts, His 297 and Lys 166 provide electrostatic

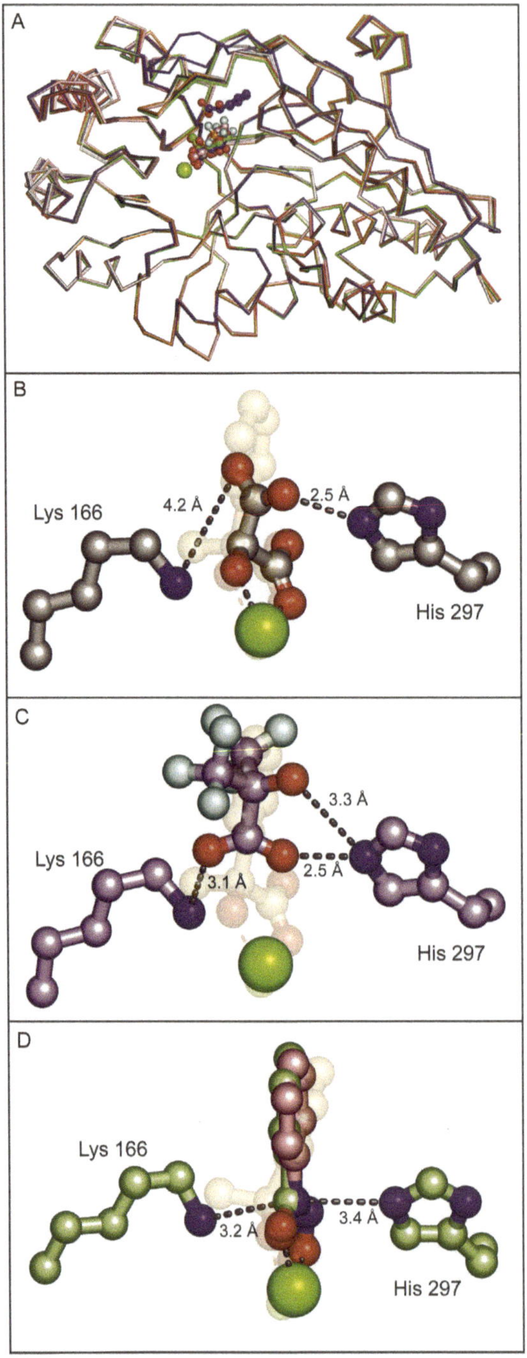

Fig. 4 See legend on opposite page.

stabilization in both the ground state and TS through cation-π/NH-π interactions with the phenyl ring, as well as stabilization through cation-π/NH-π interactions with the π-bond of the enol(ate) of the intermediate/TS (Nagar & Bearne, 2015). The X-ray crystal structures of wild-type MR with bound 3,3,3-trifluoro-2-hydroxy-2-(trifluoromethyl)propanoic acid (TFHTP) and tartronate further support the notion that electrostatic interactions from the two Brønsted acid–base catalysts serve as binding determinants for the ligands of MR and that this phenomenon may extend throughout the MR subgroup (Nagar et al., 2014). It was rationalized that TFHTP would serve as an inhibitor of MR based on the observation that benzilate, which is a substrate–product analogue of mandelate, is a reasonably potent inhibitor of MR (Siddiqi et al., 2005) and that the hydrophobic pocket binds both the phenyl and trifluoromethyl groups with approximately equal affinities (Nagar, Narmandakh, Khalak, & Bearne, 2011). Somewhat surprisingly, TFHTP, which is a substrate–product analogue of the alternative substrate trifluorolactate, was bound with an affinity that was ~20-fold greater than that exhibited for benzilate, making it the most potent ground state analogue inhibitor of MR reported to date, with a K_i value of 27 μM (Nagar et al., 2011). Structural analysis of wild-type MR

Fig. 4 Ligand binding modes in MR. There are many similarities and some differences in the binding modes occupied by various ligands of MR. (A) A structural superposition of the MR peptide backbone bound with sulfate (*gray*, PDB 2MNR), (*S*)-atrolactate (*red*, PDB 1MDR), (*S*)- and (*R*)-mandelate (*blue*, K166R mutation, PDB 1MDL), TFHTP (*orange*, PDB 4FP1), tartronate (*white*, PDB 4M6U), BzH (*pink*, PDB 3UXK), and inactivated with (*R*)-α-phenylglycidate (*green*, PDB 1MNS) illustrates that the overall conformation of MR does not change in the presence of bound (*S*)-mandelate, (*S*)-mandelate analogues, or intermediate analogues. With the exception of the nonproductive binding mode of (*R*)-mandelate within the hydrophobic cavity of K166R MR (CPK-colored *blue ball* and *sticks*), all other ligands are bound in similar orientations within the active site. (B) A structural superposition of tartonate-bound MR in monomer B (CPK-colored white ball and sticks, PDB 4M6U) with (*S*)-atrolactate-bound MR (semitransparent, CPK-colored *orange ball* and *sticks*; PDB 1MDR). The distances between the Brønsted acid–base catalysts and the bridging carboxylate oxygens of tartronate are indicated. (C) A structural superposition of TFHTP-bound MR (CPK-colored *blue ball* and *sticks*, PDB 4FP1) with (*S*)-atrolactate-bound MR (semitransparent, CPK-colored *orange ball* and *sticks*, PDB 1MDR). The distances between the Brønsted acid–base catalysts and the bridging carboxylate oxygens of TFHTP are indicated. (D) A structural superposition of BzH-bound MR (CPK-colored *green ball* and *sticks*, PDB 3UXK) with cupferron-bound MR (CPK-colored *pink ball* and *sticks*, PDB 3UXL) and (*S*)-atrolactate-bound MR (semitransparent, CPK-colored *orange ball* and *sticks*, PDB 1MDR). The distances between the Brønsted acid–base catalysts and the α-carbon of BzH are indicated.

with bound TFHTP revealed that this surprisingly potent competitive inhibition arose because of an unexpected binding mode for TFHTP at the active site (Fig. 4C). Rather than adopting the catalytically relevant conformation observed in the structure of all other ground state analogues, TFHTP did not directly chelate the active site Mg^{2+}. Instead, the carboxylate group of TFHTP formed a bridge between the two Brønsted acid–base catalysts while the trifluoromethyl groups packed intimately within the hydrophobic cavity. Thus, TFHTP takes advantage of the binding potential afforded by the interactions with the two Brønsted acid–base catalysts.

Based on the unexpected binding mode of TFHTP, it was reasoned that tartronate, a substrate analogue of mandelate with a carboxylate in place of the phenyl group, might also be an inhibitor of MR. Indeed, MR bound tartronate with an affinity approximately equal to that of mandelate and with nearly 20-fold greater affinity than that exhibited for lactate, revealing that the carboxylate group of tartronate contributes an additional 1.7 kcal/mol of ground state binding energy relative to the methyl group of lactate (Nagar et al., 2011). The crystal structure determined for tartronate bound at the active site of wild-type MR revealed that tartronate binds in a conformation that chelates Mg^{2+}, similar to other substrate analogues, but with its distal carboxylate group bridging Lys 166 and His 297 to take advantage of the additional electrostatic interactions afforded by these binding determinants (Fig. 4B).

5.1.2 Structures of MR With Bound Intermediate Analogues

As mentioned earlier, MR preferentially stabilizes the altered substrate in the TS by 26 kcal/mol. To investigate the structural basis for this extraordinary discrimination between binding of the ground state and the TS, the structures of MR bound to the intermediate/TS analogues BzH (PDB ID 3UXK) and cupferron (PDB ID 3UXL) were determined (Lietzan et al., 2012). These two intermediate/TS analogues are bound by MR with ~100–370-fold higher affinity than the substrate: $K_i = 9.3$ μM for BzH (St. Maurice & Bearne, 2000), and $K_i = 2.7$ μM for cupferron (Bourque et al., 2007). No gross conformational changes were observed relative to the substrate-bound structures for either of the two structures with bound intermediate/TS analogues (Fig. 4A), suggesting that the enhanced binding affinity of intermediate analogues results from a subtle tuning of multiple interactions at the enzyme active site. With the exception of a slight pivot about the α-carbon, both BzH and cupferron are bound in nearly equivalent positions and conformations at the enzyme active site as the ground-state ligands (Fig. 4D). Nevertheless, the nearly equidistant positioning of the two Brønsted acid–base catalysts from the α-carbon, the planar nature of these ligands, and the sp^2 character of the α-carbon, argue strongly that BzH and cupferron accurately mimic the conformation of the bound *aci*-carboxylate intermediate. The contributions that various binding determinants make to BzH binding are summarized in Fig. 2. Consistent with the contribution of the hydrophobic cavity to TS stabilization in MR (St. Maurice & Bearne, 2004), the binding of BzH is accompanied by a modest constriction of the hydrophobic pocket, exemplified by the movement of residues Phe 52 and Tyr 54 into closer proximity to the phenyl rings of both BzH and cupferron (Lietzan et al., 2012). The structures also suggest that the intermediate/TS analogues chelate the Mg^{2+} at the active site more strongly than do ground state analogues, offering an additional mechanism for preferential stabilization of the intermediate/TS over the ground state by MR (cf. Guthrie & Kluger, 1993).

5.2 The Hydrophobic Cavity and the Active Site Flap

The hydrophobic cavity within the active site of MR is remote from the site of proton abstraction but, nevertheless, makes important contributions to both ground state binding and TS stabilization (Lietzan et al., 2012; St. Maurice & Bearne, 2004). This cavity also contributes to substrate selectivity (Bourque & Bearne, 2008) and its mobility may enable translocation of the phenyl group of the substrate during catalysis (Siddiqi et al., 2005).

The relatively large hydrophobic cavity within the active site of MR is composed of residues from the 20s, 50s, and interdigitating loops and can be partitioned into two pockets: (1) an *S*-specific pocket that encloses the phenyl group of (*S*)-mandelate, with contributions from Leu 93′ of the neighboring interdigitating loop along with residues Phe 52 and Tyr 54 from the 50s loop and Val 29 from the C-terminal end of the 20s loop, and (2) an *R*-specific pocket that encloses the phenyl group of (*R*)-mandelate with contributions from Val 22, Thr 24, Ala 25, and Val 26 of the 20s loop (Fig. 5) (Siddiqi et al., 2005). The separation of the *R*- and *S*-specific pockets is exemplified by the binding of two substrate–product analogues, benzilate and (*S*)-cyclohexylphenylglycolate (Siddiqi et al., 2005), and the substrate–intermediate–product analogue 1,1-diphenyl-1-hydroxymethylphosphonate (Burley & Bearne, 2005) with binding affinities that are similar to those that MR exhibits for the substrate. The crystal structure of a triply mutated MR (C92S/C264S/K166C) with bound benzilate

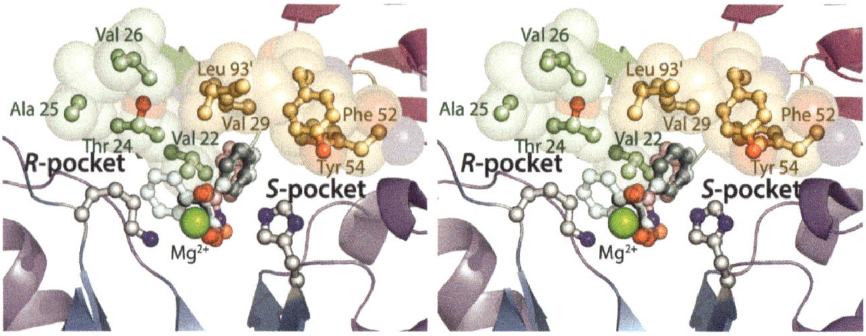

Fig. 5 A close-up stereoview of the MR hydrophobic cavity. The *R*- and *S*-specific pockets from the structure of MR bound with (*S*)-atrolactate (CPK-colored *white ball* and *sticks*, PDB 1MDR) are shown. The side chains from the α + β N-terminal capping domain of 1MDR that contribute to the *S*-specific pocket are illustrated as CPK-colored *orange ball* and *sticks* (Leu 93′ is contributed from the interdigitating loop of a neighboring protomer), while the side chains from the α + β N-terminal capping domain of 1MDR that contribute to the *R*-specific pocket are illustrated as CPK-colored *green ball* and *sticks*. The relative ligand binding positions of BzH (CPK-colored *pink ball* and *sticks*) and benzilate (semitransparent CPK-colored *blue ball* and *sticks*) from a superposition with PDB 3UXK (wild-type MR cocrystallized with BzH) and PDB 4HNC (C92S/K166C/C264S MR cocrystallized with benzilate) illustrate how these ligands are accommodated within the hydrophobic cavity.

(PDB ID 4HNC) offers an initial perspective on how the phenyl group of the substrate might be accommodated in both the *R*- and *S*-specific hydrophobic pockets (Fig. 5) (Nagar et al., 2014). Interestingly, the 20s loop, which makes all of the contributions to the *R*-specific cavity, also serves as the active site flap, which is presumed to close over the active site and form the catalytic cage that excludes solvent and promotes stabilizing interactions with the substrate in the ground state and the altered substrate in the TS. Given the absence of any MR structures with (*R*)-mandelate or (*R*)-mandelate analogues bound at the active site, it is not currently clear whether the conformation of the 20s loop and the *R*-specific pocket is altered by the presence of (*R*)-mandelate.

The binding interactions with the substrate within the hydrophobic cavity are nonspecific, leading to considerable plasticity in the substrate selectivity of MR. In general, MR catalyzes the racemization of a variety of aryl- and heteroaryl-substituted mandelate derivatives, wherein the β,γ-unsaturated or aromatic ring can efficiently delocalize the negative charge of the intermediate (Felfer et al., 2005, 2001; St. Maurice & Bearne, 2004). As alluded to above, (*R*)- and (*S*)-trifluorolactate are also substrates of MR. Although trifluorolactate lacks β,γ-unsaturation, the strongly electron withdrawing CF_3 group adjacent to the α-carbon appears to reduce the C—H acidity sufficiently to permit proton abstraction, albeit MR catalyzes the interconversion of (*R*)- and (*S*)-trifluorolactate with an efficiency (k_{cat}/K_m) that is ~400-fold less than that observed when mandelate is the substrate (Nagar et al., 2011).

The general contributions to substrate selectivity and TS stabilization from the hydrophobic cavity have been probed through site-directed mutagenesis and the kinetic analysis of the resulting series of MR variants. A particularly extensive study was performed on the residues contributed by the 20s loop of MR (Bourque & Bearne, 2008). In most cases, substitutions to bulkier hydrophobic residues in the 20s loop were well tolerated by the enzyme, with minimal impact on catalytic efficiency in either the $R \rightarrow S$ or $S \rightarrow R$ reaction directions. This tolerance can be attributed to the conformational flexibility of the 20s loop/active site flap such that it is able to accommodate an increase in steric bulk without significantly compromising the ability of the hydrophobic cavity to interact with the substrate, leaving catalytic efficiency relatively unperturbed. Conversely, bulkier mutations in the more rigid 50s loop (F52W, Y54W, F52W/Y54W) restricted the size of the *S*-specific pocket without any ability to compensate for the added steric bulk, leading to a significant reduction in the catalytic efficiency in both reaction directions and, when compared to the wild-type enzyme, a greater reduction in the binding affinity for (*S*)-mandelate than for (*R*)-mandelate (Siddiqi et al., 2005). The kinetic parameters of the MR variants bearing these mutations of the 50s loop are fully consistent with the notion that the phenyl group of the substrate translocates between the *R*- and *S*-specific pockets during catalysis, while the carboxylate and α-OH groups remain fixed in place through numerous polar interactions at the active site (discussed in detail later). Such translocation of groups through an active site has also been described for α-methylacyl-coenzyme A racemase. Structural studies by Wierenga and colleagues revealed that the aromatic and fatty acid side chains of the substrates ibuprofenoyl-coenzyme A and 2-methylmyristoyl-coenzyme A, respectively, migrate through a commodious, hydrophobic active site during the epimerization reaction catalyzed by this enzyme (Bhaumik et al., 2007).

5.3 Substrate and Enzyme Motion During Catalysis

MR binds and efficiently interconverts the enantiomers of mandelic acid despite retaining multiple polar binding interactions with the substrate carboxylate and α-OH groups, and a fixed coordination geometry around the active site Mg^{2+}. The number and strength of polar interactions with the substrate strongly implies that the substrate carboxylate and α-OH groups remain fixed in place during the reaction, while the α-carbon and phenyl ring move within the active site to enable inversion of stereochemistry at the α-carbon (Fig. 6A) (Lietzan et al., 2012; Siddiqi et al., 2005). This binding configuration follows the four-contact point model (Hanson, 1981; Hanson & Rose, 1975; Mesecar & Koshland, 2000a, 2000b; Topiol & Sabio, 1989), in which three different groups on the ligand interact specifically with the coplanar binding determinants A, B, and C on the enzyme (Fig. 6A). A fourth binding determinant, D or D′, interacts with group H on a stereocenter such that D would interact with one enantiomer and D′ would interact with the other enantiomer. This binding orientation has also been referred to as mirror-image packing (Hanson, 1981) wherein the coplanar binding determinants A, B, and C define the mirror plane, and the stereogenic α-carbon atoms of the enantiomers reside on either side of the mirror plane.

This hypothesis is supported by several observations from kinetic, computational, and structural studies. "Bulky" mutations in the *S*-specific pocket alter the binding affinity of (*S*)-mandelate to a greater degree than (*R*)-mandelate, indicating that there are two discrete pockets for binding the phenyl ring of the (*R*)- and (*S*)-specific enantiomers of mandelic acid (Siddiqi et al., 2005). In addition, inhibition studies showed that the substrate–product analogue benzilate (Siddiqi et al., 2005) and the substrate–intermediate–product analogue 1,1-diphenyl-α-hydroxymethylphosphonate (Burley & Bearne, 2005) are bound by MR with affinities equivalent to or slightly greater than the substrate, consistent with an active site that is structured with nonoverlapping pockets to accommodate phenyl rings in multiple conformations. Computational studies on the substrates vinylglycolate (Garcia-Viloca, Gonzalez-Lafont, & Lluch, 2001) and mandelate (Prat-Resina, Gonzalez-Lafont, & Lluch, 2005) have suggested that the α-carbon of the substrate moves such that it is located closer to the Brønsted base catalyzing the proton abstraction reaction. Indeed, in the X-ray crystal structures of MR with the bound intermediate/TS analogues BzH and cupferron, the positions of both the phenyl ring and the α-carbon are altered relative to their positions in the structures of bound (*S*)-mandelate and (*S*)-atrolactate, while the positions of the α-OH and carboxylate groups are unchanged (Fig. 4D). With the notable exception of the crystal structures of MR with bound TFHTP and (*R*)-mandelate

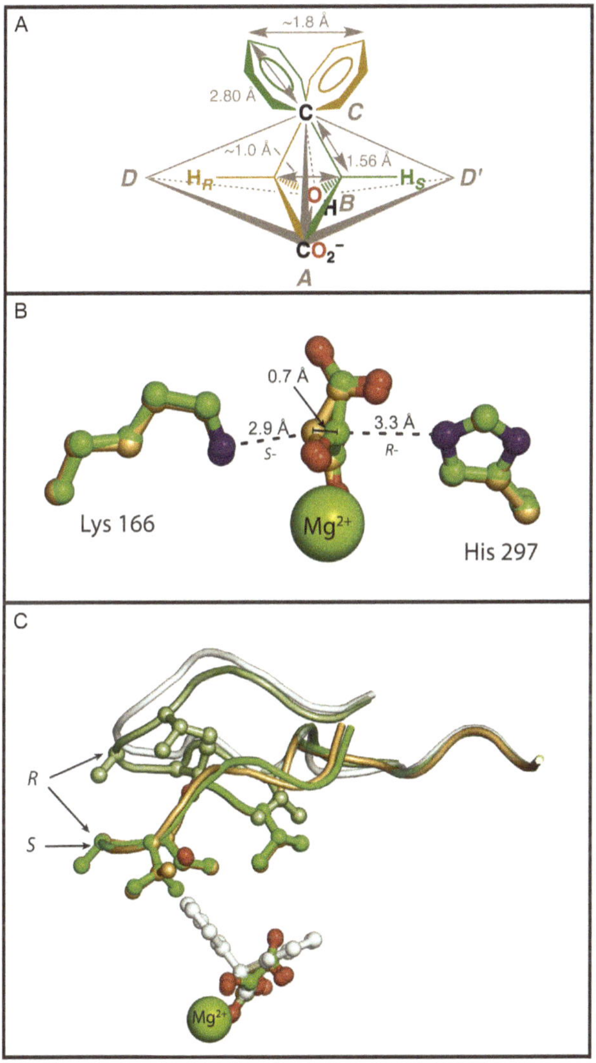

Fig. 6 Substrate and enzyme motions associated with catalysis. Catalytic turnover in MR is expected to involve significant motions of both the substrate and the enzyme. (A) To illustrate substrate motion during catalysis, (*R*)- and (*S*)-mandelate are diagrammed at the active site of MR using the four-location model (Mesecar & Koshland, 2000a, 2000b). The carboxylate, hydroxyl, and phenyl groups are bound within the same plane (sites *A*–*C*), while the α-proton can occupy either of two sites (*D* or *D*′) depending on which enantiomer is bound. Bond lengths from the crystal structure of MR with bound (*S*)-atrolactate (PDB 1MDR) were used to estimate the distance traveled by the *para* carbon atom of the phenyl ring during catalysis. If the locations of the carboxylate and hydroxyl groups are invariant and the β-carbons of (*R*)- and (*S*)-mandelate are within the same plane, then the α-carbon is predicted to move ~1.0 Å as it undergoes

(Continued)

described earlier, cocrystal structures of MR with of all other bound ligands reveal an identical, inflexible binding mode centered around Mg^{2+} for the α-OH and carboxylate groups of the ligand (Fig. 4). The structure of the substrate–product analogue benzilate, bound in the active site of a MR triple mutant (C92S/C264S/K166C), demonstrates that two phenyl rings can simultaneously be accommodated in both the *R*- and *S*-specific hydrophobic pockets (Fig. 5). It should be noted, however, that obtaining the X-ray crystal structure with benzilate required mutation of the Brønsted acid–base catalyst, Lys 166, suggesting that the true *R*-specific cavity may undergo substantial conformational rearrangements, particularly at Lys 166, in order to accommodate the phenyl ring of (*R*)-mandelate. Finally, while tartronate does not contain a stereogenic center, the X-ray crystal structure of MR with bound tartronate revealed two different orientations for the α-carbon of tartronate in opposing active sites that follows the mirror-image packing model: one tartronate binding orientation closely resembled the orientation of (*S*)-mandelate, with the α-carbon occupying a nearly identical position as it does for (*S*)-mandelate. In the other active site, however, tartronate displayed a different binding orientation, with the α-carbon ~1 Å closer to the (*R*)-specific Brønsted acid–base catalyst His297, consistent with this binding

Fig. 6—Cont'd a Walden inversion while the *para*-position of the phenyl ring may move ~1.8 Å. (B) A structural superposition of bound tartronate in monomers A (*green*) and B (*gold*) of MR (PDB 4M6U), centered on the active site, shows the distances between the α-carbon and His 297 and Lys 166. The orientation of tartronate in monomer A may mimic that of (*R*)-mandelate, while the orientation of tartronate in monomer B mimics that of (*S*)-mandelate. The distance between the overlaid α-carbon positions (~0.7 Å) correlates well with the predictions in the substrate motion diagram. (C) A superposition of the structures of the MR–tartronate monomer A (*green*), MR–tartronate monomer B (*gold*), and the triple mutant MR–benzilate complex (*cyan*). The 20s loop of the triple mutant MR–benzilate complex adopts an open conformation, while it adopts a closed conformation in the MR–tartonate complex in monomer B. The 20s loop in monomer A of the MR–tartronate complex, which mimics the α-carbon position of (*R*)-mandelate, occupies both an open (*light green*) and closed conformation (*dark green*). These structures suggest that a significant conformational change in the 20s loop may accompany catalysis in MR. *Panel (A) adapted with permission from Siddiqi, F., Bourque, J. R., Jiang, H., Gardner, M., St. Maurice, M., Blouin C., et al. (2005). Perturbing the hydrophobic pocket of mandelate racemase to probe phenyl motion during catalysis.* Biochemistry, *44(25), 9013–9021. Copyright 2005 American Chemical Society; Panel (C) adapted with permission from Nagar, M., Lietzan, A. D., St. Maurice, M., & Bearne, S. L. (2014). Potent inhibition of mandelate racemase by a fluorinated substrate-product analogue with a novel binding mode.* Biochemistry, *53(7), 1169–1178. Copyright 2014 American Chemical Society.*

orientation mimicking the position hypothesized to be occupied by the α-carbon in (*R*)-mandelate (Fig. 6B) (Nagar et al., 2014). Taken together, these kinetic, computational, and structural data consistently support a model where the α-carbon and phenyl group of the substrate migrate through space during the course of the reaction. Certainly, to discriminate between the substrate in the ground state and the altered substrate in the TS, MR must respond to the change in the physical positions of these groups during the course of the reaction and, at the appropriate point, enhance binding interactions to effect TS stabilization.

The X-ray crystal structures of MR with bound tartronate and benzilate both indicate that the enzyme itself may undergo a conformational transition during substrate racemization. Intriguingly, these structures indicate that when either the α-carbon of tartronate or the phenyl group of benzilate occupies a position that mimics a conformation analogous to (*R*)-mandelate, the 20s loop adopts an "open" conformation (Fig. 6C). It is clear from the structure of the triple mutant–benzilate complex that the loop must adopt an alternative conformation in order to accommodate the added steric bulk of the phenyl group binding in a position equivalent to (*R*)-mandelate. The occurrence of a conformational change in the enzyme during the conversion of (*S*)- to (*R*)-mandelate may account for our observations that MR crystals grown in the presence of (*S*)-atrolactate quickly crack when soaked with (*R*)-atrolactate. Increased conformational entropy for the active site flap in the presence of (*R*)-mandelate may also explain why attempts to cocrystallize or soak MR with analogues of the (*R*)-enantiomer have, to date, proven unsuccessful. These observations imply that a conformational change and/or increased conformational freedom in the 20s loop is likely to accompany the conversion of (*S*)- to (*R*)-mandelate.

It is noteworthy that truncation of Lys 166 through site-directed mutagenesis was required in order to permit the structural determination of the MR–benzilate complex. Lys 166 extends into the *R*-specific pocket and its "normal" conformation observed in the structure of the MR–BzH complex clashes with the position of the phenyl ring of benzilate (Fig. 5). Thus, we expect that the position of Lys 166 might change dramatically in the presence of (*R*)-mandelate. The observed pK_a values determined from k_{cat}-pH profiles in *both* the $R \rightarrow S$ and $S \rightarrow R$ reaction directions are ~6 and ~10 for the ascending and descending limbs, respectively (Landro et al., 1991). Considering that the K166R mutation of the *S*-specific base only altered the apparent pK_a of the ascending limb in the $S \rightarrow R$ reaction direction, with no alteration of the two observed pK_a values in the $R \rightarrow S$ reaction direction, or the descending limb in the $S \rightarrow R$ reaction direction (Kallarakal et al., 1995), Gerlt suggested that the simplest interpretation of these data is that the apparent pK_a values estimated from the ascending limbs correspond to the ionization of the conjugate acid of the Brønsted base that abstracts the α-proton from the substrate (Gerlt, 1998). The implication of this result is that Lys must alter its pK_a value, such that it exists as its conjugate acid when the (*R*)-enantiomer is bound, and as its deprotonated basic form when the (*S*)-enantiomer is bound. In order for the difference in chirality of the bound substrate to effect these perturbations in the pK_a, a significant change is expected in the local environment of Lys 166. With this in mind, it is reasonable to anticipate that conformational changes in the hydrophobic cavity and an altered orientation of Lys 166 will accompany the transition from (*S*)- to (*R*)-mandelate.

Despite the wealth of structural information available, several questions about MR catalysis remain unresolved owing, in part, to a structural picture that remains incomplete. For example, what is the conformation of the enzyme in the presence of (*R*)-mandelate and does it differ substantially from the conformation when either (*S*)-mandelate or (*S*)-atrolactate is bound? What is the structure of the metal-free enzyme and how does the presence of a metal ion alter the architecture of the active site? The answers to these questions will furnish insights into how enzymes discriminate between enantiomers and alter acid–base chemistry at active sites, and also delineate the structural role of the divalent cation present at the active sites of other ENS enzymes.

6. EVOLUTIONARY RELATIONSHIPS: A MECHANISTIC LINK TO ALDOLASES?

6.1 Inactivation of MR by 3-Hydroxypyruvate

While exploring the inhibition of MR by a variety of three-carbon acids, it was discovered that 3-hydroxypyruvate (3-HP) irreversibly inactivates the enzyme (Nagar et al., 2015). Trypsinolysis of the inactivated enzyme followed by MS/MS analysis of the peptides revealed that Lys 166 had been modified, forming an 86-Da adduct. Considering that nucleophilic attack at the 3-position was unlikely and that Schiff-base formation had not previously been reported for members of the ENS, this finding was quite surprising. Structural analysis was required to determine if covalent modification occurred at the 2-position or 3-position (via isomerization of 3-HP to tartronate semialdehyde, possibly catalyzed by magnesium; Hedrick &

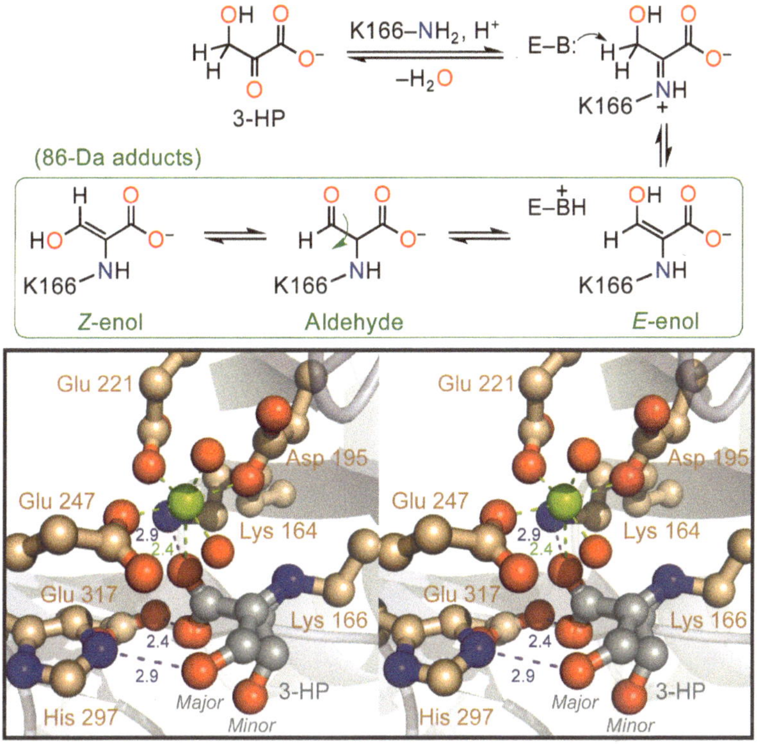

Fig. 7 Formation and structure of the 3-HP-derived adduct. MR is inactivated by 3-HP via formation of a Schiff-base that is subsequently deprotonated (possibly by His 297) to yield an 86-Da enol(ate)/aldehyde adduct. This adduct is stabilized at the active site through the multiple interactions indicated (PDB 4X2P) (Nagar et al., 2015). Distances (Å) between the heteroatoms of active site residues and the adduct are shown. Mg^{2+} (*green*) is coordinated by two water molecules.

Sallach, 1961). As shown in Fig. 7, Lys 166 added at the 2-position of 3-HP to yield an aldehyde/enol(ate)-lysine adduct. The proposed mechanism for modification involves the initial formation of a Schiff-base, which is subsequently deprotonated to yield the 86-Da adduct. Interestingly, covalent modification did not occur with the H297N mutant, suggesting that this residue may serve to deprotonate the initial Schiff-base adduct. The observed electron density was consistent with two alternative conformations for the adduct: major (~70%) and minor (~30%) conformations of either the (*E*)-enol(ate) or the aldehyde. Although it was not possible to distinguish between the enol(ate) and the aldehyde from the structural information, it is not unreasonable to expect that the bound species is stabilized as the enol(ate) since the active site of MR is optimized to stabilize such a species

(Scheme 1). Indeed, the adduct is stabilized at the active site through multiple H-bonding interactions between the carboxyl group and the side chains of Lys 164 and Glu 317, and between the 3-OH group and the side chain of His 297. In addition, one oxygen of the carboxylate is directly coordinated to the Mg^{2+}. This interaction likely facilitates proton abstraction from the Schiff-base. The protein flap (20s loop) that typically closes over the active site when either a ground state or a intermediate/TS analogue is bound was found to be in the open conformation such that the covalently modified active site is exposed to solvent. Closure of the flap to form the catalytic cage, as discussed earlier, may not occur because the hydrophobic group (i.e., phenyl ring) present on substrate and intermediate/TS analogues is not present. Alternatively, modification of Lys 166 may indirectly influence the conformation of the 20s loop (cf. Section 5.3).

6.2 Mechanistic Similarities

Although the members of the ENS are mechanistically diverse (Babbitt et al., 1995), they all share a bidomain structure: a $(\beta/\alpha)_7\beta$-barrel domain (modified $(\beta/\alpha)_8$-barrel or TIM barrel) and an $\alpha + \beta$ capping domain arising from N- and C-terminal segments of the protein (Babbitt et al., 1996; Glasner, Gerlt, & Babbitt, 2006). Indeed, the $(\alpha/\beta)_8$-barrel folding topology is highly prevalent among enzymes (Brändén, 1991; Farber, 1993; Farber & Petsko, 1990; Gerlt & Raushel, 2003; Nagano, Orengo, & Thornton, 2002; Reardon & Farber, 1995; Wierenga, 2001) and is believed to have arisen via divergent evolution from a common ancestor or limited number of progenitor enzymes (Copley & Bork, 2000; Janeček, 1995; Janeček & Baláz, 1995; Nagano et al., 2002; Raine, Scrutton, & Mathews, 1994; Reardon & Farber, 1995; Wilmanns, Hyde, Davies, Kirschner, & Jansonius, 1991). For example, many aldolases share this folding topology with members of the enolase and *N*-acetylneuraminate lyase (NAL) superfamilies. Moreover, the aldolases share a similar mechanistic theme with the ENS enzymes, i.e., the formation of a "masked" carbanionic intermediate. In the class I aldolases, formation of such an intermediate is facilitated via formation of an imine/enamine "electron-sink" with an active site Lys residue. On the other hand, the class II aldolases utilize a transition metal ion (usually Zn^{2+}) as an essential Lewis acid cofactor. It has been postulated that the ancestor of these two classes of aldolases was a Schiff-base-forming enzyme that subsequently gave rise to a lineage of metal-dependent class II aldolases (Nagano et al., 2002). Interestingly, common ancestry between the class I and II aldolases and the ENS has been suggested based on several comparisons of sequences, structures, and functions (Brändén, 1991; Copley & Bork, 2000; Nagano et al., 2002; Wilmanns et al., 1991). Certainly, the mechanistic link of the metal ion-dependence of the class II aldolases to the ENS is clear, considering the presence of a divalent cation at the active site of all ENS enzymes (Rakus et al., 2008); however, such a mechanistic link between the class I aldolases and the ENS is not obvious.

The ability of MR to form and deprotonate a Schiff-base intermediate suggested that a mechanistic relationship between the class I aldolases and the ENS may indeed exist and may be a relic of an activity possessed by a promiscuous progenitor enzyme (O'Brien & Herschlag, 1999). The irreversible inhibition of the NAL superfamily members NAL from *Clostridium fringens* (DeVries & Binkley, 1972) and *E. coli* (Lawrence et al., 1997) and dihydrodipicolinate synthase from *E. coli* (Dobson et al., 2008) by 3-HP has also been reported, and the X-ray crystal structures of these covalently modified enzymes revealed the presence of an aldehyde/enol(ate)-lysine adduct similar to the one observed for MR inactivated by 3-HP (Dobson et al., 2008; Lawrence et al., 1997).

As a caveat, however, it is possible that the inactivation of MR by 3-HP is fortuitous. Indeed, treatment of TarD and TGD, both from the MR subgroup of the ENS, with 3-HP did not produce formation of an aldehyde/enol(ate)-lysine adduct (S. Aboushawareb & S.L. Bearne, unpublished observations). This difference in reactivity could arise because the active site of MR is arranged to catalyze a racemization reaction as opposed to the β-elimination of water catalyzed by these two dehydratases.

7. CONCLUDING REMARKS

Over the past three decades, MR has been the subject of intense structural and kinetic characterization, cementing its status as a key paradigm for understanding fundamental aspects of enzyme catalysis. In this review, we have summarized a wide range of studies that have clarified the molecular description of catalysis by MR and which, by extension, have offered critical insights into a range of mechanistic strategies that enzymes may adopt to promote proton abstraction from carbon acids, a highly pervasive reaction in nature. These findings are being extended to an ever-growing number of mechanistically related enzymes, particularly those in the ENS. Past and present studies on MR further illustrate its importance as a system for understanding enzyme evolution, with recent evidence suggesting evolutionary and mechanistic connections within and beyond the enzymes of the ENS. A deeper understanding of precisely how nature utilizes the common α/β-barrel scaffold as a reaction-platform to catalyze proton abstraction, and how substrate specificity can be rationally altered will be required to develop "designer enzymes" based on the ENS. As the archetypical enzyme of the ENS, future structural and mechanistic studies of MR will play a crucial role in developing this understanding.

ACKNOWLEDGMENTS

We express thanks to the funding agencies that have supported our research, including the Natural Sciences and Engineering Research Council (NSERC) of Canada (Discovery Grants to S.L.B.) and the National Science Foundation (MRI-R2 Grant DBI-0959442 to M.St.M.). We also express our gratitude to our trainees who have worked on MR in our labs over the years.

REFERENCES

Akiva, E., Brown, S., Almonacid, D. E., Barber, A. E., Custer, A. F., Hicks, M. A., et al. (2014). The structure–function linkage database. *Nucleic Acids Research*, *42*(D1), D521–D530.

Albery, W. J. (1980). The application of the Marcus relation to reactions in solution. *Annual Review of Physical Chemistry*, *31*, 227–263.

Albery, W. J. (1982). Application of the Marcus relation to concerted proton transfers. *Journal of the Chemical Society, Faraday Transactions 1*, *78*(5), 1579–1590.

Amyes, T. L., & Richard, J. P. (1996). Determination of the pK_a of ethyl acetate: Brønsted correlation for deprotonation of a simple oxygen ester in aqueous solution. *Journal of the American Chemical Society*, *118*(13), 3129–3141.

Amyes, T. L., & Richard, J. P. (2007). *Proton transfers to and from carbon in model reactions*. New York: Wiley-VCH Verlag GmbH & Co.

Babbitt, P. C., & Gerlt, J. A. (1997). Understanding enzyme superfamilies. Chemistry as the fundamental determinant in the evolution of new catalytic activities. *The Journal of Biological Chemistry*, *272*(49), 30591–30594.

Babbitt, P. C., & Gerlt, J. A. (2000). New functions from old scaffolds: How nature reengineers enzymes for new functions. *Advances in Protein Chemistry*, *55*, 1–28.

Babbitt, P. C., Hasson, M. S., Wedekind, J. E., Palmer, D. R., Barrett, W. C., Reed, G. H., et al. (1996). The enolase superfamily: A general strategy for enzyme-catalyzed abstraction of the α-protons of carboxylic acids. *Biochemistry*, *35*(51), 16489–16501.

Babbitt, P. C., Mrachko, G. T., Hasson, M. S., Huisman, G. W., Kolter, R., Ringe, D., et al. (1995). A functionally diverse enzyme superfamily that abstracts the α-protons of carboxylic acids. *Science*, *267*(5201), 1159–1161.

Bartlett, P. A., & Marlowe, C. K. (1983). Phosphonamidates as transition-state analogue inhibitors of thermolysin. *Biochemistry*, *22*(20), 4618–4624.

Bauer, C., Boy, M., Faber, K., Felfer, U., & Voss, H. (2001). Activation of mandelate racemase via immobilization in lyotropic liquid crystals for biocatalysis in organic solvents: Application and modelling. *Journal of Molecular Catalysis B: Enzymatic*, *16*(2), 91–100.

Bearne, S. L. (2017). The interdigitating loop of the enolase superfamily as a specificity binding determinant or 'flying buttress'. *Biochimica Biophysica Acta – Proteins and Proteomics*, *1865*(5), 619–630.

Bearne, S. L., & Spiteri, R. J. (2005). Reduction of intrinsic kinetic and thermodynamic barriers for enzyme-catalysed proton transfers from carbon acid substrates. *Journal of Theoretical Biology*, *233*(4), 563–571.

Bearne, S. L., St Maurice, M., & Vaughan, M. D. (1999). An assay for mandelate racemase using high-performance liquid chromatography. *Analytical Biochemistry*, *269*(2), 332–336.

Bearne, S. L., & Wolfenden, R. (1997). Mandelate racemase in pieces: Effective concentrations of enzyme functional groups in the transition state. *Biochemistry*, *36*(7), 1646–1656.

Bernasconi, C. F. (1992). The principle of non-perfect synchronization. *Advances in Physical Organic Chemistry*, *27*, 119–238.

Bhat, S. G., & Vaidyanathan, C. S. (1976). Involvement of 4-hydroxymandelic acid in the degradation of mandelic acid by *Pseudomonas convexa*. *Journal of Bacteriology*, *127*(3), 1108–1118.

Bhaumik, P., Schmitz, W., Hassinen, A., Hiltunen, J. K., Conzelmann, E., & Wierenga, R. K. (2007). The catalysis of the 1,1-proton transfer by α-methyl-acyl-CoA racemase is coupled to a movement of the fatty acyl moiety over a hydrophobic, methionine-rich surface. *Journal of Molecular Biology*, *367*(4), 1145–1161.

Bourque, J. R., & Bearne, S. L. (2008). Mutational analysis of the active site flap (20s loop) of mandelate racemase. *Biochemistry*, *47*(2), 566–578.

Bourque, J. R., Burley, R. K., & Bearne, S. L. (2007). Intermediate analogue inhibitors of mandelate racemase: *N*-Hydroxyformanilide and cupferron. *Bioorganic & Medicinal Chemistry Letters*, *17*(1), 105–108.

Brändén, C.-I. (1991). The TIM barrel—The most frequently occurring folding motif in proteins. *Current Opinion in Structural Biology*, *1*(6), 978–983.

Burley, R. K., & Bearne, S. L. (2005). Inhibition of mandelate racemase by the substrate-intermediate-product analogue 1,1-diphenyl-1-hydroxymethylphosphonate. *Bioorganic & Medicinal Chemistry Letters*, *15*(19), 4342–4344.

Burley, S. K., & Petsko, G. A. (1986). Amino-aromatic interactions in proteins. *FEBS Letters*, *203*(2), 139–143.

Chiang, Y., Kresge, A. J., Pruszynski, P., Schepp, N. P., & Wirz, J. (1990). The enol of mandelic acid, detection, acidity in aqueous solution, and estimation of the keto-enol equilibrium constant and carbon acidity of mandelic acid. *Angewandte Chemie (International Ed. in English)*, *29*(7), 792–794.

Chiang, Y., Kresge, A. J., Santaballa, J. A., & Wirz, J. (1988). Ketonization of acetophenone enol in aqueous buffer solutions. Rate-equilibrium relations and mechanism of the "uncatalyzed" reaction. *Journal of the American Chemical Society*, *110*(16), 5506–5510.

Chiang, Y., Kresge, A. J., Schepp, N. P., & Xie, R.-Q. (2000). Generation of the enol of methyl mandelate by flash photolysis of methyl phenyldiazoacetate in aqueous solution and study of rates of ketonization of this enol in that medium. *The Journal of Organic Chemistry*, *65*(4), 1175–1180.

Cohen, A. O., & Marcus, R. A. (1968). Slope of free energy plots in chemical kinetics. *The Journal of Physical Chemistry*, *72*(12), 4249–4256.

Copley, R. R., & Bork, P. (2000). Homology among $(\beta\alpha)_8$ barrels: Implications for the evolution of metabolic pathways. *Journal of Molecular Biology*, *303*(4), 627–641.

DeVries, G. H., & Binkley, S. B. (1972). 3-Hydroxy-*N*-acetylneuraminic acid: Synthesis and inhibitory properties. *Archives of Biochemistry and Biophysics*, *151*(1), 243–250.

Dobson, R. C., Griffin, M. D., Devenish, S. R., Pearce, F. G., Hutton, C. A., Gerrard, J. A., et al. (2008). Conserved main-chain peptide distortions: A proposed role for Ile 203 in catalysis by dihydrodipicolinate synthase. *Protein Science*, *17*(12), 2080–2090.

Durham, D. R. (1984). Initial reactions involved in the dissimilation of mandelate by *Rhodotorula graminis*. *Journal of Bacteriology*, *160*(2), 778–780.

Eigen, M. (1964). Proton transfer, acid-base catalysis, and enzymatic hydrolysis. *Angewandte Chemie (International Ed. in English)*, *3*(1), 1–72.

El Gihani, M. T., & Williams, J. M. (1999). Dynamic kinetic resolution. *Current Opinion in Chemical Biology*, *3*(1), 11–15.

Fan, C. W., Xu, G. C., Ma, B. D., Bai, Y. P., Zhang, J., & Xu, J. H. (2015). A novel D-mandelate dehydrogenase used in three-enzyme cascade reaction for highly efficient synthesis of non-natural chiral amino acids. *Journal of Biotechnology*, *195*, 67–71.

Farber, G. K. (1993). An α/β-barrel full of evolutionary trouble. *Current Opinion in Structural Biology*, *3*(3), 409–412.

Farber, G. K., & Petsko, G. A. (1990). The evolution of α/β barrel enzymes. *Trends in Biochemical Sciences*, *15*(6), 228–234.

Fee, J. A., Hegeman, G. D., & Kenyon, G. L. (1974). Mandelate racemase from *Pseudomonas putida*. Subunit composition and absolute divalent metal ion requirement. *Biochemistry*, *13*(12), 2528–2532.

Felfer, U., Goriup, M., Koegl, M. F., Wagner, U., Larissegger-Schnell, B., Faber, K., et al. (2005). The substrate spectrum of mandelate racemase: Minimum structural requirements for substrates and substrate model. *Advanced Synthesis and Catalysis*, *347*(7–8), 951–961.

Felfer, U., Strauss, U. T., Kroutil, W., Fabian, W. M. F., & Faber, K. (2001). Substrate spectrum of mandelate racemase: Part 2. (hetero)-aryl-substituted mandelate derivatives and modulation of activity. *Journal of Molecular Catalysis B: Enzymatic*, *15*(4–6), 213–222.

Gadler, P., Glueck, S. M., Kroutil, W., Nestl, B. M., Larissegger-Schnell, B., Ueberbacher, B. T., et al. (2006). Biocatalytic approaches for the quantitative production of single stereoisomers from racemates. *Biochemical Society Transactions*, *34*(2), 296–300.

Garcia-Viloca, M., Gonzalez-Lafont, A., & Lluch, J. M. (2001). A QM/MM study of the racemization of vinylglycolate catalyzed by mandelate racemase enzyme. *Journal of the American Chemical Society*, *123*(4), 709–721.

Gerlt, J. A. (1998). Enzyme-catalyzed proton transfer reactions to and from carbon. In S. M. Hecht (Ed.), *Bioorganic chemistry: Peptides and proteins* (pp. 279–311). New York: Oxford University Press.

Gerlt, J. A., & Babbitt, P. C. (1998). Mechanistically diverse enzyme superfamilies: The importance of chemistry in the evolution of catalysis. *Current Opinion in Chemical Biology*, *2*(5), 607–612.

Gerlt, J. A., & Babbitt, P. C. (2001). Divergent evolution of enzymatic function: Mechanistically diverse superfamilies and functionally distinct suprafamilies. *Annual Review of Biochemistry*, *70*, 209–246.

Gerlt, J. A., Babbitt, P. C., Jacobson, M. P., & Almo, S. C. (2012). Divergent evolution in the enolase superfamily: Strategies for assigning functions. *The Journal of Biological Chemistry*, *287*(1), 29–34.

Gerlt, J. A., Babbitt, P. C., & Rayment, I. (2005). Divergent evolution in the enolase superfamily: The interplay of mechanism and specificity. *Archives of Biochemistry and Biophysics*, *433*(1), 59–70. http://dx.doi.org/10.1016/j.abb.2004.07.034.

Gerlt, J. A., & Gassman, P. G. (1992). Understanding enzyme-catalyzed proton abstraction from carbon acids: Details of stepwise mechanisms for β-elimination reactions. *Journal of the American Chemical Society*, *114*(15), 5928–5934.

Gerlt, J. A., & Gassman, P. G. (1993a). An explanation for rapid enzyme-catalyzed proton abstraction from carbon acids: Importance of late transition states in concerted mechanisms. *Journal of the American Chemical Society*, *115*(24), 11552–11568.

Gerlt, J. A., & Gassman, P. G. (1993b). Understanding the rates of certain enzyme-catalyzed reactions: Proton abstraction from carbon acids, acyl-transfer reactions, and displacement reactions of phosphodiesters. *Biochemistry*, *32*(45), 11943–11952.

Gerlt, J. A., Kenyon, G. L., Kozarich, J. W., Neidhart, D. C., & Petsko, G. A. (1992). Mandelate racemase and class-related enzymes. *Current Opinion in Structural Biology*, *2*(5), 736–742.

Gerlt, J. A., Kozarich, J. W., Kenyon, G. L., & Gassman, P. G. (1991). Electrophilic catalysis can explain the unexpected acidity of carbon acids in enzyme-catalyzed reactions. *Journal of the American Chemical Society*, *113*(25), 9667–9669.

Gerlt, J. A., Kreevoy, M. M., Cleland, W., & Frey, P. A. (1997). Understanding enzymic catalysis: The importance of short, strong hydrogen bonds. *Chemistry & Biology*, *4*(4), 259–267.

Gerlt, J. A., & Raushel, F. M. (2003). Evolution of function in $(\beta/\alpha)_8$-barrel enzymes. *Current Opinion in Chemical Biology*, 7, 252–264.

Ghasempur, S., Eswaramoorthy, S., Hillerich, B. S., Seidel, R. D., Swaminathan, S., Almo, S. C., et al. (2014). Discovery of a novel L-lyxonate degradation pathway in *Pseudomonas aeruginosa* PAO1. *Biochemistry*, *53*(20), 3357–3366.

Glasner, M. E., Gerlt, J. A., & Babbitt, P. C. (2006). Evolution of enzyme superfamilies. *Current Opinion in Chemical Biology*, *10*(5), 492–497.

Goriup, M., Strauss, U. T., Felfer, U., Kroutil, W., & Faber, K. (2001). Substrate spectrum of mandelate racemase part 1: Variation of the α-hydroxy acid moiety. *Journal of Molecular Catalysis B: Enzymatic*, *15*(4–6), 207–212.

Groninger-Poe, F. P., Bouvier, J. T., Vetting, M. W., Kalyanaraman, C., Kumar, R., Almo, S. C., et al. (2014). Evolution of enzymatic activities in the enolase superfamily: Galactarate dehydratase III from *Agrobacterium tumefaciens* C58. *Biochemistry*, *53*(25), 4192–4203.

Gu, J., Liu, M., Guo, F., Xie, W., Lu, W., Ye, L., et al. (2014). Virtual screening of mandelate racemase mutants with enhanced activity based on binding energy in the transition state. *Enzyme and Microbial Technology*, *55*, 121–127.

Gu, J., & Yu, H. (2012). The role of S139 of mandelate racemase: Synergistic effect of S139 and E317 on transition state stabilization. *Journal of Biomolecular Structure & Dynamics*, *30*(5), 585–593.

Guan, R., Ho, M. C., Brenowitz, M., Tyler, P. C., Evans, G. B., Almo, S. C., et al. (2011). Entropy-driven binding of picomolar transition state analogue inhibitors to human 5′-methylthioadenosine phosphorylase. *Biochemistry*, *50*(47), 10408–10417.

Gulick, A. M., Hubbard, B. K., Gerlt, J. A., & Rayment, I. (2000). Evolution of enzymatic activities in the enolase superfamily: Crystallographic and mutagenesis studies of the reaction catalyzed by D-glucarate dehydratase from *Escherichia coli*. *Biochemistry*, *39*(16), 4590–4602.

Gunsalus, I. C., Gunsalus, C. F., & Stanier, R. Y. (1953a). The enzymatic conversion of mandelic acid to benzoic acid. I. Gross fractionation of the system into soluble are particulate components. *Journal of Bacteriology*, *66*(5), 538–542.

Gunsalus, C. F., Stanier, R. Y., & Gunsalus, I. C. (1953b). The enzymatic conversion of mandelic acid to benzoic acid. III. Fractionation and properties of the soluble enzymes. *Journal of Bacteriology*, *66*(5), 548–553.

Guthrie, J. P. (1998). Predicting the rates of proton transfer reactions: A simple model using equilibrium constants and distortion energies. *Journal of Physical Organic Chemistry*, *11*(8–9), 632–641.

Guthrie, J. P., & Kluger, R. (1993). Electrostatic stabilization can explain the unexpected acidity of carbon acids in enzyme-catalyzed reactions. *Journal of the American Chemical Society*, *115*(24), 11569–11572.

Hanson, K. R. (1981). Phenylalanine ammonia-lyase: Mirror-image packing of D- and L-phenylalanine and D- and L-transition state analogs into the active site. *Archives of Biochemistry and Biophysics*, *211*(2), 575–588.

Hanson, K. R., & Rose, I. A. (1975). Interpretations of enzyme reaction stereospecificity. *Accounts of Chemical Research*, *8*(1), 1–10.

Hasson, M. S., Schlichting, I., Moulai, J., Taylor, K., Barrett, W., Kenyon, G. L., et al. (1998). Evolution of an enzyme active site: The structure of a new crystal form of muconate lactonizing enzyme compared with mandelate racemase and enolase. *Proceedings of the National Academy of Sciences of the United States of America*, *95*(18), 10396–10401.

Hedrick, J. L., & Sallach, H. J. (1961). The metabolism of hydroxypyruvate. I. The nonenzymatic decarboxylation and autoxidation of hydroxypyruvate. *The Journal of Biological Chemistry*, *236*(7), 1867–1871.

Hegeman, G. D., Rosenberg, E. Y., & Kenyon, G. L. (1970). Mandelic acid racemase from *Pseudomonas putida*. Purification and properties of the enzyme. *Biochemistry*, *9*(21), 4029–4036.

Houk, R. J., Anslyn, E. V., & Stanton, J. F. (2006). Carbonyl coordination chemistry from a new angle: A computational study of α-carbon acidity based on electrophile coordination geometry. *Organic Letters*, *8*(16), 3461–3463.

Houk, R. J., Monzingo, A., & Anslyn, E. V. (2008). Electrophilic coordination catalysis: A summary of previous thought and a new angle of analysis. *Accounts of Chemical Research*, *41*(3), 401–410.

Jamaluddin, M., Rao, P. V., & Vaidyanathan, C. S. (1970). Involvement of the protocatechuate pathway in the metabolism of mandelic acid by *Aspergillus niger*. *Journal of Bacteriology*, *101*(3), 786–793.

Janeček, S. (1995). Close evolutionary relatedness among functionally distantly related members of the $(\alpha/\beta)_8$-barrel glycosyl hydrolases suggested by the similarity of their fifth conserved sequence region. *FEBS Letters*, *377*(1), 6–8.

Janeček, S., & Baláz, S. (1995). Functionally essential, invariant glutamate near the C-terminus of strand β5 in various $(\alpha/\beta)_8$-barrel enzymes as a possible indicator of their evolutionary relatedness. *Protein Engineering*, *8*(8), 809–813.

Kaftzik, N., Kroutil, W., Faber, K., & Kragl, U. (2004). Mandelate racemase activity in ionic liquids: Scopes and limitations. *Journal of Molecular Catalysis A: Chemical*, *214*(1), 107–112.

Kallarakal, A. T., Mitra, B., Kozarich, J. W., Gerlt, J. A., Clifton, J. G., Petsko, G. A., et al. (1995). Mechanism of the reaction catalyzed by mandelate racemase: Structure and mechanistic properties of the K166R mutant. *Biochemistry*, *34*(9), 2788–2797.

Kenyon, G. L., Gerlt, J. A., Petsko, G. A., & Kozarich, J. W. (1995). Mandelate racemase: Structure-function studies of a pseudosymmetric enzyme. *Accounts of Chemical Research*, *28*(4), 178–186.

Kenyon, G. L., & Hegeman, G. D. (1970). Mandelic acid racemase from *Pseudomonas putida*. Evidence favoring a carbanion intermediate in the mechanism of action. *Biochemistry*, *9*(21), 4036–4043.

Kenyon, G. L., & Hegeman, G. D. (1977). Mandelate racemase. *Methods in Enzymology*, *46*, 541–548.

Kenyon, G. L., & Hegeman, G. D. (1979). Mandelate racemase. *Advances in Enzymology and Related Areas of Molecular Biology*, *50*, 325–360.

Kraut, D. A., Carroll, K. S., & Herschlag, D. (2003). Challenges in enzyme mechanism and energetics. *Annual Review of Biochemistry*, *72*, 517–571.

Kresge, A. J., & Silverman, D. N. (1999). Application of marcus rate theory to proton transfer in enzyme-catalyzed reactions. *Methods in Enzymology*, *308*, 276–297.

Landro, J. A., Gerlt, J. A., Kozarich, J. W., Koo, C. W., Shah, V. J., Kenyon, G. L., et al. (1994). The role of lysine 166 in the mechanism of mandelate racemase from *Pseudomonas putida*: Mechanistic and crystallographic evidence for stereospecific alkylation by (*R*)-α-phenylglycidate. *Biochemistry*, *33*(3), 635–643.

Landro, J. A., Kallarakal, A. T., Ransom, S. C., Gerlt, J. A., Kozarich, J. W., Neidhart, D. J., et al. (1991). Mechanism of the reaction catalyzed by mandelate racemase. 3.

Asymmetry in reactions catalyzed by the H297N mutant. *Biochemistry*, *30*(38), 9274–9281.

Larissegger-Schnell, B., Kroutil, W., & Faber, K. (2005). Chemo-enzymatic synthesis of (*R*)- and (*S*)-2-hydroxy-4-phenylbutanoic acid via enantio-complementary deracemization of (±)-2-hydroxy-4-phenyl-3-butenoic acid using a racemase-lipase two-enzyme system. *Synlett*, *12*(12), 1936–1938.

Larsen, T. M., Wedekind, J. E., Rayment, I., & Reed, G. H. (1996). A carboxylate oxygen of the substrate bridges the magnesium ions at the active site of enolase: Structure of the yeast enzyme complexed with the equilibrium mixture of 2-phosphoglycerate and phosphoenolpyruvate at 1.8-Å resolution. *Biochemistry*, *35*(14), 4349–4358.

Lawrence, M. C., Barbosa, J. A. R. G., Smith, B. J., Hall, N. E., Pilling, P. A., Ooi, H. C., et al. (1997). Structure and mechanism of a sub-family of enzymes related to *N*-acetylneuraminate lyase. *Journal of Molecular Biology*, *266*(2), 381–399.

Li, D., Zeng, Z., Yang, J., Wang, P., Jiang, L., Feng, J., et al. (2013). Mandelate racemase and mandelate dehydrogenase coexpressed recombinant *Escherichia coli* in the synthesis of benzoylformate. *Bioscience, Biotechnology, and Biochemistry*, 77(6), 1236–1239.

Lietzan, A. D., Nagar, M., Pellmann, E. A., Bourque, J. R., Bearne, S. L., & St. Maurice, M. (2012). Structure of mandelate racemase with bound intermediate analogues benzohydroxamate and Cupferron. *Biochemistry*, *51*(6), 1160–1170.

Lin, D. T., Powers, V. M., Reynolds, L. J., Whitman, C. P., Kozarich, J. W., & Kenyon, G. L. (1988). Evidence for the generation of α-carboxy-α-hydroxy-*p*-xylylene from *p*-(bromomethyl)mandelate by mandelate racemase. *Journal of the American Chemical Society*, *110*(1), 323–324.

Marcus, R. A. (1969). Unusual slopes of free energy plots in kinetics. *Journal of the American Chemical Society*, *91*(26), 7224–7225.

Maskill, H. (1985). *The physical basis of organic chemistry*. New York: Oxford University Press, pp. 437–442.

Mesecar, A. D., & Koshland, D. E., Jr. (2000a). A new model for protein stereospecificity. *Nature*, *403*(6770), 614–615.

Mesecar, A. D., & Koshland, D. E., Jr. (2000b). Sites of binding and orientation in a four-location model for protein stereospecificity. *IUBMB Life*, *49*(5), 457–466.

Mitchell, J. B., Nandi, C. L., McDonald, I. K., Thornton, J. M., & Price, S. L. (1994). Amino/aromatic interactions in proteins: Is the evidence stacked against hydrogen bonding? *Journal of Molecular Biology*, *239*(2), 315–331.

Mitra, B., Kallarakal, A. T., Kozarich, J. W., Gerlt, J. A., Clifton, J. G., Petsko, G. A., et al. (1995). Mechanism of the reaction catalyzed by mandelate racemase: Importance of electrophilic catalysis by glutamic acid 317. *Biochemistry*, *34*(9), 2777–2787.

Myslinski, J. M., DeLorbe, J. E., Clements, J. H., & Martin, S. F. (2011). Protein-ligand interactions: Thermodynamic effects associated with increasing nonpolar surface area. *Journal of the American Chemical Society*, *133*(46), 18518–18521.

Nagano, N., Orengo, C. A., & Thornton, J. M. (2002). One fold with many functions: The evolutionary relationships between TIM barrel families based on their sequences, structures and functions. *Journal of Molecular Biology*, *321*(5), 741–765.

Nagar, M., & Bearne, S. L. (2015). An additional role for the Brønsted acid-base catalysts of mandelate racemase in transition state stabilization. *Biochemistry*, *54*(44), 6743–6752.

Nagar, M., Lietzan, A. D., St. Maurice, M., & Bearne, S. L. (2014). Potent inhibition of mandelate racemase by a fluorinated substrate-product analogue with a novel binding mode. *Biochemistry*, *53*(7), 1169–1178.

Nagar, M., Narmandakh, A., Khalak, Y., & Bearne, S. L. (2011). Redefining the minimal substrate tolerance of mandelate racemase. Racemization of trifluorolactate. *Biochemistry*, *50*(41), 8846–8852.

Nagar, M., Wyatt, B. N., St. Maurice, M., & Bearne, S. L. (2015). Inactivation of mandelate racemase by 3-hydroxypyruvate reveals a potential mechanistic link between enzyme superfamilies. *Biochemistry*, *54*(17), 2747–2757.

Narmandakh, A., & Bearne, S. L. (2010). Purification of recombinant mandelate racemase: Improved catalytic activity. *Protein Expression and Purification*, *69*(1), 39–46.

Neidhart, D. J., Howell, P. L., Petsko, G. A., Powers, V. M., Li, R. S., Kenyon, G. L., et al. (1991). Mechanism of the reaction catalyzed by mandelate racemase. 2. Crystal structure of mandelate racemase at 2.5-Å resolution: Identification of the active site and possible catalytic residues. *Biochemistry*, *30*(38), 9264–9273.

Neidhart, D. J., Kenyon, G. L., Gerlt, J. A., & Petsko, G. A. (1990). Mandelate racemase and muconate lactonizing enzyme are mechanistically distinct and structurally homologous. *Nature*, *347*(6294), 692–694.

Neidhart, D. J., Powers, V. M., Kenyon, G. L., Tsou, A. Y., Ransom, S. C., Gerlt, J. A., et al. (1988). Preliminary X-ray data on crystals of mandelate racemase. *The Journal of Biological Chemistry*, *263*(19), 9268–9270.

O'Brien, P. J., & Herschlag, D. (1999). Catalytic promiscuity and the evolution of new enzymatic activities. *Chemistry & Biology*, *6*(4), R91–R105.

Parker, M. H., Lunney, E. A., Ortwine, D. F., Pavlovsky, A. G., Humblet, C., & Brouillette, C. G. (1999). Analysis of the binding of hydroxamic acid and carboxylic acid inhibitors to the stromelysin-1 (matrix metalloproteinase-3) catalytic domain by isothermal titration calorimetry. *Biochemistry*, *38*(41), 13592–13601.

Phillips, M. A., Kaplan, A. P., Rutter, W. J., & Bartlett, P. A. (1992). Transition-state characterization: A new approach combining inhibitor analogues and variation in enzyme structure. *Biochemistry*, *31*(4), 959–963.

Powers, V. M., Koo, C. W., Kenyon, G. L., Gerlt, J. A., & Kozarich, J. W. (1991). Mechanism of the reaction catalyzed by mandelate racemase. 1. Chemical and kinetic evidence for a two-base mechanism. *Biochemistry*, *30*(38), 9255–9263.

Prat-Resina, X., Gonzalez-Lafont, A., & Lluch, J. M. (2005). Reaction mechanism of the mandelate anion racemization catalyzed by mandelate racemase enzyme: A QM/MM molecular dynamics free energy study. *The Journal of Physical Chemistry. B*, *109*(44), 21089–21101.

Radzicka, A., & Wolfenden, R. (1995). Transition state and multisubstrate analog inhibitors. *Methods in Enzymology*, *249*, 284–312.

Raine, A. R., Scrutton, N. S., & Mathews, F. S. (1994). On the evolution of alternate core packing in eightfold β/α-barrels. *Protein Science*, *3*(10), 1889–1892.

Rakus, J. F., Fedorov, A. A., Fedorov, E. V., Glasner, M. E., Hubbard, B. K., Delli, J. D., et al. (2008). Evolution of enzymatic activities in the enolase superfamily: L-rhamnonate dehydratase. *Biochemistry*, *47*(38), 9944–9954.

Rakus, J. F., Fedorov, A. A., Fedorov, E. V., Glasner, M. E., Vick, J. E., Babbitt, P. C., et al. (2007). Evolution of enzymatic activities in the enolase superfamily: D-Mannonate dehydratase from *Novosphingobium aromaticivorans*. *Biochemistry*, *46*(45), 12896–12908.

Reardon, D., & Farber, G. K. (1995). The structure and evolution of α/β barrel proteins. *The FASEB Journal*, *9*(7), 497–503.

Renaud, P., & Fox, M. A. (1988). An electrochemical characterization of dianions: Dilithiated carboxylic acids. *Journal of the American Chemical Society*, *110*(17), 5705–5709.

Resch, V., Fabian, W. M. F., & Kroutil, W. (2010). Deracemization of mandelic acid to optically pure non-natural L-phenylglycine via a redox-neutral biocatalytic cascade. *Advanced Synthesis and Catalysis*, *352*(6), 993–997.

Richard, J. P., & Amyes, T. L. (2001). Proton transfer at carbon. *Current Opinion in Chemical Biology*, *5*(6), 626–633.

Richard, J. P., Amyes, T. L., Goryanova, B., & Zhai, X. (2014). Enzyme architecture: On the importance of being in a protein cage. *Current Opinion in Chemical Biology*, *21*, 1–10.

Richard, J. P., Williams, G., O'Donoghue, A. C., & Amyes, T. L. (2002). Formation and stability of enolates of acetamide and acetate anion: An eigen plot for proton transfer at α-carbonyl carbon. *Journal of the American Chemical Society*, *124*(12), 2957–2968.

Sakai, A., Fedorov, A. A., Fedorov, E. V., Schnoes, A. M., Glasner, M. E., Brown, S., et al. (2009). Evolution of enzymatic activities in the enolase superfamily: Stereochemically distinct mechanisms in two families of *cis,cis*-muconate lactonizing enzymes. *Biochemistry*, *48*(7), 1445–1453.

Schafer, S. L., Barrett, W. C., Kallarakal, A. T., Mitra, B., Kozarich, J. W., Gerlt, J. A., et al. (1996). Mechanism of the reaction catalyzed by mandelate racemase: Structure and mechanistic properties of the D270N mutant. *Biochemistry*, *35*(18), 5662–5669.

Schramm, V. L. (1998). Enzymatic transition states and transition state analog design. *Annual Review of Biochemistry*, *67*, 693–720.

Schramm, V. L. (2011). Enzymatic transition states, transition-state analogs, dynamics, thermodynamics, and lifetimes. *Annual Review of Biochemistry*, *80*, 703–732.

Scrutton, N. S., & Raine, A. R. C. (1996). Cation-π bonding and amino-aromatic interactions in the biomolecular recognition of substituted ammonium ligands. *The Biochemical Journal*, *319*(1), 1–8.

Siddiqi, F., Bourque, J. R., Jiang, H., Gardner, M., St. Maurice, M., Blouin, C., et al. (2005). Perturbing the hydrophobic pocket of mandelate racemase to probe phenyl motion during catalysis. *Biochemistry*, *44*(25), 9013–9021.

Silverman, D. N. (2000). Marcus rate theory applied to enzymatic proton transfer. *Biochimica et Biophysica Acta*, *1458*(1), 88–103.

St. Maurice, M., & Bearne, S. L. (2000). Reaction intermediate analogues for mandelate racemase: Interaction between Asn 197 and the α-hydroxyl of the substrate promotes catalysis. *Biochemistry*, *39*(44), 13324–13335.

St. Maurice, M., & Bearne, S. L. (2002). Kinetics and thermodynamics of mandelate racemase catalysis. *Biochemistry*, *41*(12), 4048–4058.

St. Maurice, M., & Bearne, S. L. (2004). Hydrophobic nature of the active site of mandelate racemase. *Biochemistry*, *43*(9), 2524–2532.

St. Maurice, M., Bearne, S. L., Lu, W., & Taylor, S. D. (2003). Inhibition of mandelate racemase by α-fluorobenzylphosphonates. *Bioorganic & Medicinal Chemistry Letters*, *13*(12), 2041–2044.

Stecher, H., Felfer, U., & Faber, K. (1997). Large-scale production of mandelate racemase by *Pseudomonas putida* ATCC 12633: Optimization of enzyme induction and development of a stable crude enzyme preparation. *Journal of Biotechnology*, *56*(1), 33–40.

Strauss, U. T., & Faber, K. (1999). Deracemization of (±)-mandelic acid using a lipase-mandelate racemase two-enzyme system. *Tetrahedron, Assymetry*, *10*(21), 4079–4081.

Strauss, U. T., Kandelbauer, A., & Faber, K. (2000). Stabilization and activity-enhancement of mandelate racemase from *Pseudomonas putia* ATCC 12336 by immobilization. *Biotechnology Letters*, *22*(6), 515–520.

Thibblin, A., & Jencks, W. P. (1979). Unstable carbanions. General acid catalysis of the cleavage of 1-phenylcyclopropanol and 1-phenyl-2-arylcyclopropanol anions. *Journal of the American Chemical Society*, *101*(17), 4963–4973.

Topiol, S., & Sabio, J. (1989). Interactions between eight centers are required for chiral recognition. *Journal of the American Chemical Society*, *111*(11), 4109–4110.

Tsou, A. Y., Ransom, S. C., Gerlt, J. A., Powers, V. M., & Kenyon, G. L. (1989). Selection and characterization of a mutant of the cloned gene for mandelate racemase that confers resistance to an affinity label by greatly enhanced production of enzyme. *Biochemistry*, *28*(3), 969–975.

Vetting, M. W., Bouvier, J. T., Gerlt, J. A., & Almo, S. C. (2016). Purification, crystallization and structural elucidation of D-galactaro-1,4-lactone cycloisomerase from *Agrobacterium tumefaciens* involved in pectin degradation. *Acta Crystallographica. Section F, Structural Biology Communications*, *72*(1), 36–41.

Wichelecki, D. J., Froese, D. S., Kopec, J., Muniz, J. R., Yue, W. W., & Gerlt, J. A. (2014). Enzymatic and structural characterization of rTSγ provides insights into the function of rTSβ. *Biochemistry*, *53*(16), 2732–2738.

Wierenga, R. K. (2001). The TIM-barrel fold: A versatile framework for efficient enzymes. *FEBS Letters*, *492*(3), 193–198.

Wilmanns, M., Hyde, C. C., Davies, D. R., Kirschner, K., & Jansonius, J. N. (1991). Structural conservation in parallel β/α-barrel enzymes that catalyze three sequential reactions in the pathway of tryptophan biosynthesis. *Biochemistry*, *30*(38), 9161–9169.

Wolfenden, R. (1972). Analog approaches to the structure of the transition state in enzyme reactions. *Accounts of Chemical Research*, *5*, 10–18.

Wolfenden, R. (1976). Transition state analog inhibitors and enzyme catalysis. *Annual Review of Biophysics and Bioengineering*, *5*(1), 271–306.

Wolfenden, R., & Frick, L. (1987). Transition state affinity and the design of enzyme inhibitors. In M. I. Page & A. Williams (Eds.), *Enzyme Mechanisms* (pp. 97–122). London: Royal Society of Chemistry.

Wrzosek, K., García Rivera, M. A., Battenbrock, K., & Seidel-Morgenstern, A. (2016). Racemization of undesired enantiomers: Immobilization of mandelate racemase and application in a fixed-bed reactor. *Biotechnology Journal*, *11*(4), 453–463.

Yang, C., Ye, L., Gu, J., Yang, X., Li, A., & Yu, H. (2017). Directed evolution of mandelate racemase by a novel high-throughput screening method. *Applied Microbiology and Biotechnology*, *101*(3), 1063–1072.

Yew, W. S., Fedorov, A. A., Fedorov, E. V., Almo, S. C., & Gerlt, J. A. (2007). Evolution of enzymatic activities in the enolase superfamily: L-talarate/galactarate dehydratase from *Salmonella typhimurium* LT2. *Biochemistry*, *46*(33), 9564–9577.

Yew, W. S., Fedorov, A. A., Fedorov, E. V., Rakus, J. F., Pierce, R. W., Almo, S. C., et al. (2006a). Evolution of enzymatic activities in the enolase superfamily: L-fuconate dehydratase from *Xanthomonas campestris*. *Biochemistry*, *45*(49), 14582–14597.

Yew, W. S., Fedorov, A. A., Fedorov, E. V., Wood, B. M., Almo, S. C., & Gerlt, J. A. (2006b). Evolution of enzymatic activities in the enolase superfamily: D-tartrate dehydratase from *Bradyrhizobium japonicum*. *Biochemistry*, *45*(49), 14598–14608.

Zacharias, N., & Dougherty, D. A. (2002). Cation-π interactions in ligand recognition and catalysis. *Trends in Pharmacological Sciences*, *23*(6), 281–287.

Zhuohang, M., Ji, L., & Hongwei, Y. (2011). Modeling of transition state by molecular dynamics. Prediction of catalytic efficiency of mutants of mandelate racemase. *Journal of Biomolecular Structure & Dynamics*, *28*(6), 871–879.

CHAPTER FIVE

Striking Diversity in Holoenzyme Architecture and Extensive Conformational Variability in Biotin-Dependent Carboxylases

Liang Tong[1]
Columbia University, New York, NY, United States
[1]Corresponding author: e-mail address: ltong@columbia.edu

Contents

1. Introduction 162
2. Structures of Conserved Components in Biotin-Dependent Carboxylases 166
   2.1 BC Component 167
   2.2 CT Component in Acyl-CoA Carboxylases 169
   2.3 CT Component in PC 170
   2.4 CT Component in UC 171
   2.5 BCCP Component 171
   2.6 BT Domain in Acyl-CoA Carboxylases 171
   2.7 PT Domain and BT-Like Domain in PC 171
   2.8 AC Domains in Single-Chain ACCs 173
3. Extensive Diversity in Holoenzyme Architectures 173
   3.1 Architectures of Acyl-CoA Carboxylase Holoenzymes 173
   3.2 Architectures of PC and UC Holoenzymes 177
4. Extensive Conformational Variability 180
   4.1 Conformational Variability in the BC Domain Dimer Interface of Single-Chain ACCs 180
   4.2 Conformational Variability in the Central Region of Single-Chain ACCs 181
   4.3 Conformational Variability in the CT Active Site of Single-Chain ACCs for Herbicide Binding 183
   4.4 Conformational Variability in Single-Chain PCs 185
   4.5 BCCP Translocation During Catalysis 187
5. Perspectives 187
Acknowledgments 188
References 189

Advances in Protein Chemistry and Structural Biology, Volume 109
ISSN 1876-1623
http://dx.doi.org/10.1016/bs.apcsb.2017.04.006

Abstract

Biotin-dependent carboxylases are widely distributed in nature and have central roles in the metabolism of fatty acids, amino acids, carbohydrates, and other compounds. The last decade has seen the accumulation of structural information on most of these large holoenzymes, including the 500-kDa dimeric yeast acetyl-CoA carboxylase, the 750-kDa $\alpha_6\beta_6$ dodecameric bacterial propionyl-CoA carboxylase, 3-methylcrotonyl-CoA carboxylase, and geranyl-CoA carboxylase, the 720-kDa hexameric bacterial long-chain acyl-CoA carboxylase, the 500-kDa tetrameric bacterial single-chain pyruvate carboxylase, the 370-kDa $\alpha_2\beta_4$ bacterial two-subunit pyruvate carboxylase, and the 130-kDa monomeric eukaryotic urea carboxylase. A common theme that has emerged from these studies is the dramatic structural flexibility of these holoenzymes despite their strong overall sequence conservation, evidenced both by the extensive diversity in the architectures of the holoenzymes and by the extensive conformational variability of their domains and subunits. This structural flexibility is crucial for the function and regulation of these enzymes and identifying compounds that can interfere with it represents an attractive approach for developing novel modulators and drugs. The extensive diversity observed in the structures so far and its biochemical and functional implications will be the focus of this review.

1. INTRODUCTION

Biotin-dependent carboxylases are widely distributed in nature, being found from bacteria, archaea, fungi, plants to humans, and have central roles in the metabolism of fatty acids, amino acids, carbohydrates, and other compounds (Alberts & Vagelos, 1972; Tong, 2013; Wakil, Stoops, & Joshi, 1983; Waldrop, Holden, & St. Maurice, 2012). The majority of these enzymes carboxylate coenzyme A (CoA) esters of organic acids, including acetyl-CoA (acetyl-CoA carboxylase, ACC), propionyl-CoA (PCC), 3-methylcrotonyl-CoA (MCC), geranyl-CoA (GCC), and long-chain acyl-CoA (LCC). Other enzymes in this family carboxylate small organic compounds, including pyruvate (pyruvate carboxylase, PC) and urea (UC).

Biotin-dependent carboxylases have two distinct catalytic activities, and the prosthetic group biotin (vitamin B_7) plays a central role in both reactions (Fig. 1A). In the first reaction, a biotin carboxylase (BC) activity carboxylates the $N_{1'}$ atom of biotin, using bicarbonate as the CO_2 donor and requiring ATP hydrolysis. The carboxylated biotin then translocates to the second active site, carboxyltransferase (CT), where the CO_2 is conveyed to the acceptor substrate.

Biotin itself is covalently attached through an amide bond to a conserved lysine residue in the biotin carboxyl carrier protein (BCCP)

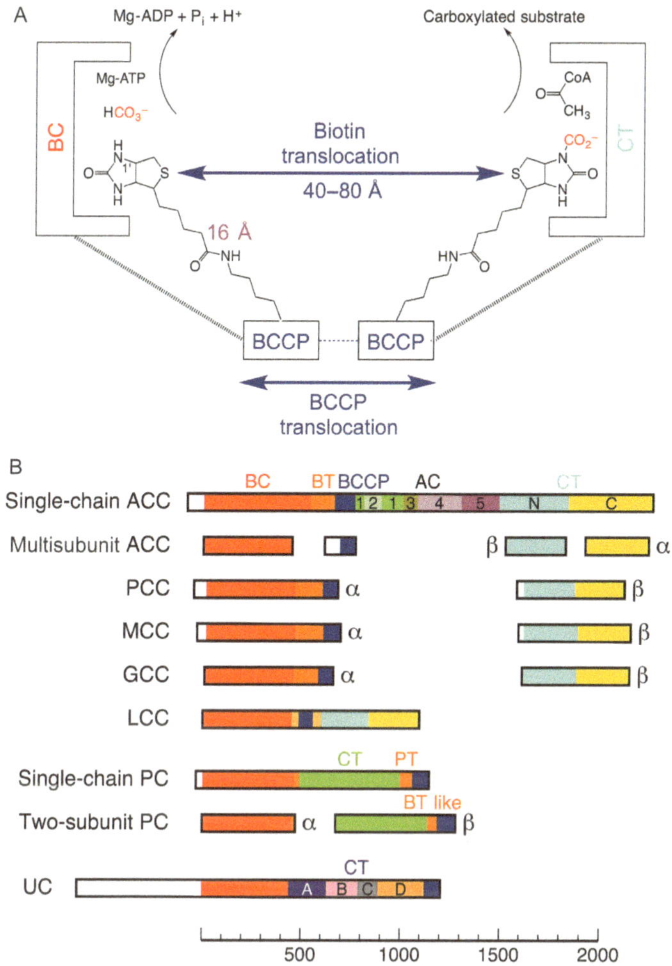

Fig. 1 Biochemical activity and domain organization of biotin-dependent carboxylases. (A) Schematic drawing showing the reactions in the biotin carboxylase (BC) and carboxyltransferase (CT) active sites. Acetyl-CoA is shown as the acceptor substrate as an example. Both biotin and the biotin carboxyl carrier protein (BCCP) component need to translocate during catalysis, known as the swinging-domain model. (B) Domain organization of representative biotin-dependent carboxylases. Domains or subunits with sequence homology to each other are shown with the same color and labeled. The central region of single-chain ACC contains five domains, AC1–AC5 (AC: ACC Central). The proteins are drawn to scale, which is indicated with the scale bar at the *bottom* (in number of residues).

component. (The term component will be used here to indicate a functional module, which can be composed of domain(s) or subunit(s).) This places its $N_{1'}$ atom at the end of a flexible arm with eight methylene groups, which could be ~16 Å long when fully extended (Fig. 1A). It was generally believed that biotin visits both active sites on this flexible arm, the so-called swinging-arm model, which could reach a distance of ~30 Å. However, recent structures of various biotin-dependent carboxylase holoenzymes have revealed that the distances between the BC and CT active sites range between 40 and 80 Å, which cannot be covered by the flexible arm alone. Therefore, the entire BCCP component also needs to translocate during catalysis, giving rise to the swinging-domain model (Fig. 1A). This is reminiscent of the movement of the acyl carrier protein during the catalysis by fatty acid synthase (Maier, Leibundgut, Boehringer, & Ban, 2010) and other enzyme systems.

The BC component (~50 kDa) is shared among all biotin-dependent carboxylases and is well conserved (Fig. 1B). The CT components for those enzymes that carboxylate CoA esters (~60–80 kDa) have a common organization, with two domains (N and C domains) or two subunits. In contrast, the CT components for PC and UC are entirely different because they recognize small organic compounds rather than CoA esters. The core of the BCCP component contains ~80 amino acid residues, although there are often linkers that connect it to the rest of the enzyme. The biotinylation motif (Met-Lys-Met) is highly conserved, although the sequences for the rest of the BCCP component are more variable.

Five biotin-dependent carboxylases exist in humans, ACC1, ACC2 (two isoforms of ACC), PCC, MCC, and PC. ACC1 and ACC2 are ~250 kDa single-chain enzymes (Fig. 1B) and function as 500 kDa dimers and higher oligomers. They contain a central region (~80 kDa) that is not found in other biotin-dependent carboxylases. ACC1 is localized in the cytoplasm of lipogenic tissues (liver and adipose), and its malonyl-CoA product is crucial for the biosynthesis of long-chain fatty acids and other compounds. ACC2 is primarily found in the heart and skeletal muscle, where its malonyl-CoA product is a potent inhibitor of carnitine palmitoyltransferase I (CPT-I), the enzyme that mediates the transport of long-chain acyl-CoAs into the mitochondria for fatty acid oxidation (McGarry & Brown, 1997; Ramsay, Gandour, & van der Leij, 2001). ACC2 is attached to the outer membrane of the mitochondria through a unique segment at its N-terminus, prior to the BC domain.

PCC, MCC, and PC are localized in the mitochondrial matrix. The holoenzymes of PCC and MCC are 750 kDa $\alpha_6\beta_6$ dodecamers, and the holoenzyme of PC is a 500-kDa homotetramer. PCC is required for the metabolism of four amino acids (valine, threonine, isoleucine, and methionine), cholesterol, and fatty acids with an odd number of carbon atoms

(Campeau et al., 2001; Chloupkova, Maclean, Alkhateeb, & Kraus, 2002; Clavero et al., 2002; Deodato, Boenzi, Santorelli, & Dionisi-Vici, 2006; Desviat et al., 2004), and MCC is necessary for the metabolism of the amino acid leucine (Baumgartner et al., 2001; Gallardo et al., 2001; Holzinger et al., 2001). PC is crucial for replenishing the tricarboxylic acid cycle intermediate oxaloacetate, removing pyruvate from tissues, and is also important for gluconeogenesis, fatty acid biosynthesis, and glucose-induced insulin secretion (Jitrapakdee et al., 2008; Wallace, 2010). A large collection of mutations in PCC, MCC, and PC are linked to serious metabolic diseases, even death in infants, underscoring their functional importance (Baumgartner et al., 2001; Campeau et al., 2001; Chloupkova et al., 2002; Clavero et al., 2002; Deodato et al., 2006; Desviat et al., 2004, 2003; Gallardo et al., 2001; Holzinger et al., 2001; Robinson, 2006).

Biotin-dependent carboxylases are attractive targets for drug discovery. Mice that are deficient in ACC2 have elevated fatty acid oxidation and reduced body fat and body weight (Abu-Elheiga, Matzuk, Abo-Hashema, & Wakil, 2001; Abu-Elheiga, Oh, Kordari, & Wakil, 2003), indicating ACC2 (and also ACC1) as targets for drug discovery against many manifestations of the metabolic syndrome, such as diabetes, cancer, and cardiovascular disease (Reaven, 2004, 2011; Tong & Harwood, 2006). A small-molecule inhibitor of human ACCs (ND-630) has recently entered phase II clinical trials against nonalcoholic fatty liver disease (also known as nonalcoholic steatohepatitis) (Harriman et al., 2016). Bacterial ACCs, with four subunits (Fig. 1B), are targets for antibiotic discovery (Cronan & Waldrop, 2002; Miller et al., 2009; Miller & Waldrop, 2010; Polyak, Abell, Wilce, Zhang, & Booker, 2012; Silvers et al., 2016). Recent studies have implicated PC in cancer development (Cardaci et al., 2015; Christen et al., 2016; Phannasil et al., 2016; Sellers et al., 2015) and insulin resistance (Perry et al., 2015).

Biotin-dependent carboxylases have been studied biochemically and functionally for several decades, but obtaining the structures of their holoenzymes have been rather challenging due to their large sizes and the presence of multiple domains/subunits (Fig. 1B). Only over the past decade have structures at ~3 Å resolution or higher become available. These include the crystal structures of the 500-kDa dimeric yeast *Saccharomyces cerevisiae* ACC (ScACC) holoenzyme (Wei & Tong, 2015), a 750-kDa $\alpha_6\beta_6$ dodecameric bacterial PCC holoenzyme (α subunit from *Ruegeria pomeroyi* and β subunit from *Roseobacter denitrificans*, 88% identical to the β subunit from *R. pomeroyi*) (Huang et al., 2010), the 750-kDa $\alpha_6\beta_6$ dodecameric *Pseudomonas aeruginosa*

MCC (PaMCC) holoenzyme (Huang, Ge, Zhou, & Tong, 2012), the 750-kDa $\alpha_6\beta_6$ dodecameric *Pseudomonas fluorescens* GCC (PfGCC) holoenzyme (Jurado, Huang, Zhang, Zhou, & Tong, 2015), the 720-kDa hexameric *Mycobacterium avium subspecies paratuberculosis* LCC (MapLCC) holoenzyme (Tran et al., 2015), the 500-kDa tetrameric single-chain PC holoenzymes from *Rhizobium etli* (RePC) (Lietzan et al., 2011; St. Maurice et al., 2007), *Staphylococcus aureus* (SaPC) (Xiang & Tong, 2008; Yu, Chou, Choi, & Tong, 2013; Yu et al., 2009), and *Listeria monocytogenes* (LmPC) (Sureka et al., 2014), the 370-kDa $\alpha_2\beta_4$ two-subunit *Methylobacillus flagellatus* PC (MfPC) holoenzyme (Choi et al., 2016), and the 130-kDa monomeric *Kluyveromyces lactis* UC (KlUC) holoenzyme (Fan, Chou, Tong, & Xiang, 2012). Because of their large sizes, three-dimensional electron microscopy (EM) reconstructions, at lower resolutions, had been obtained for several holoenzymes, including human PCC (HsPCC) (Huang et al., 2010), PaMCC (Huang et al., 2012), PfGCC (Jurado et al., 2015), and SaPC (Lasso et al., 2014, 2010; Yu et al., 2009). With the recent revolution in EM technology, it can be expected that near atomic-resolution structures of these holoenzymes by EM will be available in the near future.

A common theme that has emerged from the structural studies on these holoenzymes is their dramatic structural flexibility despite their strong sequence conservation, evidenced both by the extensive diversity in the architectures of their holoenzymes and by the extensive conformational variability of their domains and subunits. This extensive diversity and its biochemical and functional consequences will be the focus of this review.

2. STRUCTURES OF CONSERVED COMPONENTS IN BIOTIN-DEPENDENT CARBOXYLASES

Biochemical studies identified three conserved components among biotin-dependent carboxylases, BC, CT, and BCCP. Additional components were identified from structural studies on their holoenzymes, especially a BT (BC–CT interaction) domain in most acyl-CoA carboxylases (Huang et al., 2012, 2010; Jurado et al., 2015; Wei & Tong, 2015) and a PT (PC tetramerization) domain (Sureka et al., 2014; Xiang & Tong, 2008), also known as allosteric domain (St. Maurice et al., 2007), in single-chain PC enzymes (Fig. 1B). These components are located in the linker segments among BC, CT, and/or BCCP in the primary sequence.

The sequence conservation of these segments is generally weaker, and they were not recognized as separate domains prior to the structure determination of the holoenzymes.

2.1 BC Component

The structure of the BC subunit of *Escherichia coli* ACC was the first atomic structural information available for biotin-dependent carboxylases (Waldrop, Rayment, & Holden, 1994). Many other structures of the BC component have been reported since then (Broussard, Pakhomova, Neau, Bonnot, & Waldrop, 2015; Cho et al., 2010; Chou & Tong, 2011; Hagmann, Hunkeler, Stuttfeld, & Maier, 2016; Harriman et al., 2016; Kondo et al., 2007; Miller et al., 2009; Mochalkin et al., 2008; Raymer et al., 2009), especially the structure of this BC subunit in complex with Mg-ADP, biotin, and bicarbonate (Fig. 2A; Chou et al., 2009) and the structure of the BC domain of ScACC, in complex with the polyketide natural product inhibitor soraphen A (Fig. 2B; Shen et al., 2004).

The structure of BC contains three subdomains, A, B, and C (Fig. 2B). The yeast BC domain is slightly larger in size due to a linker segment between the A and B domains (AB linker, Fig. 2B). The active site is located at the interface between the three domains. Especially, domain B acts like a flap and closes over the active site upon substrate binding (Broussard et al., 2015; Chou et al., 2009; Thoden, Blanchard, Holden, & Waldrop, 2000). The open–closed conformational transition of this domain is likely important for BC catalysis.

The *E. coli* BC subunit is a dimer with an extensive, though mostly hydrophilic interface between the two monomers (Fig. 2A). The active site of BC is located 25 Å away from the dimer interface, with no contributions from the other monomer of the dimer. It has been reported that dimerization of this BC subunit is required for its activity (Janiyani, Bordelon, Waldrop, & Cronan, 2001). Mutations in the dimer interface can disrupt dimer formation, but such monomeric mutants are still catalytically active (Shen, Chou, Chang, & Tong, 2006). These mutants support growth in vivo only when overexpressed (Smith & Cronan, 2012), suggesting that a BC dimer might be important for the overall activity of the ACC holoenzyme. On the other hand, a monomeric BC domain is observed in the holoenzyme structures of PCC (Huang et al., 2010) and KlUC (Fan et al., 2012), and the structures of these BC monomers are generally similar to that of *E. coli* BC.

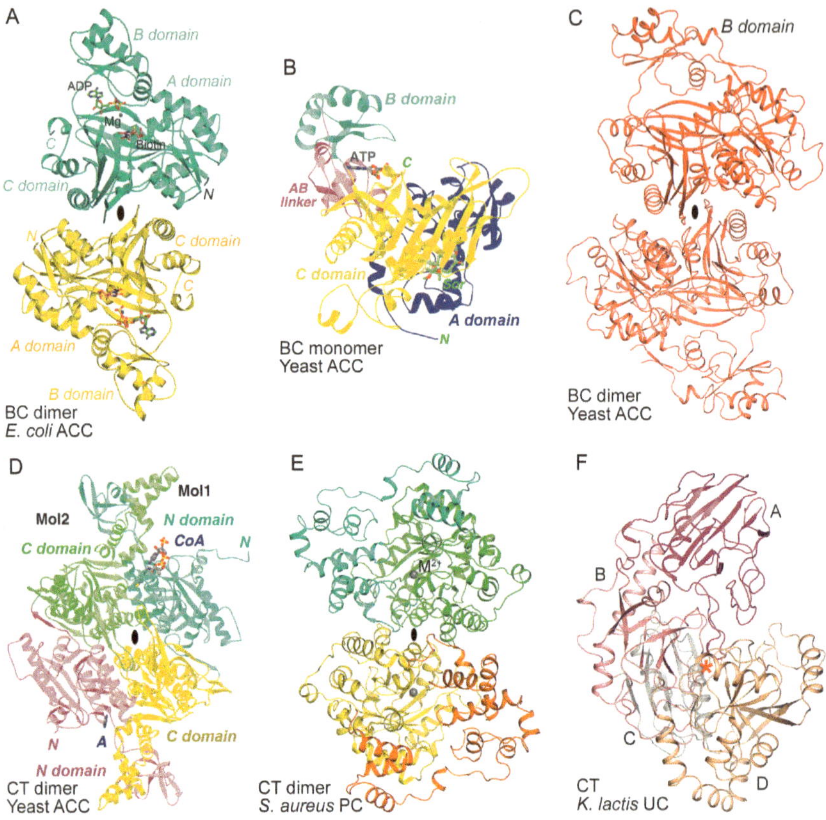

Fig. 2 Structures of the BC and CT components of biotin-dependent carboxylases. (A) Structure of the BC subunit dimer of *E. coli* ACC (one subunit in *cyan* and the other in *yellow*) in complex with Mg^{2+} (*gray sphere*), ADP (*green*), biotin (*pink*), and bicarbonate (*black*) (Chou, Yu, & Tong, 2009). The twofold symmetry axis of the dimer is indicated with the *black* oval. (B) Structure of the BC domain monomer of ScACC in complex with the polyketide inhibitor soraphen A (*green*) (Shen, Volrath, Weatherly, Elich, & Tong, 2004). The three subdomains of BC and the AB linker are given different colors and labeled. (C) Structure of the BC domain dimer in the ScACC holoenzyme (Wei & Tong, 2015). One monomer is in *red*, and the other in *pink*. (D) Structure of the CT domain dimer of ScACC (Zhang, Yang, Shen, & Tong, 2003). The N and C domains are in *cyan* and *yellow* for one monomer, and *magenta* and *green* for the other. The active site position is indicated by the bound CoA (*gray*). (E) Structure of the CT domain dimer of SaPC (Xiang & Tong, 2008). The TIM barrel domain and the C-terminal helical extension are in *green* and *cyan* for one monomer, and *yellow* and *orange* for the other. The active site position is indicated by the bound metal ion (*gray spheres*). (F) Structure of the CT domain of KlUC (Fan et al., 2012). The four domains are given different colors, and the active site is indicated with the *red asterisk*. The coloring schemes of this figure do not necessarily follow that of Fig. 1B, in order to show structural features within the components.

The BC domain of ScACC, as well as human ACCs, is a monomer (Fig. 2B) and catalytically inactive when isolated from the rest of the holoenzyme. There are large conformational differences in the region of the structure that is equivalent to the dimer interface of *E. coli* BC, and the BC domain cannot form the same dimer. The soraphen A polyketide natural product inhibitor and the small-molecule inhibitor ND-630 are bound to this region of the structure. On the other hand, this domain is a dimer in the holoenzyme (Fig. 2C; Wei & Tong, 2015). This requires large conformational changes for residues in the dimer interface region, which is one of the examples for the extensive conformational variability in these enzymes (see later). On the other hand, the overall organization of the BC domain dimer is similar to that of the BC subunit dimer of *E. coli* ACC, suggesting a conserved mode of dimerization for the BC components.

2.2 CT Component in Acyl-CoA Carboxylases

The CT component in acyl-CoA carboxylases contains two domains (N and C domains) or subunits (β and α, with weak sequence conservation to N and C domains) (Fig. 1B), as illustrated by the CT domain structure of ScACC (Fig. 2D; Zhang et al., 2003). Despite sharing very low sequence identity, the structures of the two domains have the same backbone fold, with a β–β–α superhelix core (also known as the crotonase fold or the spiral fold). Many structures of various CT components are now available (Arabolaza et al., 2010; Bilder et al., 2006; Lietzan, Lin, & St. Maurice, 2014; Lietzan & St. Maurice, 2013; Madauss et al., 2009; Silvers et al., 2016; Xiang, Callaghan, Watson, & Tong, 2009; Yu, Kim, & Tong, 2010; Zhang, Tweel, Li, & Tong, 2004; Zhang, Tweel, & Tong, 2004). This component is shared with other enzymes, for example, the transcarboxylase 12S subunit (Hall et al., 2003) and the biotin-dependent decarboxylase glutaconyl-CoA decarboxylase Na^+ pump (Wendt, Schall, Huber, Buckel, & Jacob, 2003).

CT is a head-to-tail dimer, with the N domain of one monomer in contact with the C domain of the other (Fig. 2D). The dimer interface is extensive and hydrophobic in nature, suggesting that this dimer is likely very stable. The CT domain of eukaryotic ACCs contains a small subdomain that further stabilizes the dimer (Fig. 2D), which is absent in the structures of bacterial ACC CTs (Bilder et al., 2006; Silvers et al., 2016) and other CT components. The β subunit of bacterial ACCs contains a zinc-binding domain prior to the crotonase fold, which may have a separate functional role (Meades, Benson, Grove, & Waldrop, 2010). In PCC (Huang et al., 2010), MCC (Huang et al., 2012), GCC (Jurado et al., 2015), LCC (Hagmann et al., 2016; Tran et al., 2015), and several other enzymes (Diacovich et al., 2004, 2002; Lin et al., 2006; Reddy et al., 2014), three CT dimers assemble into a hexamer, forming the central core of these holoenzymes.

The CT active site is located at the dimer interface, and therefore CT (and acyl-CoA carboxylases) must at least dimerize to be catalytically active. The crotonase fold is well known for recognizing CoA esters, and the N domain in CT mediates the binding of the acyl-CoA substrate (Fig. 2D), which explains why this CT component is shared among all acyl-CoA carboxylases. On the other hand, the crotonase fold of the C domain recognizes the biotin cofactor, indicating another function for this fold.

Potent, small-molecule inhibitors are known against the CT active site of eukaryotic ACCs, such as herbicides (haloxyfop, tepraloxydim, and pinoxaden) that target plant ACCs (Xiang et al., 2009; Yu et al., 2010; Zhang, Tweel, & Tong, 2004) and CP-640186 that targets mammalian ACCs (Harwood et al., 2003; Zhang, Tweel, Li, et al., 2004). A large collection of resistance mutations have been identified in plant ACCs (Kaundun, 2014), which are clustered around the CT active site region. The natural product moiramide b inhibits bacterial CT in the active site (Silvers et al., 2016).

2.3 CT Component in PC

The CT component in PC contains a TIM (triosephosphate isomerase) barrel core, followed by a long, helical extension at the C-terminus, as illustrated by the structure of the CT domain of SaPC (Fig. 2E; Xiang & Tong, 2008). This CT is a dimer in all the structures determined so far, and the dimer interface primarily involves the TIM barrel. This component is shared with other enzymes, for example, the transcarboxylase 5S subunit (Hall et al., 2004) and the biotin-dependent decarboxylase oxaloacetate decarboxylase Na^+ pump (Balsera, Buey, & Li, 2011; Studer et al., 2007).

The active site of CT is located in the center of the TIM barrel, far away from the dimer interface (Fig. 2E). The C-terminal helical extension contributes to the recognition of BCCP–biotin. CT requires a divalent metal ion (oftentimes Mn^{2+}) for catalysis, and the substrate pyruvate is directly coordinated to the metal ion.

2.4 CT Component in UC

The CT component of UC consists of four subdomains, named A, B, C, and D, as illustrated by the structure of KlUC (Fig. 2F; Fan et al., 2012). The active site is located in a cleft at the interface between the B and D domains. This component is similar to the *Bacillus subtilis* KipA–KipI complex (Jacques, Langley, Hynson, et al., 2011; Jacques, Langley, Kuramitsu, et al., 2011), with ~20%–35% sequence identity, but residues that are important for CT catalysis are not conserved in KipA and KipI. In addition, domains B and D have the cyclophilin fold, although they do not share any recognizable sequence conservation with cyclophilin. In contrast to the other CT components, the CT component of UC functions as a monomer.

2.5 BCCP Component

The structure of the BCCP core was first reported in 1995 (Athappilly & Hendrickson, 1995). It can be thought of as a long β-hairpin structure, with four pairs of antiparallel β-strands that wrap around a central hydrophobic core (Fig. 3A). The biotinylation motif Met-Lys-Met is located at the tip of the β-hairpin structure. Rotations around the Cα—Cβ bond of this Lys residue contribute to the swinging-arm model. The connection to the rest of the enzyme at the N-terminus of BCCP core is located at the opposite end of the structure from the biotin moiety. Rotations around this region contribute to the swinging-domain model, and the $N_{1'}$ atom of biotin is ~40 Å from this pivot point. This gives a range of ~80 Å for the swinging-domain model, and the BC–CT active site distances observed so far are between 40 and 80 Å. In addition, the linker before the BCCP core in the holoenzyme could also be flexible, which would give further reach for the biotin $N_{1'}$ atom.

2.6 BT Domain in Acyl-CoA Carboxylases

The BT domain was first observed in the structure of the PCC holoenzyme (Huang et al., 2010), with a structure that contains a central helix surrounded by an eight-stranded antiparallel β-barrel (Fig. 3B). This domain has also been observed in the structures of PaMCC, PfGCC, and ScACC holoenzymes, although this backbone fold is rarely observed in other proteins.

2.7 PT Domain and BT-Like Domain in PC

The PT domain was first observed in the structure of RePC and was named the allosteric domain (St. Maurice et al., 2007), and it has since been

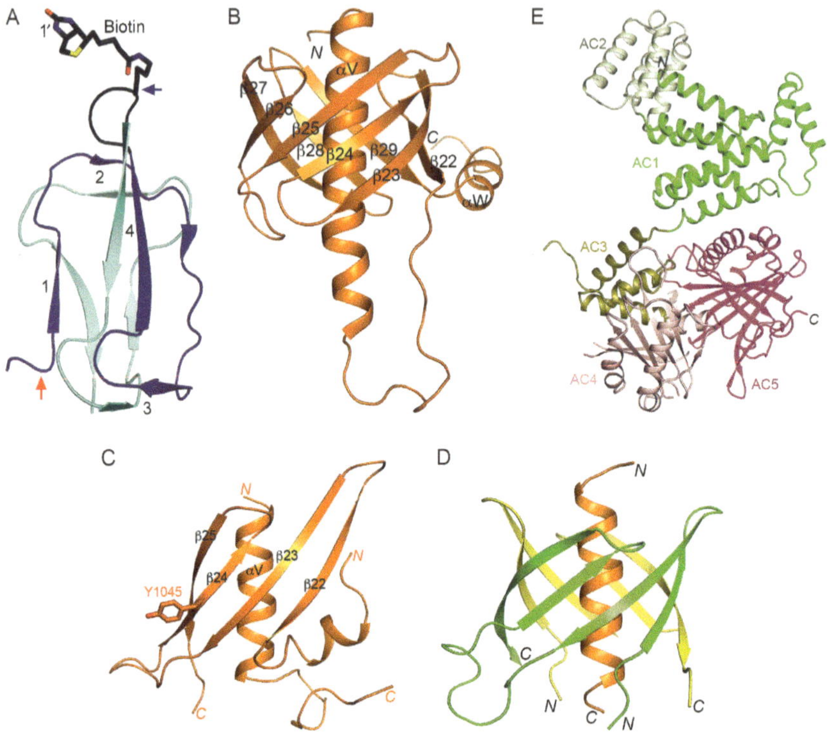

Fig. 3 Structures of the BCCP, BT, and PT components of biotin-dependent carboxylases. (A) Structure of the BCCP domain core (*blue* for the first half, *cyan* for the second, and *black* for the tip of the β-hairpin) in the PCC holoenzyme (Huang et al., 2010). The biotin cofactor (*black*) is located in the CT active site. The *blue arrow* indicates the pivot point for the swinging-arm model, and the *red arrow* for the swinging-domain model. (B) Structure of the BT domain in the bacterial PCC holoenzyme. (C) Structure of the PT domain in the SaPC holoenzyme (Xiang & Tong, 2008). (D) Structure of the BT-like domain in the two-subunit MfPC holoenzyme (Choi et al., 2016). The α-helix at the C-terminus of the α subunit is shown in *orange*, and the four-stranded β-sheets from the two β subunits are in *green* and *yellow*, respectively. (E) Structure of the central region of ScACC holoenzyme, domains AC1–AC5 (Wei & Tong, 2015). The coloring schemes of this figure do not necessarily follow that of Fig. 1B, in order to show structural features within the components.

observed in other single-chain PC enzymes, including HsPC and SaPC (where it is named the PT domain; Fig. 3C) (Xiang & Tong, 2008), and LmPC (Sureka et al., 2014). The structure consists of a short helix flanked on one side by a four-stranded antiparallel β-sheet. However, the PT domain is formed from two noncontiguous segments of the enzyme, with the helix being located between the BC and CT domains and the β-sheet between the CT and BCCP domains in the primary sequence (Fig. 1B). In this regard, the CT domain could be considered as an insert to the PT domain in single-chain PCs.

The PT domain bears weak structural similarity to the BT domain. In the structure of the two-subunit MfPC, the helix at the end of the α subunit (containing BC) is surrounded by two four-stranded β-sheets from two β subunits (containing CT and BCCP), which form an eight-stranded antiparallel β-sheet (Fig. 3D) (Choi et al., 2016). The structure of this domain is remarkably similar to that of the BT domain, and it has been named the BT-like domain. The structural similarities among the PT, BT-like, and BT domains may provide information in terms of their evolution.

2.8 AC Domains in Single-Chain ACCs

The central region in eukaryotic, single-chain ACC enzymes (Fig. 1B), contains five domains, and they are named AC1–AC5 (AC: ACC Central) (Fig. 3E; Hunkeler, Stuttfeld, Hagmann, Imseng, & Maier, 2016; Wei & Tong, 2015). Domains AC1, AC2, and AC3 are all helical, with domains AC1 and AC3 being four-helical bundles. Domain AC2 is an insert in domain AC1, while domain AC3 is positioned between AC4 and AC5, mediating interactions between them. Domains AC4 and AC5 have similar backbone folds, consisting of a twisted β-sheet flanked on one face by helices, despite sharing very low sequence identity (12%). This backbone fold has weak similarity to several other proteins, but the similarity is unlikely to have any functional significance for ACC.

3. EXTENSIVE DIVERSITY IN HOLOENZYME ARCHITECTURES

3.1 Architectures of Acyl-CoA Carboxylase Holoenzymes

Among the acyl-CoA carboxylases, PCC, MCC, and GCC have similar domain organization and good sequence conservation, and LCC can be considered as a single-chain version of these enzymes (Fig. 1B). Therefore, it was anticipated that the overall structures of these enzymes are similar to each other, but striking differences are observed in their actual structures (Fig. 4A–K).

The structure of a bacterial $\alpha_6\beta_6$ PCC holoenzyme possesses 32 symmetry and contains a central β_6 core (Fig. 4A; Huang et al., 2010), formed by three β_2 (CT) dimers. There are three monomeric α subunits at each end of the core, with their BC domains splayed far away from each other (Fig. 4B).

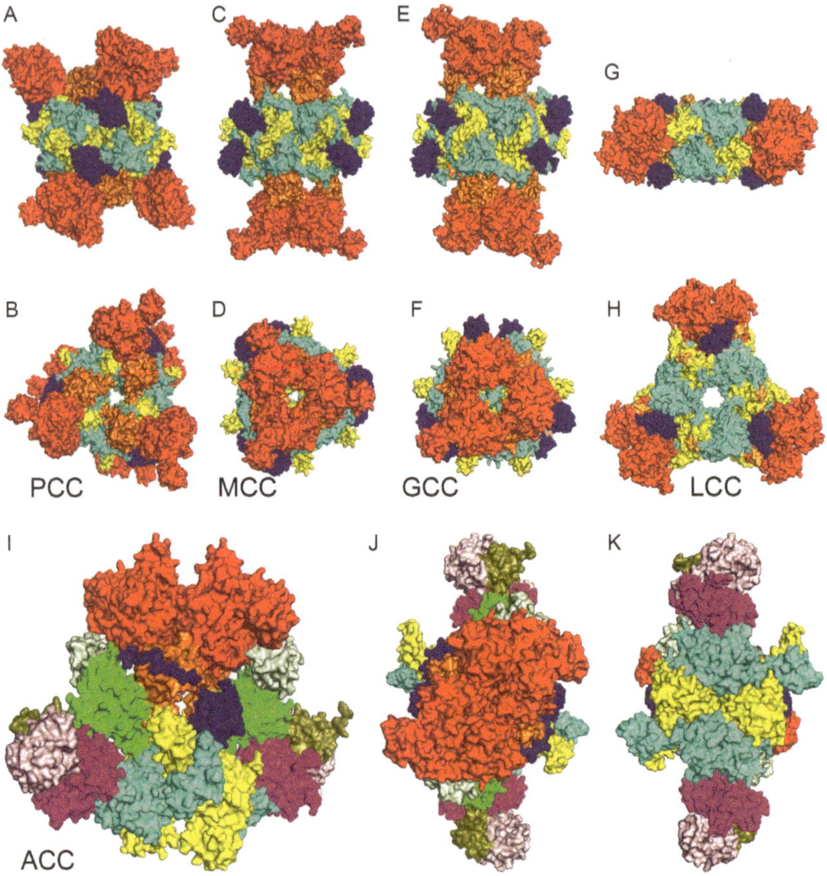

Fig. 4 Diverse architectures of acyl-CoA carboxylase holoenzymes. (A) Structure of a bacterial PCC holoenzyme (Huang et al., 2010), viewed down a twofold symmetry axis. The molecular surface of the structure is shown, and the domains are colored as in Fig. 1B in all the panels. (B) Structure of the PCC holoenzyme viewed down the threefold symmetry axis, related to panel (A) by a 90 degree rotation around the horizontal axis. (C) Structure of the PaMCC holoenzyme (Huang et al., 2012). (D) Structure of PaMCC viewed down the threefold symmetry axis. (E) Structure of the PfGCC holoenzyme (Jurado et al., 2015). (F) Structure of PfGCC viewed down the threefold symmetry axis. (G) Structure of the MapLCC holoenzyme (Tran et al., 2015). (H) Structure of MapLCC viewed down the threefold symmetry axis. (I) Structure of the ScACC holoenzyme (Wei & Tong, 2015). (J) Structure of ScACC viewed down the twofold symmetry axis at the BC domain dimer. (K) Structure of ScACC viewed down the twofold symmetry axis at the CT domain dimer.

The BT domain is essential for mediating interactions between the α (BC) and β (CT) subunits, as there are few direct contacts between BC and CT in the holoenzyme. BCCP–biotin is bound in the active site of the β subunit, providing molecular insights into CT catalysis. The BC and CT active sites are separated by 55 Å in the holoenzyme, indicating the swinging-domain model (Fig. 1A). EM studies of HsPCC suggest that it has a similar overall structure as the bacterial PCC (Huang et al., 2010).

The structure of the $\alpha_6\beta_6$ PaMCC holoenzyme also contains a central β_6 core (Fig. 4C; Huang et al., 2012), but the three α subunits at either end of the β_6 core are arranged close to each other, in fact showing trimeric association for the BC domains (Fig. 4D). As a consequence, the overall shape of PaMCC is strikingly different from that of PCC. The active sites of BC and CT are separated by 80 Å, also indicating the swinging-domain model.

In addition, while the overall structure of the MCCβ hexamer is similar to that of PCCβ, the positions of the two domains in each subunit are swapped (Fig. 5A and B), due to a difference in the conformation of the linker segment between the two domains in each enzyme (Fig. 5C and D) rather than a swap of the two domains in the primary sequence (Fig. 1B). Both domains have the crotonase fold, and the β_6 core has pseudo-sixfold symmetry. The swap of the positions of the two domains is equivalent to a 60 degree rotation around this axis. The structural differences may be linked to the differences in the biochemical activities of these enzymes. While PCC, ACC, and LCC carboxylate the α carbon of the acceptor substrate (Fig. 5E), MCC and GCC carboxylate the γ carbon of an α,β-unsaturated acid (Fig. 5F). Therefore, the structural and biochemical observations suggest that there may be two distinct lineages of acyl-CoA carboxylases (Tong, 2013).

The overall structure of the $\alpha_6\beta_6$ PfGCC holoenzyme is similar to that of PaMCC (Fig. 4E and F; Jurado et al., 2015), consistent with them being in the same lineage of acyl-CoA carboxylases. However, there are significant conformational differences for the linker between the two domains as well as the C-terminal segment of the β subunit, which are examples of the extensive conformational variability among these enzymes.

The structure of the hexameric MapLCC holoenzyme contains a hexameric core of the CT domains (Fig. 4G) that is similar to that of PCC (Fig. 4A; Tran et al., 2015). However, the overall architecture of the holoenzyme is remarkably different from that of PCC and MCC. The BC domains form a dimer and are located at the side of the CT domain core, and the holoenzyme is much shorter (Fig. 4G) but wider (Fig. 4H). The exact location of the BC domains relative to the CT core is rather variable, and this variability is also observed in studies on the homologous enzyme from *Deinococcus radiodurans* (Hagmann et al., 2016). A BT domain does not exist in MapLCC, and the linker between BC and BCCP is partly

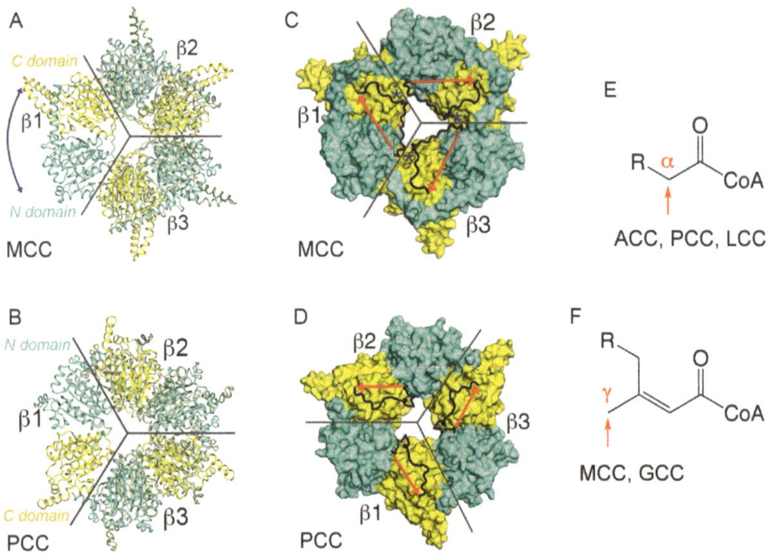

Fig. 5 Structural differences in the β_6 hexamer cores of PCC and MCC. (A) Structure of the β_6 hexamer of PaMCC. The N and C domains are colored in *cyan* and *yellow*, respectively, for the three subunits in the *top layer*, and the subunits in the *bottom layer* are colored in *green*. The subunit beneath β1 is omitted for clarity. *Gray lines* mark the boundaries of the subunits, and the *blue arrow* indicates the swapping of the positions of the N and C domains relative to PCCβ. (B) Structure of the β_6 hexamer of *Roseobacter denitrificans* PCCβ. (C) The N and C domains in MCCβ are shown as a molecular surface, and the linker between them is shown as a trace in *black*, with the *red arrow* indicating the direction of the linker. This linker runs in the opposite direction compared to that in PCC and connects to a different C domain, leading to the swap of the positions of the N and C domains in each subunit between the two enzymes. The two neighboring linkers approach each other closely at one point (*blue star*) in MCCβ, and a change in connectivity at that position will lead to the PCC organization. (D) Structure of PCCβ showing the linker between the N and C domains. (E) The substrates of PCC, ACC, and LCC. The site of carboxylation is indicated with the *red arrow*. (F) The substrates of MCC and GCC.

disordered. The BC and CT active sites are separated by 40 Å, again consistent with the swinging-domain model.

The structure of the dimeric ScACC holoenzyme is strikingly different from those of the other acyl-CoA carboxylase holoenzymes (Fig. 4I). The structure contains dimers of BC (Fig. 4J) and CT (Fig. 4K) domains, and their active sites are separated by 80 Å. The central region is located far away from the active sites and has no direct involvement in the catalysis. This region and the BT domain function as a scaffold and are important for positioning the BC and CT domains for catalysis.

3.2 Architectures of PC and UC Holoenzymes

The structures of the tetrameric single-chain PC holoenzymes RePC (Fig. 6A and B; St. Maurice et al., 2007), SaPC (Fig. 6C and D; Xiang & Tong, 2008; Yu et al., 2009), and LmPC (Fig. 6E and F; Sureka et al., 2014)

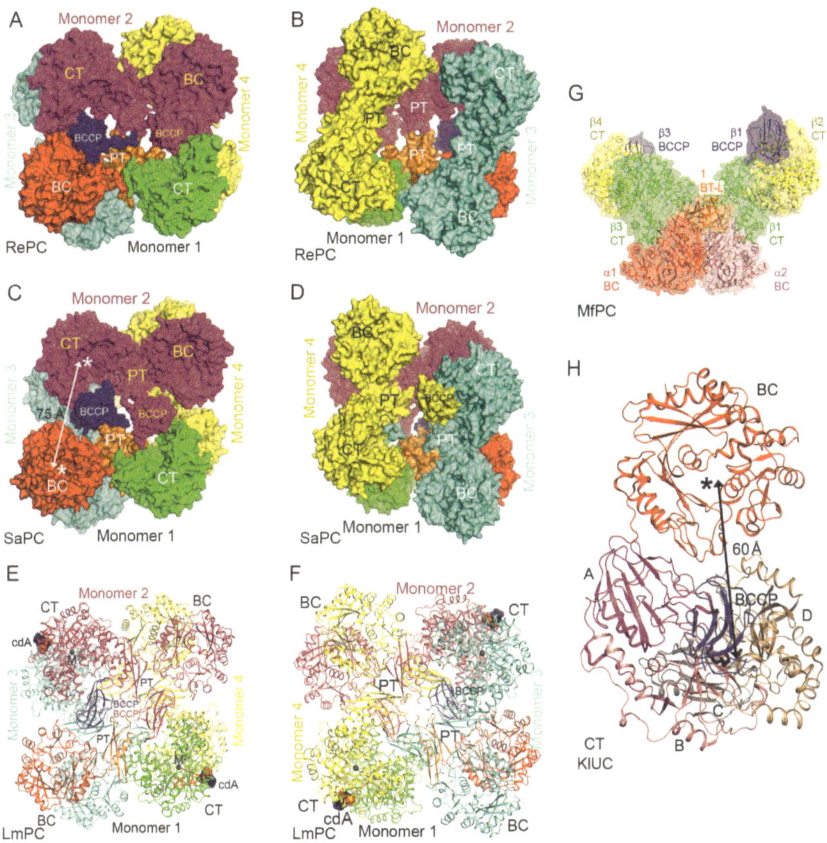

Fig. 6 Diverse architectures of the PC and UC holoenzymes. (A) Structure of the tetrameric RePC holoenzyme (St. Maurice et al., 2007). Monomer 1 is colored according to Fig. 1B, and monomers 2, 3, and 4 are in *magenta*, *yellow* and *cyan*, respectively. (B) Structure of the *bottom layer* of RePC tetramer, viewed after 180 degree rotation around the vertical axis from panel (A). (C) Structure of the tetrameric SaPC holoenzyme (Xiang & Tong, 2008). The BC and CT active sites are indicated with the *asterisks*, and the *arrow* indicates a possible route for BCCP translocation during catalysis. (D) Structure of the *bottom layer* of SaPC tetramer. (E) Structure of the tetrameric LmPC holoenzyme in complex with cyclic di-AMP (sphere model, labeled cdA) (Sureka et al., 2014). (F) Structure of the *bottom layer* of LmPC tetramer. (G) Structure of the $\alpha_2\beta_4$ holoenzyme of MfPC (Choi et al., 2016). (H) Structure of the monomeric KlUC holoenzyme (Fan et al., 2012).

consist of two layers, each with two monomers, and are in the shape of a diamond. Interactions between the layers are mediated by BC and CT dimers, located at alternate corners of the diamond, as well as a dimer of the PT domain in some of the holoenzymes. BCCP–biotin is observed in the CT active site, or in an exosite near the CT (Xiang & Tong, 2008). In many monomers the BCCP domain is disordered, consistent with its dynamic translocation during catalysis.

The distance between the BC and CT active sites is ~75 Å in the tetramer, indicating BCCP translocation during catalysis. Moreover, the structure suggests that BCCP migrates between the BC and CT active sites of different monomers for catalysis (Fig. 6C), which provides an explanation why PC is only active as a tetramer. In addition, the domains in each monomer have little contact with each other, and the PC holoenzyme can only be stabilized in the context of the tetramer, which is another reason why PC is only active as a tetramer.

While the overall structures of the single-chain PC holoenzymes appear similar, there are significant differences among them, both in the conformations of the monomers and in the relative positions of the monomers. The RePC tetramer is highly asymmetrical in the positioning of its four monomers, which can be visualized by comparing the top (Fig. 6A) and bottom (Fig. 6B) layers. In comparison, the SaPC tetramer is more symmetrical (Xiang & Tong, 2008), especially in the complex with CoA (Yu et al., 2009). The LmPC structure in the complex with cyclic di-AMP (c-di-AMP) is even more symmetrical (Sureka et al., 2014). This asymmetry is also reflected by the differences in the conformations of the four monomers in the tetramer. When the monomers are superposed based on their CT domains, the positions of the BC domains are significantly different in RePC (Fig. 7A). The differences are smaller for SaPC (Fig. 7B), while the four monomers of LmPC in the complex with c-di-AMP are essentially identical (Fig. 7C). Studies on RePC suggest that its asymmetric structure may be an inherent property (Lietzan et al., 2011), despite sharing strong sequence conservation with other single-chain PC enzymes.

Another evidence for the structural differences among the PC holoenzymes is the relative positions of the PT + CT domains of their four monomers. There are very few contacts between the two PT + CT dimers in RePC (Fig. 7D). In contrast, there are close contacts among the PT domains in SaPC (Fig. 7E) and especially HsPC (Fig. 7F), mediated by the exposed surface of the β-sheet in the PT domain. In fact, the HsPC structure lacks the BC domain, and the contacts among the PT domains are the primary

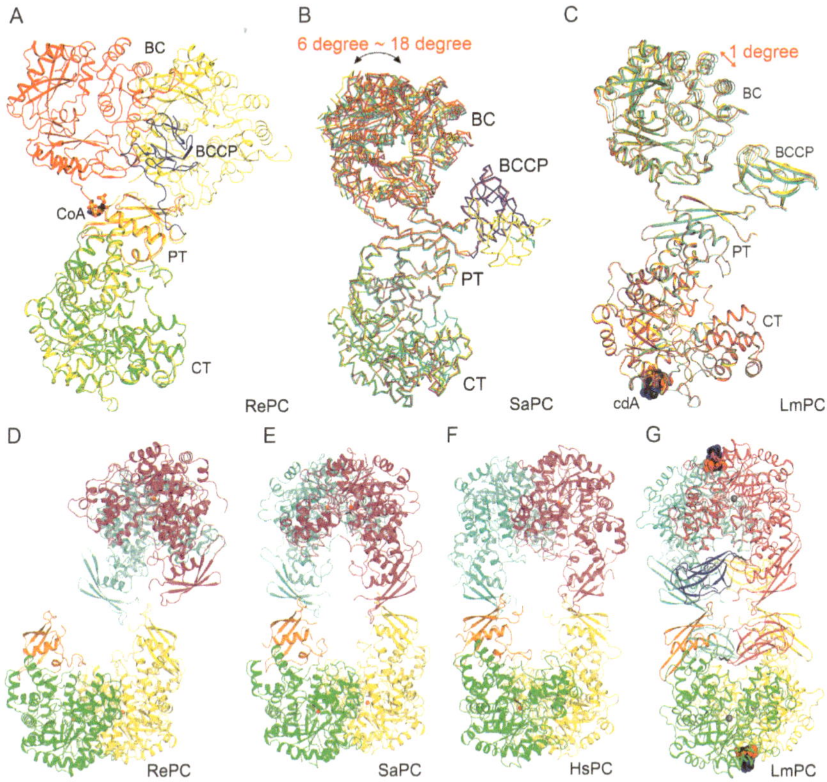

Fig. 7 Structural differences among the PC holoenzymes. (A) Overlay of the monomer structures in the RePC holoenzyme (St. Maurice et al., 2007). The overlay is based on the CT domain only. Only two monomers are shown because the tetramer obeys crystallographic twofold symmetry. (B) Overlay of the monomer structures in the SaPC holoenzyme (Xiang & Tong, 2008). (C) Overlay of the monomer structures in the LmPC holoenzyme in complex with c-di-AMP (Sureka et al., 2014). (D) Structure of the PT+CT domains of the RePC tetramer. (E) Structure of the PT+CT domains of the SaPC tetramer. (F) Structure of the PT+CT domains of the HsPC tetramer. (G) Structure of the PT+CT domains of the LmPC tetramer.

interactions that enable the formation of this tetramer, hence the domain was named the PT domain (Xiang & Tong, 2008). A highly conserved aromatic residue on the surface of the β-sheet of this domain (Tyr1045 in SaPC, Fig. 3C; Phe1077 in HsPC) is crucial for this interaction, and the F1077A mutation in HsPC disrupts the holoenzyme tetramer and inactivates the enzyme. Interestingly, acetyl-CoA can rescue this mutant, enabling tetramer formation and catalytic activity. The overall organization of the

PT + CT domains in LmPC (Fig. 7G) is similar to that in SaPC and HsPC, although the PT domains are not in contact.

The two-subunit PC was believed to have $\alpha_4\beta_4$ stoichiometry, which would be equivalent to the tetramer of single-chain PC (Fig. 1B). The structure of MfPC showed that it has $\alpha_2\beta_4$ stoichiometry (Fig. 6G; Choi et al., 2016), and the overall structure of this holoenzyme is dramatically different from the tetrameric single-chain PCs. Further biophysical experiments confirm this stoichiometry, and the $\alpha_1\beta_2$ organization of the BT-like domain (Fig. 3D) may be the molecular determinant for this stoichiometry. A BC dimer forms the base of the structure, with two CT dimers extending out from it. Such a stoichiometry contains excess BCCP–biotin and CT active sites, and it is not known whether there is coordination in the use of the BC and CT active sites during catalysis.

The structure of monomeric KlUC has a distance of 60 Å between the BC and CT active sites (Fig. 6H; Fan et al., 2012). The BC domain directly contacts the A and D domains of CT, and BCCP–biotin is bound in the active site of CT.

4. EXTENSIVE CONFORMATIONAL VARIABILITY

Structural information currently available on the biotin-dependent carboxylases indicates that they are highly dynamic and can exhibit extensive conformational variability, especially upon the binding of substrates and/or (allosteric) regulators. Some of the types of variability has been observed frequently with other enzymes as well, such as the open–closed transition for the B domain of BC during catalysis and conformational changes in the CT active site, including the C-terminal helical extension of the CT in PC (Xiang & Tong, 2008). These will not be described further here. This section will instead focus on the conformational variability that is more unique to biotin-dependent carboxylases.

4.1 Conformational Variability in the BC Domain Dimer Interface of Single-Chain ACCs

While a BC domain dimer is observed in the structure of ScACC holoenzyme (Fig. 2C; Wei & Tong, 2015), BC domain alone of yeast (Fig. 2B) and human ACCs is monomeric (Cho et al., 2010; Harriman et al., 2016; Raymer et al., 2009; Shen et al., 2004) and catalytically inactive (Weatherly, Volrath, & Elich, 2004). There are extensive differences between the structure of BC domain in the holoenzyme and that of BC domain alone, both in the dimer interface region (Fig. 8A) and in the biotin-binding site of the active site (Fig. 8B). As a consequence of these conformational changes, BC domain alone assumes a conformation in the dimer interface that is incompatible with dimerization (Fig. 8A and C). In addition, BC domain alone cannot bind biotin in the active site because a loop in this state would clash with biotin, which explains why BC domain alone is catalytically inactive. Finally, BC domain alone has a binding site in the dimer interface region for the natural product inhibitor soraphen A (Fig. 8A; Shen et al., 2004) and the small-molecule inhibitor ND-630 (Harriman et al., 2016). This indicates that these compounds allosterically inhibit ACC by stabilizing a conformation of the BC domain that is incompatible with catalysis (and dimerization) (Fig. 8C).

Eukaryotic ACCs are regulated by phosphorylation, for example, by AMP-activated protein kinase (AMPK, known as SNF1 in yeast) (Carling & Viollet, 2015; Hardie, Schaffer, & Brunet, 2016; Hedbacker & Carlson, 2008). An AMPK phosphorylation site is located just prior to the BC domain core in the primary sequence of mammalian ACCs (Ser80 in human ACC1 and Ser222 in human ACC2). The phosphorylated segment containing Ser222 of ACC2 is recognized by the soraphen A-binding site (Fig. 8D; Cho et al., 2010), suggesting that the physiological function of the soraphen A-binding site is to recognize this phosphorylated segment, thereby mediating regulation by AMPK (Fig. 8C). Interestingly, ScACC does not have a phosphorylation site that is equivalent to Ser222 of human ACC2, but it does contain a soraphen A-binding site. There may be another physiological ligand to this site in ScACC, which remains to be identified.

4.2 Conformational Variability in the Central Region of Single-Chain ACCs

AMPK has a second phosphorylation site in mammalian ACCs, Ser1216 in human ACC1 and Ser1340 in human ACC2. This site is equivalent to Ser1157 in ScACC, and SNF1 was suggested to be the protein kinase for this site (Choi & Da Silva, 2014; Hofbauer et al., 2014; Hunkeler et al., 2016; Shi, Chen, Siewers, & Nielsen, 2014). It was recently demonstrated that SNF1 can directly phosphorylate this site in vitro (Wei et al., 2016). Ser1157 is located in domain AC4, more than 70 Å away from the CT and BC active sites in the ScACC holoenzyme structure (Wei & Tong, 2015). Phosphorylated Ser1157 is recognized by a pocket at the AC4–AC5 interface (Fig. 8E), and mutations of the two Arg residues in this pocket

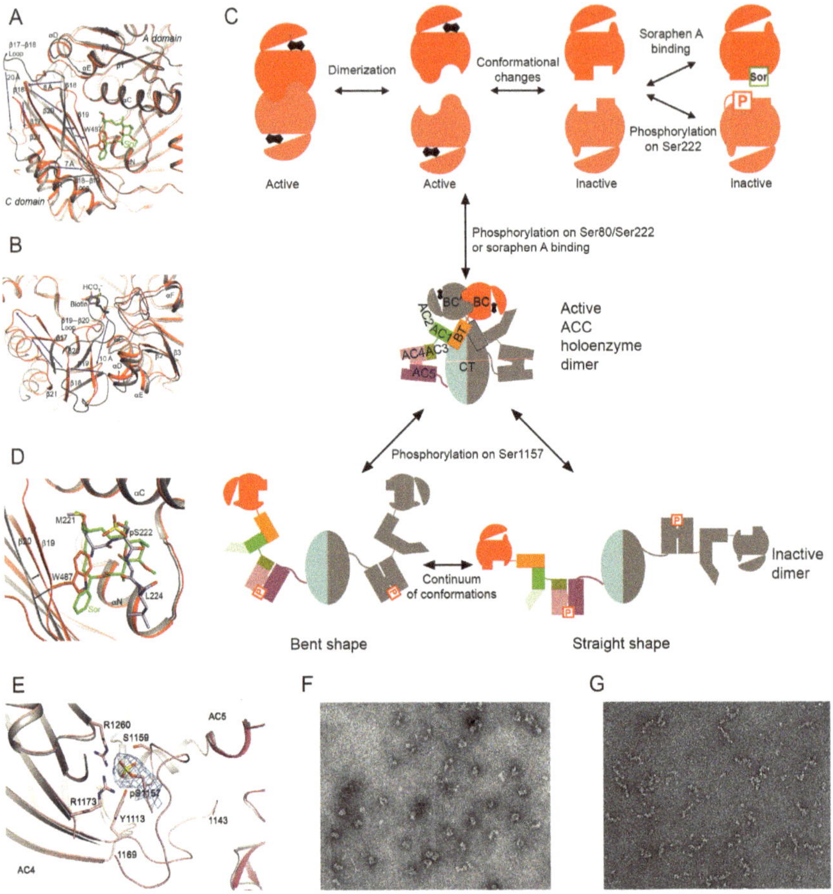

Fig. 8 Conformational variability of single-chain ACCs and a unified mechanism for their regulation by phosphorylation. (A) Overlay of the structure of BC domain in the ScACC holoenzyme (*red*) (Wei & Tong, 2015) with that of BC domain alone (*gray*) (Shen et al., 2004) in the dimer interface region. *Blue arrows* highlight large differences. The binding mode of soraphen A (*green*) is also shown (Shen et al., 2004). (B) Overlay of the structure of BC domain in the ScACC holoenzyme (*red*) with that of BC domain alone (*gray*) in the active site region. The bound position of biotin in *E. coli* BC is shown (*black*) (Chou et al., 2009). (C) A unified mechanism for the regulation of single-chain ACCs by AMPK/SNF1 phosphorylation (Wei et al., 2016). Only the BC domains are shown in the *top half*. (D) Close-up of the BC domain dimer interface region. The bound position of the pSer222 peptide in human ACC2 (*light blue*) is also shown (Cho et al., 2010). (E) Binding site for pSer1157 in the structure of domains AC3–AC5 of ScACC (in *color*), overlaid with that of these domains alone (*gray*) (Wei et al., 2016). (F) Negative stain EM image of ScACC in the presence of citrate (Wei et al., 2016). Primarily globular shapes are observed. (G) Negative stain EM image of phosphorylated ScACC. A continuum of bent and straight shapes is observed.

make ScACC essentially insensitive to SNF1 (Wei et al., 2016). The structure of the phosphorylated domains AC3–AC5 is essentially the same as that of the domains alone (Fig. 8E), but there are substantial differences to the structure of these three domains in the holoenzyme. Therefore, phosphorylation of Ser1157 may stabilize the structure of domains AC3–AC5 alone, which is incompatible with the holoenzyme. This situation is somewhat reminiscent of that for the BC domain, in terms of the large conformational differences between the domain alone and in the holoenzyme.

EM studies showed that ScACC primarily adopts a globular structure in the presence of citrate (Fig. 8F), highly similar to that observed in the crystal structure (Wei et al., 2016). In contrast, primarily elongated structures, in a continuum of straight and bent shapes, were observed when Ser1157 is phosphorylated (Fig. 8G), and these shapes were also observed for ScACC in the presence of soraphen A. Overall, these observations suggest that the BC domain and the central region of single-chain ACCs are highly dynamic, with BC being the weak link in the holoenzyme dimer, thereby giving rise to a unified mechanism for how AMPK/SNF1 regulates ACC (Fig. 8C; Wei et al., 2016). Phosphorylation of Ser80 or Ser1216 both stabilize a monomeric form of BC, leading to a continuum of elongated, inactive conformations for ACC. This conformational variability and the recognition of phosphorylated Ser1157 were also observed in studies on the central region of yeast and other ACCs (Hunkeler et al., 2016). On the other hand, the mechanism of how the conformational change in AC3–AC5 upon phosphorylation of Ser1157 is transmitted to the BC dimer interface in the holoenzyme is still not understood.

4.3 Conformational Variability in the CT Active Site of Single-Chain ACCs for Herbicide Binding

Three classes of compounds are used as herbicides and function through potent inhibition of plant ACCs, including haloxyfop (FOP series), tepraloxydim (DIM series), and pinoxaden (Kaundun, 2014). The binding of haloxyfop and other FOP compounds to the CT active site of ScACC requires an extensive conformational change in the dimer interface (Fig. 9A and B; Zhang, Tweel, & Tong, 2004). A part of the binding pocket extends deep into the hydrophobic dimer interface (Fig. 9C), which does not exist in the structure of CT domain alone (Fig. 9D). Smaller changes are needed for the binding of DIM compounds (Xiang et al., 2009) and pinoxaden (Yu et al., 2010).

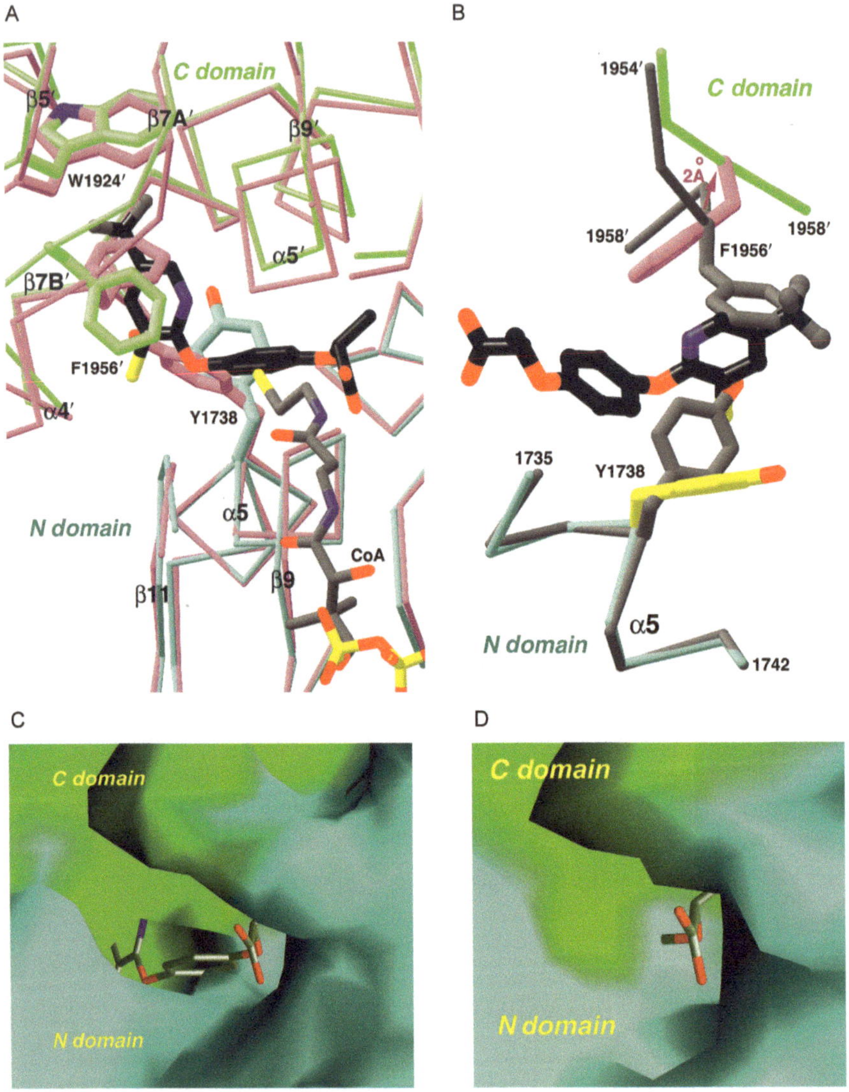

Fig. 9 Conformational variability in the CT active site of single-chain ACCs for herbicide binding. (A) Overlay of the structure of CT domain of ScACC (*cyan* and *green* for the N and C domains, respectively) in complex with haloxyfop (*black*) (Zhang, Tweel, & Tong, 2004) with that of CT domain alone (*magenta*) in complex with CoA (*gray*) (Zhang et al., 2003). (B) Close-up of the haloxyfop-binding site highlighting the crucial structural differences. The structure of CT domain alone is shown in *gray*. (C) Molecular surface of the CT domain near the haloxyfop-binding site, which involves a pocket that extends deep into the dimer interface. (D) Molecular surface of the CT domain alone. The pocket does not exist, and haloxyfop would clash into the protein.

The conformational variability of the CT domain is therefore crucial for herbicide binding, which may also contribute to the selectivity of these compounds. They are much less potent against ScACC and human ACCs, probably because the plant ACCs can more readily undergo this conformational change to accommodate the binding of the herbicides.

4.4 Conformational Variability in Single-Chain PCs

EM studies on SaPC suggest that conformational dynamics is important for its catalysis (Lasso et al., 2014, 2010). In addition, conformational variability in single-chain PCs is crucial for their regulation by acetyl-CoA and c-di-AMP. Most single-chain PCs are activated by acetyl-CoA (Jitrapakdee et al., 2008). Its nucleotide portion is bound in a pocket between the BC and PT domains (Fig. 10A), with several Arg residues recognizing the phosphate groups (Fig. 10B; St. Maurice et al., 2007; Xiang & Tong, 2008; Yu et al., 2009). One of the disease-causing mutations of HsPC, R451C, is located here. The mutant has significantly reduced sensitivity to acetyl-CoA, which is likely the molecular basis for its disease-causing effects (Xiang & Tong, 2008). On the other hand, this portion of acetyl-CoA can only activate PC with weak potency (Xiang & Tong, 2008). The pantotheine group of acetyl-CoA is bound at the BC dimer interface and causes a reorganization of the BC dimer in SaPC (Fig. 10C), while the CT dimer is essentially not affected in the holoenzyme (Yu et al., 2009). Therefore, acetyl-CoA may stimulate PC by promoting a more active organization of the BC dimer. This change in BC dimer organization also makes the conformations of the four monomers of SaPC more similar to each other, which makes the tetramer more symmetric as well. In contrast, RePC is highly asymmetrical, with only two acetyl-CoA-binding site in each tetramer (St. Maurice et al., 2007).

The acetyl-CoA-binding site does not exist in two-subunit PC, although the interactions that are observed between the nucleotide portion of acetyl-CoA and single-chain PC are mimicked by those among highly conserved residues in two-subunit PCs (Choi et al., 2016).

c-di-AMP (cdA) is a recently identified, widely conserved second messenger in bacteria and has important roles in stress response, antibiotic resistance, cell wall homeostasis, virulence, host immune response, and other processes (Commichau, Dickmanns, Gundlach, Ficner, & Stulke, 2015; Corrigan & Grundling, 2013; Woodward, Iavarone, & Portnoy, 2010).

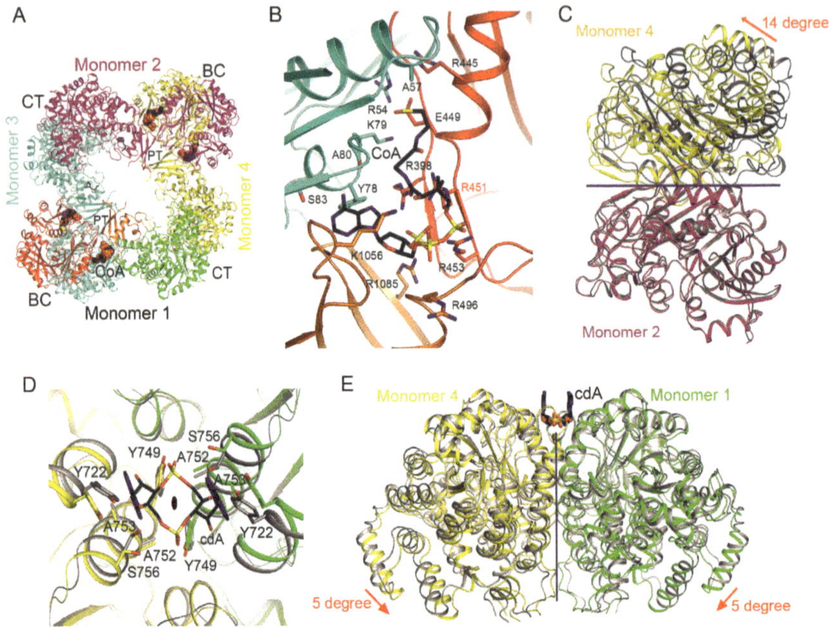

Fig. 10 Conformational variability in the single-chain PCs and their allosteric regulation. (A) Overall structure of SaPC in complex with CoA (*black*, sphere model). All four BCCP domains are disordered (Yu et al., 2009). (B) The binding site for CoA in SaPC. The nucleotide portion is located between the BC and PT domains, and the pantotheine portion in the BC dimer interface. (C) Reorganization of the BC dimer in the CoA complex of SaPC (in *color*) compared to that of SaPC alone (*gray*). The *blue line* indicates the twofold symmetry axis of the dimer. Monomer 2 is used for the overlay. (D) Overlay of the binding site of c-di-AMP (*black*) in LmPC (*green* and *yellow*) with that of LmPC alone (*gray*) (Sureka et al., 2014). Tyr722 π-stacks with the adenine base in the c-di-AMP complex, but would clash with this base in the free LmPC. (E) Reorganization of the CT dimer in the c-di-AMP complex of LmPC (in *color*) compared to that of LmPC alone (*gray*). Both monomers are used for the overlay.

LmPC was identified as a target for c-di-AMP by a chemical proteomics study, indicating a role for this compound in regulating central metabolism (Sureka et al., 2014). c-di-AMP is bound at the CT dimer interface of LmPC, 25 Å from the CT active site (Fig. 6E), aligning its twofold symmetry axis with that of the CT dimer (Fig. 10D; Sureka et al., 2014). c-di-AMP causes a reorganization of the CT dimer (Fig. 10E) and the holoenzyme tetramer of LmPC. The structure of LmPC alone is incompatible with c-di-AMP binding (Fig. 10D). In addition, the conformations of the four monomers of LmPC are different in the free enzyme, but are essentially the same in the complex with c-di-AMP (Fig. 7C). The structure suggests that c-di-AMP allosterically inhibits LmPC by causing a change in the CT dimer and/or by "freezing" the overall structure of the enzyme into a state that is incompatible with catalysis.

4.5 BCCP Translocation During Catalysis

The distances between the BC and CT active sites observed in the holoenzyme structures so far are consistent with the swinging-domain model (Fig. 1A). While a productive binding mode for BCCP and its biotin in the CT active site has been observed in many of the holoenzyme structures, productive binding of BCCP–biotin to the BC active site has not been observed yet. In fact, BCCP is rarely located near the BC domain in the holoenzyme structures so far. A nonproductive binding mode was observed for BCCP in the structure of the *E. coli* BC–BCCP complex (Broussard et al., 2013). It is not clear why it has been so difficult to trap the BCCP–biotin in the BC active site, although EM studies may offer another approach for observing productive binding of BCCP in BC. Having such structures would provide direct evidence for the swinging-domain model.

5. PERSPECTIVES

Biotin-dependent carboxylases exhibit extensive diversity in the architectures of their holoenzymes, despite sharing significant overall sequence conservation. At first glance, this might seem to contradict the general understanding in terms of the relationship between sequence conservation and structural similarity. However, the sequence conservation among these enzymes is mostly limited to their BC and CT components, and to a weaker extent BCCP and the other components, and these homologous components do have similar structures in the holoenzymes. On the other hand, how these components are assembled into the holoenzymes can be strikingly different. An analogy would be the use of the same collection of bricks (components) to build entirely different buildings (holoenzymes), and biotin-dependent carboxylases are a beautiful example of nature's versatility. The sequence conservation for the linker segments between these components is much weaker, which may also contribute the extensive diversity in the holoenzyme architectures.

The biotin-dependent carboxylase holoenzymes are large, complex systems. The divide-and-conquer approach has been commonly used to study such systems and can be very powerful. This approach has led to the structures of most of the components of these enzymes, many of which are directly relevant for their structures in the holoenzymes. However, the structures of the BC domain and the central region alone of eukaryotic ACCs exhibit substantial conformational differences to those in the holoenzyme. In addition, protein samples for the central region are dimeric, which would appear to be consistent with the dimeric association of the ACC holoenzyme. However, the central region is actually monomeric in the holoenzyme. These observations could represent a cautionary tale for interpreting results from the divide-and-conquer approach. On the other hand, the structures of these components alone represent another state for them and the conformational differences to the holoenzyme are related to the regulation of the eukaryotic ACCs. Therefore, the structural information from the divide-and-conquer approach helps reveal the conformational variability of these components and is still useful for a complete understanding of these enzymes.

The extensive conformational variability has been primarily observed for ACC and PC so far, because of the large amount of structural information available for them. As more structures are determined for the other biotin-dependent carboxylases, it might be possible that additional mechanisms of conformational variability will be found.

The conformational variability is crucial both for the catalysis by biotin-dependent carboxylases and for the regulation of these enzymes. Interfering with these conformational transitions represents an attractive approach for developing novel modulators of these enzymes. For example, soraphen A and ND-630 allosterically inhibit ACC by stabilizing (freezing) a conformation of the BC domain that is incompatible with catalysis (and dimerization). Therefore, further elucidation of the molecular mechanisms of this extensive conformational variability will greatly enhance our understanding of these enzymes and could also have a significant impact on drug discovery.

ACKNOWLEDGMENTS

I thank Jia Wei, Philip Choi, Timothy Tran, Matthew Callaghan, Chi-Yuan Chou, Daniel Floyd, Christine Huang, Ashley Jurado, Yi Seul Kim, Jiang Li, Svetlana Novoseletskaya, Kianoush Sadre-Bazzaz, Yang Shen, Benjamin Tweel, Song Xiang, Zhiru Yang, Linda

Yu, and Hailong Zhang for contributions to the studies on biotin-dependent carboxylases in my laboratory and for helpful discussions. This research was supported in part by grants from the NIH to L.T. (R01DK067238 and S10OD012018).

REFERENCES

Abu-Elheiga, L., Matzuk, M. M., Abo-Hashema, K. A. H., & Wakil, S. J. (2001). Continuous fatty acid oxidation and reduced fat storage in mice lacking acetyl-CoA carboxylase 2. *Science*, *291*, 2613–2616.

Abu-Elheiga, L., Oh, W., Kordari, P., & Wakil, S. J. (2003). Acetyl-CoA carboxylase 2 mutant mice are protected against obesity and diabetes induced by high-fat/high-carbohydrate diets. *Proceedings of the National Academy of Sciences of the United States of America*, *100*, 10207–10212.

Alberts, A. W., & Vagelos, P. R. (1972). Acyl-CoA carboxylases. In P. D. Boyer (Ed.), *Vol. 6. The enzymes* (pp. 37–82). New York: Academic Press.

Arabolaza, A., Shillito, M. E., Lin, T. W., Diacovich, L., Melgar, M. M., Pham, H., et al. (2010). Crystal structures and mutational analyses of acyl-CoA carboxylase β subunit of Streptomyces coelicolor. *The Biochemist*, *49*, 7367–7376.

Athappilly, F. K., & Hendrickson, W. A. (1995). Structure of the biotinyl domain of acetyl-coenzyme A carboxylase determined by MAD phasing. *Structure*, *3*, 1407–1419.

Balsera, M., Buey, R. M., & Li, X.-D. (2011). Quaternary structure of the oxaloacetate decarboxylase membrane complex and mechanistic relationships to pyruvate carboxylase. *The Journal of Biological Chemistry*, *286*, 9457–9467.

Baumgartner, M. R., Almashanu, S., Suormala, T., Obie, C., Cole, R. N., Packman, S., et al. (2001). The molecular basis of human 3-methylcrotonyl-CoA carboxylase deficiency. *Journal of Clinical Investigation*, *107*, 495–504.

Bilder, P., Lightle, S., Bainbridge, G., Ohren, J., Finzel, B., Sun, F., et al. (2006). The structure of the carboxyltransferase component of acetyl-CoA carboxylase reveals a zinc-binding motif unique to the bacterial enzyme. *The Biochemist*, *45*, 1712–1722.

Broussard, T. C., Kobe, M. J., Pakhomova, S., Neau, D. B., Price, A. E., Champion, T. S., et al. (2013). The three-dimensional structure of the biotin carboxylase-biotin carboxyl carrier protein complex of E. coli acetyl-CoA carboxylase. *Structure*, *21*, 650–657.

Broussard, T. C., Pakhomova, S., Neau, D. B., Bonnot, R., & Waldrop, G. L. (2015). Structural analysis of substrate, reaction intermediate, and product binding in Haemophilus influenzae biotin carboxylase. *The Biochemist*, *54*, 3860–3870.

Campeau, E., Desviat, L. R., Leclerc, D., Wu, X., Perez, B., Ugarte, M., et al. (2001). Structure of the PCCA gene and distribution of mutations causing propionic acidemia. *Molecular Genetics and Metabolism*, *74*, 238–247.

Cardaci, S., Zheng, L., MacKay, G., van den Broek, N. J., MacKenzie, E. D., Nixon, C., et al. (2015). Pyruvate carboxylation enables growth of SDH-deficient cells by supporting aspartate biosynthesis. *Nature Cell Biology*, *17*, 1317–1326.

Carling, D., & Viollet, B. (2015). Beyond energy homeostasis: The expanding role of AMP-activated protein kinase in regulating metabolism. *Cell Metabolism*, *21*, 799–804.

Chloupkova, M., Maclean, K. N., Alkhateeb, A., & Kraus, J. P. (2002). Propionic acidemia: Analysis of mutant propionyl-CoA carboxylase enzymes expressed in Escherichia coli. *Human Mutation*, *19*, 629–640.

Cho, Y. S., Lee, J. I., Shin, D., Kim, H. T., Jung, H. Y., Lee, T. G., et al. (2010). Molecular mechanism for the regulation of human ACC2 through phosphorylation by AMPK. *Biochemical and Biophysical Research Communications*, *391*, 187–192.

Choi, J. W., & Da Silva, N. A. (2014). Improving polyketide and fatty acid synthesis by engineering of the yeast acetyl-CoA carboxylase. *Journal of Biotechnology*, *187*, 56–59.

Choi, P. H., Jo, J., Lin, Y. C., Lin, M. H., Chou, C.-Y., Dietrich, L. E. P., et al. (2016). A distinct holoenzyme organization for two-subunit pyruvate carboxylase. *Nature Communications*, 7, 12713.

Chou, C.-Y., & Tong, L. (2011). Structural and biochemical studies on the regulation of biotin carboxylase by substrate inhibition and dimerization. *The Journal of Biological Chemistry*, *286*, 24417–24425.

Chou, C.-Y., Yu, L. P. C., & Tong, L. (2009). Crystal structure of biotin carboxylase in complex with substrates and implications for its catalytic mechanism. *The Journal of Biological Chemistry*, *284*, 11690–11697.

Christen, S., Lorendeau, D., Schmieder, R., Broekaert, D., Metzger, K., Veys, K., et al. (2016). Breast cancer-derived lung metastases show increased pyruvate carboxylase-dependent anaplerosis. *Cell Reports*, *17*, 837–848.

Clavero, S., Martinez, M. A., Perez, B., Perez-Cerda, C., Ugarte, M., & Desviat, L. R. (2002). Functional characterization of PCCA mutations causing propionic acidemia. *Biochimica et Biophysica Acta*, *1588*, 119–125.

Commichau, F. M., Dickmanns, A., Gundlach, J., Ficner, R., & Stulke, J. (2015). A jack of all trades: The multiple roles of the unique essential second messenger cyclic di-AMP. *Molecular Microbiology*, *97*, 189–204.

Corrigan, R. M., & Grundling, A. (2013). Cyclic di-AMP: Another second messenger enters the fray. *Nature Reviews. Microbiology*, *11*, 513–524.

Cronan, J. E., Jr., & Waldrop, G. L. (2002). Multi-subunit acetyl-CoA carboxylases. *Progress in Lipid Research*, *41*, 407–435.

Deodato, F., Boenzi, S., Santorelli, F. M., & Dionisi-Vici, C. (2006). Methylmalonic and propionic aciduria. *American Journal of Medical Genetics. Part C, Seminars in Medical Genetics*, *142*, 104–112.

Desviat, L. R., Perez, B., Perez-Cerda, C., Rodriguez-Pombo, P., Clavero, S., & Ugarte, M. (2004). Propionic acidemia: Mutation update and functional and structural effects of the variant alleles. *Molecular Genetics and Metabolism*, *83*, 28–37.

Desviat, L. R., Perez-Cerda, C., Perez, B., Esparza-Gordillo, J., Rodriguez-Pombo, P., Penalva, M. A., et al. (2003). Functional analysis of MCCA and MCCB mutations causing methylcrotonylglycinuria. *Molecular Genetics and Metabolism*, *80*, 315–320.

Diacovich, L., Mitchell, D. L., Pham, H., Gago, G., Melgar, M. M., Khosla, C., et al. (2004). Crystal structure of the β-subunit of acyl-CoA carboxylase: Structure-based engineering of substrate specificity. *The Biochemist*, *43*, 14027–14036.

Diacovich, L., Peiru, S., Kurth, D., Rodriguez, E., Podesta, F., Khosla, C., et al. (2002). Kinetic and structural analysis of a new group of acyl-CoA carboxylases found in Streptomyces coelicolor A3(2). *The Journal of Biological Chemistry*, *277*, 31228–31236.

Fan, C., Chou, C.-Y., Tong, L., & Xiang, S. (2012). Crystal structure of urea carboxylase provides insights into the carboxyltransfer reaction. *The Journal of Biological Chemistry*, *287*, 9389–9398.

Gallardo, M. E., Desviat, L. R., Rodriguez, J. M., Esparza-Gordillo, J., Perez-Cerda, C., Perez, B., et al. (2001). The molecular basis of 3-methylcrotonylglycinuria, a disorder of leucine metabolism. *American Journal of Human Genetics*, *68*, 334–346.

Hagmann, A., Hunkeler, M., Stuttfeld, E., & Maier, T. (2016). Hybrid structure of a dynamic single-chain carboxylase from Deinococcus radiodurans. *Structure*, *24*, 1227–1236.

Hall, P. R., Wang, Y.-F., Rivera-Hainaj, R. E., Zheng, X., Pustai-Carey, M., Carey, P. R., et al. (2003). Transcarboxylase 12S crystal structure: Hexamer assembly and substrate binding to a multienzyme core. *The EMBO Journal*, *22*, 2334–2347.

Hall, P. R., Zheng, R., Antony, L., Pustai-Carey, M., Carey, P. R., & Yee, V. C. (2004). Transcarboxylase 5S structures: Assembly and catalytic mechanism of a multienzyme complex subunit. *The EMBO Journal*, *23*, 3621–3631.

Hardie, D. G., Schaffer, B. E., & Brunet, A. (2016). AMPK: An energy-sensing pathway with multiple inputs and outputs. *Trends in Cell Biology, 26*, 190–201.

Harriman, G., Greenwood, J., Bhat, S., Huang, X., Wang, R., Paul, D., et al. (2016). Acetyl-CoA carboxylase inhibition by ND-630 reduces hepatic steatosis, improves insulin sensitivity, and modulates dyslipidemia in rats. *Proceedings of the National Academy of Sciences of the United States of America, 113*, E1796–E1805.

Harwood, H. J., Jr., Petras, S. F., Shelly, L. D., Zaccaro, L. M., Perry, D. A., Makowski, M. R., et al. (2003). Isozyme-nonselective N-substituted bipiperidylcarboxamide acetyl-CoA carboxylase inhibitors reduce tissue malonyl-CoA concentrations, inhibit fatty acid synthesis, and increase fatty acid oxidation in cultured cells and in experimental animals. *The Journal of Biological Chemistry, 278*, 37099–37111.

Hedbacker, K., & Carlson, M. (2008). SNF1/AMPK pathways in yeast. *Frontiers in Bioscience, 13*, 2408–2420.

Hofbauer, H. F., Schopf, F. H., Schleifer, H., Knittelfelder, O. L., Pieber, B., Rechberger, G. N., et al. (2014). Regulation of gene expression through a transcriptional repressor that senses acyl-chain length in membrane phospholipids. *Developmental Cell, 29*, 729–739.

Holzinger, A., Roschinger, W., Lagler, F., Mayerhofer, P. U., Lichtner, P., Kattenfeld, T., et al. (2001). Cloning of the human MCCA and MCCB genes and mutations therein reveal the molecular cause of 3-methylcrotonyl-CoA carboxylase deficiency. *Human Molecular Genetics, 10*, 1299–1306.

Huang, C. S., Ge, P., Zhou, Z. H., & Tong, L. (2012). An unanticipated architecture of the 750-kDa α6β6 holoenzyme of 3-methylcrotonyl-CoA carboxylase. *Nature, 481*, 219–223.

Huang, C. S., Sadre-Bazzaz, K., Shen, Y., Deng, B., Zhou, Z. H., & Tong, L. (2010). Crystal structure of the α6β6 holoenzyme of propionyl-coenzyme A carboxylase. *Nature, 466*, 1001–1005.

Hunkeler, M., Stuttfeld, E., Hagmann, A., Imseng, S., & Maier, T. (2016). The dynamic organization of fungal acetyl-CoA carboxylase. *Nature Communications*, 7, 11196.

Jacques, D. A., Langley, D. B., Hynson, R. M. G., Whitten, A. E., Kwan, A., Guss, J. M., et al. (2011a). A novel structure of an antikinase and its inhibitor. *Journal of Molecular Biology, 405*, 214–226.

Jacques, D. A., Langley, D. B., Kuramitsu, S., Yokoyama, S., Trewhella, J., & Guss, J. M. (2011b). The structure of TTHA0988 from Thermus thermophilus, a KipI-KipA homologue incorrectly annotated as an allophanate hydrolase. *Acta Crystallographica. Section D, Biological Crystallography*, 67, 105–111.

Janiyani, K., Bordelon, T., Waldrop, G. L., & Cronan, J. E., Jr. (2001). Function of Escherichia coli biotin carboxylase requires catalytic activity of both subunits of the homodimer. *The Journal of Biological Chemistry, 276*, 29864–29870.

Jitrapakdee, S., St. Maurice, M., Rayment, I., Cleland, W. W., Wallace, J. C., & Attwood, P. V. (2008). Structure, mechanism and regulation of pyruvate carboxylase. *The Biochemical Journal, 413*, 369–387.

Jurado, A. R., Huang, C. S., Zhang, X., Zhou, Z. H., & Tong, L. (2015). Structure and substrate selectivity of the 750-kDa α6β6 holoenzyme of geranyl-CoA carboxylase. *Nature Communications, 6*, 8986.

Kaundun, S. S. (2014). Resistance to acetyl-CoA carboxylase-inhibiting herbicides. *Pest Management Science, 70*, 1405–1417.

Kondo, S., Nakajima, Y., Sugio, S., Sueda, S., Islam, M. N., & Kondo, H. (2007). Structure of the biotin carboxylase domain of pyruvate carboxylase from Bacillus thermodenitrificans. *Acta Crystallographica. Section D, Biological Crystallography, 63*, 885–890.

Lasso, G., Yu, L. P. C., Gil, D., Lazaro, M., Tong, L., & Valle, M. (2014). Functional conformations for pyruvate carboxylase during catalysis explored by cryoelectron microscopy. *Structure, 22*, 911–922.

Lasso, G., Yu, L. P. C., Gil, D., Xiang, S., Tong, L., & Valle, M. (2010). Cryo-EM analysis reveals new insights into the mechanism of action of pyruvate carboxylase. *Structure, 18*, 1300–1310.

Lietzan, A. D., Lin, Y., & St. Maurice, M. (2014). The role of biotin and oxamate in the carboxyltransferase reaction of pyruvate carboxylase. *Archives of Biochemistry and Biophysics, 562*, 70–79.

Lietzan, A. D., Menefee, A. L., Zeczycki, T. N., Kumar, S., Attwood, P. V., Wallace, J. C., et al. (2011). Interaction between the biotin carboxyl carrier domain and the biotin carboxylase domain in pyruvate carboxylase from Rhizobium etli. *The Biochemist, 50*, 9708–9723.

Lietzan, A. D., & St. Maurice, M. (2013). A substrate-induced biotin binding pocket in the carboxyltransferase domain of pyruvate carboxylase. *The Journal of Biological Chemistry, 288*, 19915–19925.

Lin, T. W., Melgar, M. M., Kurth, D., Swamidass, S. J., Purdon, J., Tseng, T., et al. (2006). Structure-based inhibitor design of AccD5, an essential acyl-CoA carboxylase carboxyltransferase domain of Mycobacterium tuberculosis. *Proceedings of the National Academy of Sciences of the United States of America, 103*, 3072–3077.

Madauss, K. P., Burkhart, W. A., Consler, T. G., Cowan, D. J., Gottschalk, W. K., Miller, A. B., et al. (2009). The human ACC2 CT-domain C-terminus is required for full functionality and has a novel twist. *Acta Crystallographica. Section D, Biological Crystallography, 65*, 449–461.

Maier, T., Leibundgut, M., Boehringer, D., & Ban, N. (2010). Structure and function of eukaryotic fatty acid synthases. *Quarterly Reviews of Biophysics, 43*, 373–422.

McGarry, J. D., & Brown, N. F. (1997). The mitochondrial carnitine palmitoyltransferase system. From concept to molecular analysis. *European Journal of Biochemistry, 244*, 1–14.

Meades, G., Jr., Benson, B. K., Grove, A., & Waldrop, G. L. (2010). A tale of two functions: Enzymatic activity and translational repression by carboxyltransferase. *Nucleic Acids Research, 38*, 1217–1227.

Miller, J. R., Dunham, S., Mochalkin, I., Banotai, C., Bowman, M., Buist, S., et al. (2009). A class of selective antibacterials derived from a protein kinase inhibitor pharmacophore. *Proceedings of the National Academy of Sciences of the United States of America, 106*, 1737–1742.

Miller, J. R., & Waldrop, G. L. (2010). Discovery of novel antibacterials. *Expert Oinion on Drug Discovery, 5*, 145–154.

Mochalkin, I., Miller, J. R., Evdokimov, A., Lightle, S., Yan, C., Stover, C. K., et al. (2008). Structural evidence for substrate-induced synergism and half-sites reactivity in biotin carboxylase. *Protein Science, 17*, 1706–1718.

Perry, R. J., Camporez, J. P. G., Kursawe, R., Titchenell, P. M., Zhang, D., Perry, C. J., et al. (2015). Hepatic acetyl CoA links adipose tissue inflammation to hepatic insulin resistance and type 2 diabetes. *Cell, 160*, 745–758.

Phannasil, P., Ansari, I. H., El Azzouny, M., Longacre, M. J., Rattanapornsompong, K., Burant, C. F., et al. (2016). Mass spectrometry analysis shows the biosynthetic pathways supported by pyruvate carboxylase in highly invasive breast cancer cells. *Biochimica et Biophysica Acta, 1863*, 537–551.

Polyak, S. W., Abell, A. D., Wilce, M. C. J., Zhang, L., & Booker, G. W. (2012). Structure, function and selective inhibition of bacterial acetyl-CoA carboxylase. *Applied Microbiology and Biotechnology, 93*, 983–992.

Ramsay, R. R., Gandour, R. D., & van der Leij, F. R. (2001). Molecular enzymology of carnitine transfer and transport. *Biochimica et Biophysica Acta, 1546*, 21–43.

Raymer, B., Kavana, M., Price, A., Wang, B., Corcoran, L., Kulathila, R., et al. (2009). Synthesis and characterization of a BODIPY-labeled derivative of soraphen A that binds to acetyl-CoA carboxylase. *Bioorganic & Medicinal Chemistry Letters*, *19*, 2804–2807.

Reaven, G. M. (2004). The metabolic syndrome or the insulin resistance syndrome? Different names, different concepts, and different goals. *Endocrinology and Metabolism Clinics of North America*, *33*, 283–303.

Reaven, G. M. (2011). Insulin resistance: The link between obesity and cardiovascular disease. *The Medical Clinics of North America*, *95*, 875–892.

Reddy, M. C., Breda, A., Bruning, J. B., Sherekar, M., Valluru, S., Thurman, C., et al. (2014). Structure, activity, and inhibition of the carboxyltransferase β-subunit of acetyl coenzyme A carboxylase (AccD6) from Mycobacterium tuberculosis. *Antimicrobial Agents and Chemotherapy*, *58*, 6122–6132.

Robinson, B. H. (2006). Lactic acidemia and mitochondrial disease. *Molecular Genetics and Metabolism*, *89*, 3–13.

Sellers, K., Fox, M. P., Bousamra, M., 2nd, Slone, S. P., Higashi, R. M., Miller, D. M., et al. (2015). Pyruvate carboxylase is critical for non-small-cell lung cancer proliferation. *Journal of Clinical Investigation*, *125*, 687–698.

Shen, Y., Chou, C.-Y., Chang, G.-G., & Tong, L. (2006). Is dimerization required for the catalytic activity of bacterial biotin carboxylase? *Molecular Cell*, *22*, 807–818.

Shen, Y., Volrath, S. L., Weatherly, S. C., Elich, T. D., & Tong, L. (2004). A mechanism for the potent inhibition of eukaryotic acetyl-coenzyme A carboxylase by soraphen A, a macrocyclic polyketide natural product. *Molecular Cell*, *16*, 881–891.

Shi, S., Chen, Y., Siewers, V., & Nielsen, J. (2014). Improving production of malonyl coenzyme A-derived metabolites by abolishing Snf1-dependent regulation of Acc1. *mBio*, *5*, e01130-14.

Silvers, M. A., Pakhomova, S., Neau, D. B., Silvers, W. C., Anzalone, N., Taylor, C. M., et al. (2016). Crystal structure of carboxyltransferase from Staphylococcus aureus bound to the antibacterial agent moiramide B. *The Biochemist*, *55*, 4666–4674.

Smith, A. C., & Cronan, J. E. (2012). Dimerization of the bacterial biotin carboxylase subunit is required for acetyl coenzyme A carboxylase activity in vivo. *Journal of Bacteriology*, *194*, 72–78.

St. Maurice, M., Reinhardt, L., Surinya, K. H., Attwood, P. V., Wallace, J. C., Cleland, W. W., et al. (2007). Domain architecture of pyruvate carboxylase, a biotin-dependent multifunctional enzyme. *Science*, *317*, 1076–1079.

Studer, R., Dahinden, P., Wang, W.-W., Auchli, Y., Li, X.-D., & Dimroth, P. (2007). Crystal structure of the carboxyltransferase domain of the oxaloacetate decarboxylase Na+ pump from Vibrio cholerae. *Journal of Molecular Biology*, *367*, 547–557.

Sureka, K., Choi, P. H., Precit, M., Delince, M., Pensinger, D. A., Huynh, T. N., et al. (2014). The cyclic dinucleotide c-di-AMP is an allosteric regulator of metabolic enzyme function. *Cell*, *158*, 1389–1401.

Thoden, J. B., Blanchard, C. Z., Holden, H. M., & Waldrop, G. L. (2000). Movement of the biotin carboxylase B-domain as a result of ATP binding. *The Journal of Biological Chemistry*, *275*, 16183–16190.

Tong, L. (2013). Structure and function of biotin-dependent carboxylases. *Cellular and Molecular Life Sciences*, *70*, 863–891.

Tong, L., & Harwood, H. J., Jr. (2006). Acetyl-coenzyme A carboxylases: Versatile targets for drug discovery. *Journal of Cellular Biochemistry*, *99*, 1476–1488.

Tran, T. H., Hsiao, Y.-S., Jo, J., Chou, C.-Y., Dietrich, L. E. P., Walz, T., et al. (2015). Structure and function of a single-chain, multi-domain long-chain acyl-CoA carboxylase. *Nature*, *518*, 120–124.

Wakil, S. J., Stoops, J. K., & Joshi, V. C. (1983). Fatty acid synthesis and its regulation. *Annual Review of Biochemistry*, *52*, 537–579.

Waldrop, G. L., Holden, H. M., & St. Maurice, M. (2012). The enzymes of biotin dependent CO2 metabolism: What structures reveal about their reaction mechanisms. *Protein Science, 21*, 1597–1619.

Waldrop, G. L., Rayment, I., & Holden, H. M. (1994). Three-dimensional structure of the biotin carboxylase subunit of acetyl-CoA carboxylase. *The Biochemist, 33*, 10249–10256.

Wallace, J. C. (2010). My favorite pyruvate carboxylase. *IUBMB Life, 62*, 535–538.

Weatherly, S. C., Volrath, S. L., & Elich, T. D. (2004). Expression and characterization of recombinant fungal acetyl-CoA carboxylase and isolation of a soraphen-binding domain. *The Biochemical Journal, 380*, 105–110.

Wei, J., & Tong, L. (2015). Crystal structure of the 500-kDa yeast acetyl-CoA carboxylase holoenzyme dimer. *Nature, 526*, 723–727.

Wei, J., Zhang, Y., Yu, T.-Y., Sadre-Bazzaz, K., Rudolph, M. J., Amodeo, G. A., et al. (2016). A unified molecular mechanism for the regulation of acetyl-CoA carboxylase by phosphorylation. *Cell Discovery, 2*, 16044.

Wendt, K. S., Schall, I., Huber, R., Buckel, W., & Jacob, U. (2003). Crystal structure of the carboxyltransferase subunit of the bacterial sodium ion pump glutaconyl-coenzyme A decarboxylase. *The EMBO Journal, 22*, 3493–3502.

Woodward, J. J., Iavarone, A. T., & Portnoy, D. A. (2010). c-di-AMP secreted by intracellular Listeria monocytogenes activates a host type I interferon response. *Science, 328*, 1703–1705.

Xiang, S., Callaghan, M. M., Watson, K. G., & Tong, L. (2009). A different mechanism for the inhibition of the carboxyltransferase domain of acetyl-coenzyme A carboxylase by tepraloxydim. *Proceedings of the National Academy of Sciences of the United States of America, 106*, 20723–20727.

Xiang, S., & Tong, L. (2008). Crystal structures of human and Staphylococcus aureus pyruvate carboxylase and molecular insights into the carboxyltransfer reaction. *Nature Structural & Molecular Biology, 15*, 295–302.

Yu, L. P. C., Chou, C.-Y., Choi, P. H., & Tong, L. (2013). Characterizing the importance of the biotin carboxylase domain dimer for Staphylococcus aureus pyruvate carboxylase catalysis. *The Biochemist, 52*, 488–496.

Yu, L. P. C., Kim, Y. S., & Tong, L. (2010). Mechanism for the inhibition of the carboxyltransferase domain of acetyl-coenzyme A carboxylase by pinoxaden. *Proceedings of the National Academy of Sciences of the United States of America, 107*, 22072–22077.

Yu, L. P. C., Xiang, S., Lasso, G., Gil, D., Valle, M., & Tong, L. (2009). A symmetrical tetramer for S. aureus pyruvate carboxylase in complex with coenzyme A. *Structure, 17*, 823–832.

Zhang, H., Tweel, B., Li, J., & Tong, L. (2004a). Crystal structure of the carboxyltransferase domain of acetyl-coenzyme A carboxylase in complex with CP-640186. *Structure, 12*, 1683–1691.

Zhang, H., Tweel, B., & Tong, L. (2004b). Molecular basis for the inhibition of the carboxyltransferase domain of acetyl-coenzyme A carboxylase by haloxyfop and diclofop. *Proceedings of the National Academy of Sciences of the United States of America, 101*, 5910–5915.

Zhang, H., Yang, Z., Shen, Y., & Tong, L. (2003). Crystal structure of the carboxyltransferase domain of acetyl-coenzyme A carboxylase. *Science, 299*, 2064–2067.

CHAPTER SIX

Enzymology of Microbial Dimethylsulfoniopropionate Catabolism

Mishtu Dey[1]
The University of Iowa, Iowa City, IA, United States
[1]Corresponding author: e-mail address: mishtu-dey@uiowa.edu

Contents

1. Introduction 196
2. DMSP Demethylation Pathway and the Demethylase, DmdA 198
3. DMSP Cleavage Enzymes 202
4. The CoA-Transferase Pathway of DMS and 3-Hydroxypropionate Formation Using DddD 202
5. The Lyase Pathway of DMS and Acrylate Generation 203
   5.1 The Eukaryotic DMSP Lyase Alma1 203
   5.2 Ddd Families of Marine Bacterial DMSP Lyases 204
6. Summary and Perspective 218
Acknowledgments 219
References 219
Further Reading 222

Abstract

The biochemistry of dimethylsulfoniopropionate (DMSP) catabolism is reviewed. The microbes that catalyze the reactions central to DMSP catabolic pathways are described, and the focus is on the enzymology of the process. Approximately 10^9 tons of DMSP is released annually by marine eukaryotes as an osmolyte. A vast majority of DMSP is assimilated by bacteria through either a demethylation or lyase pathways, producing either the methane thiol or the volatile dimethylsulfide (DMS), respectively. Enzymatic breakdown of DMSP generates $\sim 10^7$ tons of DMS annually, which may have impact on global climate. DMS also acts as a chemoattractant for zooplanktons and seabirds. Both DMSP and DMS play a key role in the global sulfur cycle and are key nutrients for marine microbial growth. Important enzymes in the biochemical pathways of DMSP catabolism are covered in this review, with a focus on the latest developments in their mechanism.

Advances in Protein Chemistry and Structural Biology, Volume 109
ISSN 1876-1623
http://dx.doi.org/10.1016/bs.apcsb.2017.05.001

1. INTRODUCTION

Dimethylsulfoniopropionate (DMSP) is one of Earth's most abundant organosulfur molecules generated in marine environments at ~10^9 tons annually and is involved in global sulfur cycling (Kiene, Linn, & Bruton, 2000; Simo, 2001). It is produced as an antistress molecule by many marine phytoplankton (coccolithophores, dinoflagellates, and diatoms), macroalgae, a few angiosperms, and some corals (Otte, Wilson, Morris, & Moran, 2004; Raina et al., 2013). DMSP can occur at remarkably high (~0.5 M) intracellular levels in these eukaryotic organisms (Curson, Todd, Sullivan, & Johnston, 2011; Dickschat, Rabe, & Citron, 2015; Kiene et al., 2000; Kiene, Linn, Gonzalez, Moran, & Bruton, 1999; Simo, 2001). Many of these marine eukaryotes, e.g., *Emiliania huxleyi* form blooms, and when lysed by viruses, release large amounts of DMSP into the marine milieu with important environmental consequences due to its catabolism by marine microbial communities (Keller, 1989). Although some marine eukaryotic phytoplankton has been shown to catabolize DMSP, it is an important solute for marine bacteria. A vast majority of DMSP is imported and catabolized by bacteria via either demethylation or lyase pathways and supplies 3%–10% of the carbon and 30%–100% of their sulfur (Curson, Todd, et al., 2011; Reisch, Moran, & Whitman, 2011; Fig. 1).

Microbial DMSP catabolism is responsible for liberating the most abundant biogenic sulfur gas, dimethylsulfide (DMS), emitted to the atmosphere. Approximately 10^7 tons of DMS is generated annually (Curson, Todd, et al., 2011; Kettle & Andreae, 2000). DMSP biodegradation pathways exist in a wide range of organisms, including bacteria, eukaryotic phytoplankton, macroalgae, and fungi. Regardless of the microbe and the pathway, DMS serves as a key nutrient for marine bacteria (Kiene & Bates, 1990), as a chemoattractant for zooplankton and seabirds (Malin, 2006; Raina, Dinsdale, Willis, & Bourne, 2010; Raina, Tapiolas, Willis, & Bourne, 2009; Seymour, Simo, Ahmed, & Stocker, 2010), has wide industrial uses (Roy, 2000), and at pathologically high concentrations it is associated with blood-borne halitosis (Tangerman, 2009; Tangerman & Winkel, 2007). More importantly DMS is thought to influence the global climate, because DMS oxidation products are acidic sulfate aerosols that act as cloud condensation nuclei and reflect sunlight (Curson, Todd, et al., 2011; Kiene et al., 2000, 1999; Simo, 2001; Fig. 1). These products are delivered

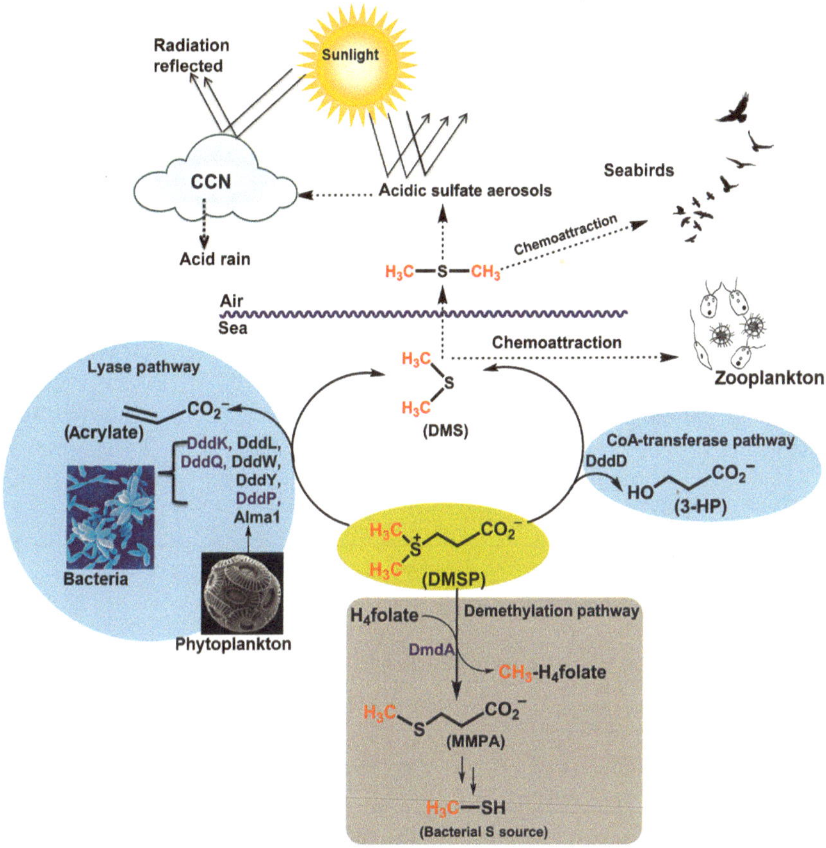

Fig. 1 DMSP catabolic pathways. *Bottom*, the demethylation pathway: the demethylase DmdA catalyzes the conversion of DMSP generating MMPA that is further degraded to form methane thiol, a bacterial C and S source. *Right*, the CoA-transferase pathway: the CoA transferase, DddD generates DMS and 3-hydroxypropionate (3-HP). The lyase pathway (*left*): the eukaryotic enzyme Alma1 and various bacterial enzymes (DddY, DddP, DddL, DddQ, DddW, and DddK) cleave DMSP to produce acrylate and DMS. The enzymes for which crystal structures are available are highlighted in *blue*.

back to Earth as acid rain, representing a key component of the global sulfur cycle. Furthermore, the other DMSP cleavage products, acrylate and 3-hydroxypropionate (3-HP), are industrially important compounds (Dickschat et al., 2015). In addition, acrylate is believed to be utilized in cellular metabolism (Sun et al., 2016).

Despite the environmental and industrial significance of DMSP degradation products, only in the past decade genetic analysis led to the identification of genes and enzymes involved in DMSP catabolism (Curson, Todd, et al., 2011; Reisch, Moran, et al., 2011). These gene-based studies in concert with recent biochemical developments have already uncovered a remarkable diversity in different aspects of DMSP catabolic processes, from the types of enzymes involved in the initial biotransformation, to the range of microbial lineages that can accomplish it. However, biochemical and structural studies elucidating mechanistic details of the DMSP catabolic enzymes are at its infancy.

Studies related to the enzymatic cleavage of DMSP have been covered in Johnston and Todd's chapter (Johnston, Green, & Todd, 2016). This review summarizes recent advances on DMSP catabolism, with an emphasis on the structural enzymology and mechanism of relevant enzymes in different pathways.

2. DMSP DEMETHYLATION PATHWAY AND THE DEMETHYLASE, DmdA

It is believed that the global majority of DMSP in seawater is catabolized via the demethylation pathway accounting for up to 80% of DMSP catabolism (Curson, Todd, et al., 2011; Moran, Reisch, Kiene, & Whitman, 2012; Reisch, Moran, et al., 2011). The demethylation pathway is found in marine α-proteobacteria, specifically the *Roseobacters* and the ubiquitous SAR11 bacteria that contain the *dmdA* gene encoding a DMSP demethylase (Kiene & Taylor, 1988a; Moran et al., 2012). The demethylation pathway involves a series of enzymes producing methylmercaptopropionate (MMPA) and major downstream catabolites, acetaldehyde and methanethiol (MeSH), an important microbial reduced sulfur source (Moran et al., 2012; Reisch, Moran, et al., 2011; Reisch, Stoudemayer, et al., 2011; Fig. 1, bottom). The demethylase DmdA (~38 kDa) transfers a methyl group from DMSP to the coenzyme tetrahydrofolate (THF) forming methyl-THF and MMPA. Subsequent demethiolation of MMPA through a fatty acid β-oxidation-like pathway involving three coenzyme-A-mediated reactions leads to the production of MeSH (Reisch, Moran, et al., 2011; Reisch, Stoudemayer, et al., 2011). Biochemical and kinetic studies with purified DmdA from *Pelagibacter ubique* (*P. ubique*) and *Ruegeria pomeroyi* demonstrated that these enzymes had low affinities for DMSP, with K_m values of 13.2 and 5.4 mM, respectively (Reisch, Moran, & Whitman, 2008; Table 1).

Table 1 Kinetic Parameters of DMSP Catabolic Enzymes, DmdA, DddD, Alma1, and the DMSP Lyases

Enzyme	Products	Organism	Apparent K_m (mM)	k_{cat}/K_m (M^{-1} s^{-1})	PDB ID
DmdA	MMPA	*P. ubique*	13.2 ± 2.0	620	3TFH, 3TFI, 3TFJ
	Me-THF	*R. pomeroyi*	5.4 ± 2.3 (Reisch et al., 2008)	450	—
DddD	3-HP	*Marinomonas* sp.	—	318 (Alcolombri, Laurino, Lara-Astiaso, Vardi, & Tawfik, 2014)	—
	DMS				
Alma1	Acrylate	*Emiliania huxleyi*	9.0 ± 0.9 (Alcolombri et al., 2015)	0.8×10^5	—
	DMS				
DddY	Acrylate	*Alcaligenes faecalis*	1.4 (de Souza & Yoch, 1995)	139×10^3	—
	DMS	*Desulfovibrio acrylicus*	0.4	—	
DddP	Acrylate	*R. pomeroyi*	21.6 ± 5.8	39	
	DMS	*R. nubinhibens*	13.8 ± 5.5 (Kirkwood, Le Brun, Todd, & Johnston, 2010)	19	
		R. denitrificans	—		4B28
		R. lacuscaerulensi	17.1 ± 0.98 (Wang et al., 2015)	—	4RZY, 4RZZ, 4S00, 4S01
DddL	Acrylate		—	—	
	DMS				

Continued

Table 1 Kinetic Parameters of DMSP Catabolic Enzymes, DmdA, DddD, Alma1, and the DMSP Lyases—cont'd

Enzyme	Products	Organism	Apparent K_m (mM)	k_{cat}/K_m (M^{-1} s^{-1})	PDB ID
DddQ	Acrylate	*R. pomeroyi*	18.1 ± 1.0	30	—
	DMS	*R. lacuscaerulensi*	21.5 ± 6.8, Zn(II) (Li, Wei, et al., 2014)	47	4LA2, 4LA3
		R. lacuscaerulensi	14.3 ± 3.3, Fe(II) (Brummett & Dey, 2016)	0.70	5JSO, 5JSP, 5JSR
			39.1 ± 6.1, Fe(III) (Brummett & Dey, 2016)	0.27	
DddW	Acrylate	*R. pomeroyi*	8.7 ± 0.7, Fe(II) (Brummett, Schnicker, Crider, Todd, & Dey, 2015)	2.1×10^3	—
	DMS		4.5 ± 0.8, Mn(II) (Brummett et al., 2015)	3.85×10^3	
DddK	Acrylate	*P. ubique*	81.9 ± 17.2 (Sun et al., 2016)	0.012	5TFZ
	DMS		5.1 ± 0.4, Ni(II) (Schnicker, De Silva, Todd, & Dey, 2017)	608 ± 0.1	5TFO

The crystal structure of DmdA from *P. ubique* has been determined in the apo-form and in the presence of either THF or DMSP (Schuller, Reisch, Moran, Whitman, & Lanzilotta, 2012). The dimeric structure of DmdA in crystals is consistent with it being a dimer in solution (Reisch et al., 2008; Schuller et al., 2012). Based on its sequence and structure similarities, DmdA belongs to the glycine cleavage T-protein (GcvT) family, one of the four proteins in the glycine cleavage system. It has a triple-domain structure similar to GcvTs and sarcosine oxidase and consists of a conserved THF-binding fold. Despite the overall structural similarity of DmdA with GcvTs, relatively few amino acids in the THF- and substrate-binding sites are conserved. The active site of DmdA is located in a cleft between domains 1 and 2, and THF is bound in the cleft (Fig. 2). In contrast to the T-proteins, various acidic side chains interact with THF

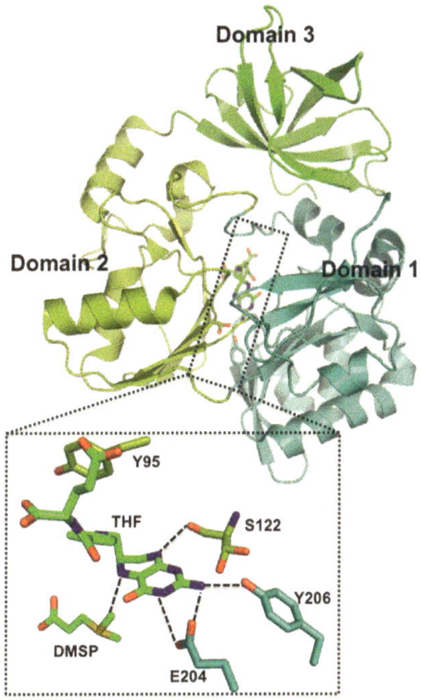

Fig. 2 Crystal structure of DmdA monomer (PDB 3TFJ), showing the tridomain architecture. The active site is located at the interface of domains 1 and 2 where THF is bound. Inset: A close-up view of active site showing key hydrogen bonding interactions (*dashed lines*) between THF and the hydroxyl group of Y206 (domain 1), carboxylate of E204 (domain 1), and the carbonyl oxygen atom of S122 (domain 2). An overlay of THF- and DMSP-bound (PDB 3TFI) structures shows that a methyl group on DMSP is located 3.3 Å away from N-5 nitrogen of THF and is within close proximity for methyl transfer. Protein residues are shown as *teal* (domain 1) and *limon* (domain 2) *sticks*: carbon in protein *backbone color*, oxygen in *red*, nitrogen in *blue*, and sulfur in *yellow*. THF and DMSP are shown as *green sticks*. The figure was generated in PYMOL using the PDBs.

and line the active site around the folate moiety in DmdA (Schuller et al., 2012). The folate ring of THF is also located in an electrostatically negative binding pocket. The structure represents a reduced form of THF with both N-5 and N-8 nitrogen atoms of the folate ring adopting sp^3 hybridization. The presence of strong hydrogen bonding interaction between N-8 nitrogen of THF and the protein backbone carbonyl is consistent with a mechanism making N-5 nitrogen available for nucleophilic attack on the methyl group of DMSP (Fig. 2, inset). An overlay of the THF- and DMSP-bound structures revealed that a methyl group on DMSP is located only 3.3 Å away from N-5 nitrogen of THF and is within close proximity for redox-neutral methyl transfer producing 5-methyl-THF (Schuller et al.,

2012; Fig. 2, inset). These structural observations were in good agreement with the biochemical data showing the formation of 5-methyl-THF with DmdA. The proposed reaction mechanism of DmdA is significantly different from the GcvTs, dimethylglycine oxidase, and sarcosine oxidase with a similar THF-binding fold and tridomain structure in that all these enzymes produce 5,10-methylene-THF (Huang et al., 2000; Schubert, Blumenthal, & Cheng, 2003; Schuller et al., 2012). Based on the crystal structures of DmdA, a concerted reaction mechanism has been proposed similar to that of *S*-adenosylmethione (SAM)-dependent *N*-methyltransferases. Like SAM, DMSP contains a sulfonium atom, making the methyl a fairly good leaving group. Thus, the DMSP demethylation reaction catalyzed by DmdA is thought to proceed via an SN^2 intermediate in which a methyl transfer is coupled to proton transfer that is probably mediated by a water molecule in the active site located within 2.6 Å of the N-5 nitrogen atom of THF (Schuller et al., 2012).

3. DMSP CLEAVAGE ENZYMES

In comparison to the DMSP demethylase DmdA that is found only in two clades of marine bacteria, the cleavage pathways of DMSP catabolism function in a far broader range of microorganisms, including bacteria, eukaryotic phytoplankton, and fungi. In this pathway DMSP is cleaved by enzymes known as "DMSP lyases" to generate the volatile DMS together with the catabolites, acrylate, or 3-HP through elimination reactions. There are seven distinct functional DMSP lyases identified in proteobacteria and some ascomycete fungi that are encoded by different *ddd* (DMSP-dependent DMS) genes (DddD, DddY, DddP, DddL, DddQ, DddW, and DddK), and, Alma1, only currently found in phytoplankton (Alcolombri et al., 2015; Johnston et al., 2016; Sun et al., 2016). We summarize recent progress of the Ddd enzymes and Alma1.

4. THE CoA-TRANSFERASE PATHWAY OF DMS AND 3-HYDROXYPROPIONATE FORMATION USING DddD

The first identified DMSP lyase was DddD found mainly in marine γ-proteobacteria and in some α- and β-proteobacteria likely through horizontal gene transfer (Todd et al., 2010, 2007). DddD (~92 kDa) is unique among the Ddd family of DMSP lyases in that it forms 3-HP instead of acrylate, in addition to acetate, and DMS as major products, when grown with DMSP and acetyl-CoA (Alcolombri et al., 2014; Fig. 1). Initially, DddD lacked lyase activity and did not form DMS in the presence of DMSP (Alcolombri et al., 2014). Because the sequence of DddD is similar to the class III CoA-transferase CaiB, a homodimeric enzyme that transfers CoA to L-carnitine, addition of acetyl-CoA triggered DMS release (Alcolombri et al., 2014). Recombinantly isolated *Marinomonas* DddD had a specific activity of $5 \pm 0.2\,\mu$mol of DMS min^{-1} (mg of DddD)$^{-1}$ (at 20 mM DMSP and 0.5 mM acetyl-CoA). With kinetic studies using variable concentrations of both DMSP and acetyl-CoA it was demonstrated that DddD followed Michaelis–Menten law with respect to acetyl-CoA ($K_m \sim 67\,\mu$M and k_{cat}/K_m of $1.6 \times 10^5\ \text{M}^{-1}\ \text{s}^{-1}$), but saturation was not achieved with varying DMSP concentrations (k_{cat} at 40 mM DMSP was $13\,\text{s}^{-1}$ and k_{cat}/K_m of $318\,\text{M}^{-1}\ \text{s}^{-1}$) (Alcolombri et al., 2014; Table 1).

In the absence of a crystal structure of DddD, homology model structure indicated that DddD is composed of two domains (see supplementary fig. S5B of Alcolombri et al., 2014). The active site and the CoA-binding site were located at the domain interface. Site-directed mutagenesis studies together with the homology model structure showed that an aspartate residue in the C-terminal domain of DddD is responsible for mediating CoA transfer prior to DMSP lyase activity (Alcolombri et al., 2014). Although the detailed catalytic mechanism of DddD has not been elucidated, biochemical data nicely demonstrated the bifunctionality of DddD, i.e., CoA transferase and lyase forming 3-HP-CoA and DMS, respectively.

5. THE LYASE PATHWAY OF DMS AND ACRYLATE GENERATION

5.1 The Eukaryotic DMSP Lyase Alma1

Alma1 (38 kDa) is the only algal DMSP lyase identified from the bloom-forming algae *E. huxleyi* in 2015 (Alcolombri et al., 2015). The *Alma1* gene shares no homology to any of the known bacterial DMSP lyases discussed later. Sequence homology searches suggested that Alma1 represents a gene family present in many marine phytoplankton. Alma1 is a homotetrameric redox-sensitive enzyme that belongs to the aspartate/glutamate/hydantoin racemase family (Alcolombri et al., 2015). Enzymes in this superfamily do not seem to require a cofactor for their activity. Instead two conserved cysteines present in these enzymes play a role in the catalytic activity. Likewise, recombinant Alma1 was found to be metal independent and exhibited redox sensitivity in a pH-dependent manner. Alcolombri et al. demonstrated that thiol oxidization and alkylation, with cysteine modifiers such as H_2O_2 and iodoacetamide, inhibited the DMSP lyase activity of Alma1. Since *E. huxleyi* possesses a DddD homolog, the effect of acetyl-CoA was examined and the DMSP lyase activity of Alma1 remained unaffected (see fig. 2A of Alcolombri et al., 2015). Furthermore, using ^{13}C-labeled DMSP, it was demonstrated that Alma1 catalyzed the formation of acrylate instead of 3-HP. Purified recombinant Alma1 exhibited a high-specific activity (450×10^3 μmol min^{-1} mg^{-1} of pure protein), with a k_{cat}/K_m of 0.8×10^5 M^{-1} s^{-1}, which is 10-fold higher than the bacterial DMSP lyase DddW (Brummett et al., 2015), but comparable to that measured for DddY (de Souza & Yoch, 1995) (see Table 1).

The mechanism of DMS production catalyzed by Alma1 is primarily based on the crystal structures of various racemases along with sequence and mutagenic analysis. Enzymes of the racemase superfamily catalyze proton addition and/or removal from a carbon center adjacent to a carboxylate group. It is known from crystal structures that these enzymes contain one or two conserved cysteine residues in the active site that act as base in proton abstraction from the substrate (Alcolombri et al., 2015). Alma1 possesses two cysteine residues similar to its relatives and site-directed mutagenesis studies confirmed that Cys265 is essential for catalysis. Thus, it has been proposed that Cys265 acts as the catalytic base in Alma1 and initiates the abstraction of a proton from the α-carbon of DMSP next to the carboxylate moiety resulting in a β-elimination, releasing DMS and acrylate.

5.2 Ddd Families of Marine Bacterial DMSP Lyases

The production of DMS via breakdown of DMSP is long known to occur in marine bacteria (Kiene & Bates, 1990), and in the past decade various bacterial DMSP lyases have been identified that belong to three different protein families outlined later.

5.2.1 The Periplasmic Lyase, DddY

DddY (46 kDa) was the first bacterial DMSP lyase to be purified and characterized in 1995 from the facultative anaerobic β-proteobacterium *Alcaligenes faecalis* (de Souza & Yoch, 1995), although the *dddY* gene encoding this protein was identified only in 2011 (Curson, Sullivan, Todd, & Johnston, 2011). DddY functions in some β-, γ-, (*Shewanella* sp.), and ε-(*Arcobacter nitrofigilis*) proteobacteria and favors anaerobic environments (Curson, Sullivan, et al., 2011; Moran et al., 2012). In *Shewanella* and *A. nitrofigilis*, since the *dddY* gene is located adjacent to membrane-bound cytochrome genes, it was thought that DddY in these bacteria may be involved in anaerobic respiration, rather than in food supply, as in *A. faecalis* (de Souza & Yoch, 1995). Interestingly, *Desulfovibrio acrylicus* uses exogenous acrylate as an electron acceptor for anaerobic respiration and has a DddY-like DMSP lyase, which may play a role in supplying acrylate when DMSP is available (Curson, Sullivan, et al., 2011).

The *A. faecalis* DddY had a K_m for DMSP of 1.41 mM and a specific activity of 402 μmol min^{-1} of DMS (mg of protein)$^{-1}$ (de Souza & Yoch, 1995) (see Table 1). Interestingly, DMSP lyases from anaerobic sediment populations have a 40-fold lower K_m (30 μM), suggesting that microbes from anaerobic environment can metabolize much lower levels of DMSP than aerobes. Unlike other bacterial DMSP lyases that are cytoplasmic, DddY exists in the periplasmic space. DddY is rare in that, apart from a short N-terminus leader sequence, it has no resemblance to any enzyme with known function and does not have any familiar conserved protein domains (Curson, Sullivan, et al., 2011). Hence, it is believed that DddY must involve novel mechanisms to generate DMS and acrylate, compared with the other lyases that catabolize DMSP, namely DddP, DddL, DddQ, DddW, and DddK described later.

5.2.2 The M24B Metallopeptidase, DddP

DddP (~110 kDa) belongs to the M24B metallopeptidase family of enzymes and functions in many strains of marine α-proteobacteria, certain γ-proteobacteria, and in some eukaryotic ascomycete fungi (Curson, Sullivan, Todd, & Johnston, 2010; Kirkwood et al., 2010; Todd, Curson, Kirkwood, Nicholson, & Johnston, 2009). The *dddP* gene is abundant and widespread, both geographically and taxonomically. Initial characterization of purified DddP from *Roseovarius nubinhibens* ISM (*Rn*DddP) indicated that it did not contain metallocofactors as in some members of the M24B family, such as creatinase, which is not a metalloenzyme (Kirkwood et al., 2010). Apo-DddP had a K_m of ~14 mM for DMSP and a specific activity of 0.3 μmol of DMS min^{-1} (mg of DddP)$^{-1}$ (Kirkwood et al., 2010), which is about three orders of magnitude lower than the DMSP lyase activity of DddY (Table 1). Subsequent purification of DddP from *Roseobacter denitrificans* (*Rd*DddP) suggested that it is a metalloprotein (Hehemann, Law, Redecke, & Boraston, 2014). Total reflection X-ray fluorescence together with inductively coupled plasma mass spectrometric data revealed that various divalent transition metal ions were tightly bound to *Rd*DddP, but iron was the most abundant ion present in the protein

(Hehemann et al., 2014). Likewise DddP from *Ruegeria lacuscaerulensis* ITI_1157 (*Rl*DddP) was dimeric both in solution and in crystal form, and had two irons per monomer (Wang et al., 2015). The reddish-brown color of concentrated *Rl*DddP suggested that the irons were in the oxidized ferric state. Furthermore, the chelator, 1,10-phenanthrolineion, almost completely inhibited the DMSP lyase activity of *Rl*DddP. The measured K_m of *Rl*DddP for DMSP was 17 mM, which is similar to that originally measured for apo-*Rn*DddP.

The crystal structure of *Rd*DddP determined at 2.15 Å resolution exposed a two-domain architecture with a C-terminal "pita bread" domain representing the conserved catalytic domain of M24B metallopeptidases (Hehemann et al., 2014). *Rd*DddP is a homodimeric enzyme, with two substrate-binding sites per homodimer. Each active site consists of a dinuclear metal center and the two metal ions are at 2.7 Å apart (Hehemann et al., 2014). It should be noted that because iron was identified to be the most abundant ion present in purified *Rd*DddP, the two metal centers were modeled as iron in the crystal structure of *Rd*DddP. While diiron centers in enzymes are usually 3–3.2 Å, the relatively short distance of 2.7 Å in *Rd*DddP, though uncommon, is not unprecedented. As reported for the iron hydrogenase from *Desulfovibrio desulfuricans*, the distance between two iron centers is only 2.6 Å (Nicolet, Piras, Legrand, Hatchikian, & Fontecilla-Camps, 1999). However, in *Rd*DddP as low levels of Ni, Cu, and Zn were identified in solution, it was speculated that while one metal site is fully occupied by iron, the second site contains potentially different metal ions (Hehemann et al., 2014).

Crystal structures of *Rl*DddP determined with bound 2-(*N*-morpholino) ethanesulfonic acid (MES) or PO_4^{3-}, and of two *Rl*DddP mutants bound to acrylate, were all dimeric as observed in solution and had a bimetallic site similar to *Rd*DddP discussed earlier (Wang et al., 2015). The two metal ions in both *Rd*DddP and *Rl*DddP are coordinated by conserved residues present in DddP enzymes from other organisms: three Asp (2 mono-, 1 bidentate), two Glu (1 mono-, 1 bidentate), and one His (Hehemann et al., 2014; Wang et al., 2015). One of the glutamines (Glu421) acts as a bridging ligand while coordinating the two iron centers (Fig. 3). Not surprisingly, mutation of any of these metal-coordinating residues to Ala resulted in complete loss of *Rl*DDP lyase activity (Wang et al., 2015). In the absence of a substrate-bound structure of DddP, the *Rl*DddP–MES complex aided in identifying possible location of DMSP. This structure together with site-directed mutagenesis showed other conserved residues (Trp95, Tyr117, Tyr366, and Asp377) in

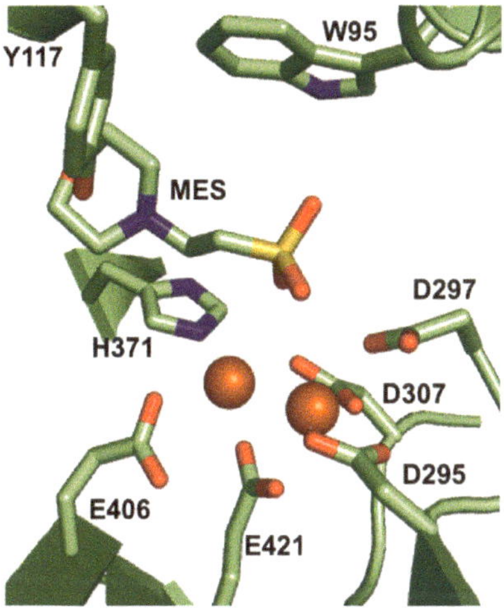

Fig. 3 The active site of *Rl*DddP in complex with MES showing a bimetallic iron center consisting of conserved metal-coordinating residues. The bound MES is located in the active site, presumably in the substrate-binding pocket. It is suggested to mimic DMSP binding, with possible cation–π interactions with Trp95 and Tyr117. The structure was generated using PDB 4RZY. Conserved residues and MES are in *light green sticks*: carbon in protein *backbone color*, oxygen in *red*, nitrogen in *blue*, and sulfur in *yellow*. The two iron ions are in *rust spheres*.

the active site of DddP that were important for its activity. Notably, Y117A, W95A mutations reduced the activity of *Rl*DddP significantly, which suggested that these residues maybe responsible for substrate positioning by interacting with the S^+ end of DMSP through cation–π interactions (Fig. 3).

Based on the structures of both *Rd*DddP and *Rl*DddP, the cleavage of DMSP by DddP enzymes is proposed to proceed via a β-elimination reaction, where Tyr366 or Asp377 are initially proposed to act as the catalytic base for abstracting the Cα-H of DMSP, as described later for DddQ (Brummett & Dey, 2016; Li, Wei, et al., 2014) and DddW (Brummett et al., 2015), respectively. However, on the basis of activity studies with *Rl*DddP variants Y366A, D377A/F/N, Asp377 has been proposed to act as the attacking base. A structural comparison of *Rl*DddP–MES with PO_4^{3-}-bound *Rl*DddP and of two *Rl*DddP variants bound to acrylate showed a shift in one of the iron centers by ~0.9 Å (see fig. 4 of Wang et al., 2015). This shift would presumably place DMSP near Asp377, which would function as the nucleophilic base to abstract a proton leading to β-elimination reaction. Thus, DddP is unique in that it is the only DMSP lyase known to use an ion shift mechanism involving a diiron active site for DMSP cleavage (Wang et al., 2015).

5.2.3 The Cupin Domain-Containing DMSP Lyases

5.2.3.1 A Brief Introduction

The polypeptide sequence of the DMSP lyases DddL, DddQ, DddW from marine *Roseobacters* and the recently identified DddK from *Pelagibacter ubique* HTCC1062, all contain a C-terminal Greek-key (or, cupin) domain characterized by a double-stranded β-helix (DSBH) structure (Fig. 4). The cupin superfamily is considered one of the most functionally diverse protein superfamilies known to exist, and all members contain two conserved sequence motifs, $GX_5HXHX_{3,4}E$ (or D)X_6G and $GX_5PXGX_2HX_3N$ (Dunwell, Culham, Carter, Sosa-Aguirre, & Goodenough, 2001; Khuri, Bakker, &

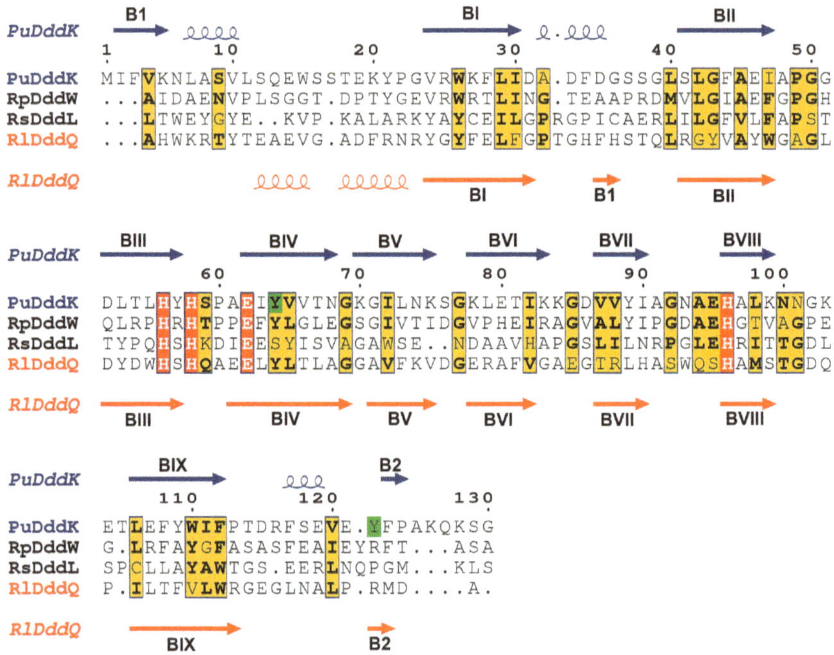

Fig. 4 Sequence alignment of *P. ubique* DddK (*Pu*DddK), *R. pomeroyi* DddW (*Rp*DddW), *Rhodobacter sphaeroides* DddL (*Rs*DddL), and *R. lacuscaerulensis* DddQ (*Rl*DddQ) with secondary structures defined above the sequence. α-Helices are shown by helical lines, β-strands are indicated by *arrows* with B labels, and a roman numeral after the B indicates a β-strand in the cupin fold. Active site residues are highlighted in *red*. Catalytically relevant tyrosines in DddK are highlighted in *green* (see Schnicker, De Silva, Todd, & Dey, 2017). Similar residues are in *black bold* characters and boxed in *yellow*. The secondary structure elements are from Fe(III)-DddQ (5JSO) and Ni-DddK (5TFZ). *Adapted from Schnicker, N. J., De Silva, S. M., Todd, J. D., Dey, M., 2017. Structural and biochemical insights into dimethylsulfoniopropionate cleavage by cofactor-bound DddK from the prolific marine bacterium Pelagibacter. http://dx.doi.org/10.1021/acs.biochem.7b00099.*

Dunwell, 2001). These conserved regions contain residues in the active site that are involved in binding diverse metal ions and play a functional role. Various combinations of ligand coordination exist, with the most common being 2/3-His and 1-carboxylate (Asp/Glu), and less frequently 2-His/1-Glu/1-Tyr, 2-His/1-Glu/1-Gln, or 3-His/1-Gln (Chavez, Banerjee, & Sljivic, 2011; Dunwell et al., 2001; Khuri et al., 2001). While transition metal ions such as iron, manganese, nickel, cobalt, and zinc typically bind at the active site in a cupin domain-containing protein, often these enzymes are promiscuous and accommodate different divalent metal ions. However, commonly there is only one preferred catalytic metal ion (Fe(II) in many cases) that is involved in substrate positioning, and therefore, contributes to the overall activity (Chavez et al., 2011; Dunwell et al., 2001; Khuri et al., 2001). All four DMSP lyases containing the cupin fold, DddL, DddQ, DddW, and DddK, are small proteins, with molecular masses of ~16–26 kDa. The metallocofactor specificity of these DMSP lyases is unclear, and the mechanism of DMSP cleavage is not fully established, which are primarily due to the lack of sufficient biochemical and structural data. Here we discuss the mechanism of DMS and acrylate formation by the cupin domain-containing DMSP lyases.

5.2.3.2 DddL Expression Is Regulated by Acrylate

The first cupin-containing DMSP lyase identified was DddL that is found mainly in marine α-proteobacteria of the *Rhodobacter* family and in bacteria that make DMSP (Curson, Rogers, Todd, Brearley, & Johnston, 2008). An unusual feature of DddL is that the catabolite acrylate is the primary coinducer for its expression (Sullivan et al., 2011). Genetic analysis showed that in *Rhodobacter sphaeroides*, the *dddL* gene is preceded by two acrylate responsive genes, *acuI* and *acuR*. The genes of *acuR*, *acuI*, and *dddL* are cotranscribed, and AcuR has been shown to downregulate the expression of *acuR–acuI–dddL* operon, which can be reversed and induced by the presence of acrylate (Sullivan et al., 2011). Interestingly, DMS production rate was enhanced by ninefold when *R. sphaeroides* cells were grown in the presence of acrylate (Sullivan et al., 2011).

The DddL-dependent DMSP cleavage induced by a product is not unprecedented as was illustrated previously that pregrowth of different bacteria in presence of acrylate or of 3-HP enhances their rates of DMSP-dependent DMS production (Yoch, 2002). The global ocean survey (GOS) data indicate that the *dddL* gene is present in ~0.5% of bacteria found in the surface oceans (Moran et al., 2012; Rusch et al., 2007). Acrylate is known to occur in natural environments where DMSP concentrations are higher (Raina et al., 2010). While acrylate is responsible for activating the *dddL* gene involved in DMSP cleavage, as of yet the enzyme has not been isolated and biochemically characterized. However, studies with *Escherichia coli* whole-cell extracts containing cloned *dddL* generated large amounts of acrylate when growth media was supplemented with 5 mM DMSP (Sullivan et al., 2011). Furthermore, the catabolite acrylate was shown to be the primary coinducer of *dddL* in *R. sphaeroides* (Sullivan et al., 2011). Interestingly, the expression of DddL is directly regulated by acrylate availability and only indirectly by the substrate DMSP (Sullivan et al., 2011).

Based on sequence similarities, DddL is predicted to use a divalent metal ion as a cofactor, which is evident in Fig. 4 of this review. A homology model structure of DddL based on the structure of Fe(III)-DddQ shows potential metal-binding residues (H154, E159, and H190), and the two catalytic relevant tyrosines of DddQ superimpose well on the tyrosine residues of DddL (Fig. 5A). It will be of interest to understand how metallocofactors regulate the activity of DddL in DMSP catabolism.

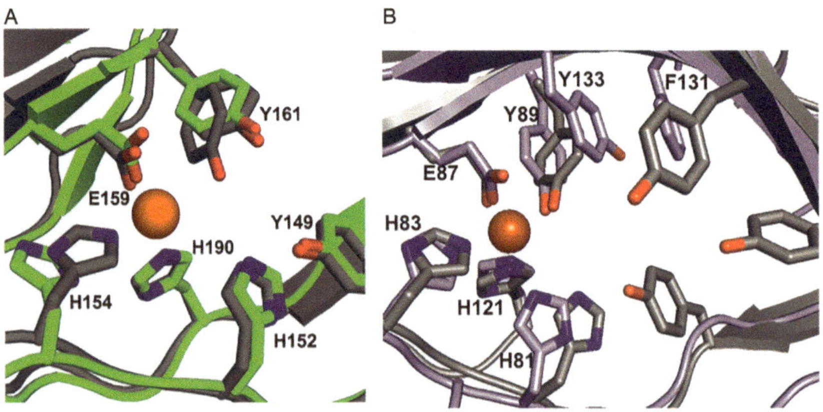

Fig. 5 Homology models of DddL and DddW. (A) The *R. sphaeroides* DddL model (*green*) shows that H154, E159, and H190 superimpose well with the metal-binding residues of Fe(III)-DddQ (*gray*, PDB 5JSP). Other relevant active site residues (Y149, H152, and Y161) that are likely important for substrate/metal ion binding or act as Lewis base are shown as *sticks*. This model was generated using SWISS-MODEL. (B) Homology model of *R. pomeroyi* DddW (*light blue*) (generated using Phyre 2) superimposed on Fe(III)-DddQ (*gray*, PDB 5JSP) showing residues (H81, H83, E87, and H121) required for DMSP lyase activity of DddW (based on Brummett et al., 2015).

5.2.3.3 DddQ Is Abundant and Promiscuous

Among the DMSP lyases, the *dddQ* and *dddP* genes are more abundant than the *dddD and dddL* genes in the GOS dataset (Curson, Todd, et al., 2011). The *dddQ* gene exists in marine bacteria of the *Roseobacter* clade and the ubiquitous SAR11 (Curson, Todd, et al., 2011; Giovannoni, 2017). DddQ is the first DMSP lyase whose crystal structure was determined in 2014 (Li, Wei, et al., 2014). Based on initial biochemical and structural studies, DddQ from *R. lacuscaerulensis* was reported to be a zinc-dependent enzyme because it was isolated with 42% zinc as detected by atomic absorption spectral analysis and contained a zinc ion at its active site in the crystal structure (Li, Wei, et al., 2014). However, the same study found that excess zinc inhibited the DMSP cleavage activity of DddQ. Furthermore, copper and alkaline earth metal ions were shown to inhibit DddQ activity, whereas the enzyme was most active in the presence of excess Co(II) and Mn(II). Other relevant metal ions, such as iron and nickel, were not investigated (Li, Wei, et al., 2014).

Recent kinetic and structural studies examining the importance of iron in DddQ-catalyzed DMSP cleavage revealed that both Fe(III)-oxidized and Fe(II)-reduced species were active (Brummett & Dey, 2016). The K_m for DMSP (14 mM with Fe(II)-DddQ and 39 mM with Fe(III)-DddQ) were similar to those reported for Zn(II)-DddQ (21.5 mM) (Li, Chen, et al., 2014; Li, Wei, et al., 2014) and DddP (13.8 mM) (Kirkwood et al., 2010) (Table 1). The specific activities of iron-DddQ in both Fe(II) ($2.7 \times 10^{-2} \pm 1.4 \times 10^{-3}$ $\mu mol\ min^{-1}\ mg^{-1}$) and Fe(III) ($2.8 \times 10^{-2} \pm 1.9 \times 10^{-3}$ $\mu mol\ min^{-1}\ mg^{-1}$) states are 10 times lower than the reported specific activity for DddP (0.31 μmol DMS $min^{-1}\ mg^{-1}$) (Kirkwood et al., 2010). The poor catalytic efficiencies for DddQ in the presence of either Fe(II) ($0.70\ M^{-1}\ s^{-1}$) or Fe(III) ($0.27\ M^{-1}\ s^{-1}$) indicated that iron may not be the most catalytically active metal ion for DddQ, and/or DMSP may not be its natural substrate (Brummett & Dey, 2016). The enzyme activities determined in the presence of other divalent transition metal ions showed that Mn(II) and Co(II) were more effective in activating DddQ than Zn(II) or iron in either redox state (Brummett & Dey, 2016; Li, Wei, et al., 2014).

The crystal structures of both zinc- and iron-bound DddQ showed that the enzyme contains a DSBH fold and uses 3-His, 1-Glu, and 1-Tyr to bind the central metal ion at its active site (Brummett & Dey, 2016; Li, Wei, et al., 2014). Site-directed mutagenesis coupled with activity studies revealed that these highly conserved hydrophilic residues of the cupin motifs located in DddQ active site are essential for catalysis (fig. 4 of Li, Wei, et al., 2014).

The slow-turnover rates of iron-bound DddQ permitted to capture a substrate-bound form of the enzyme without turnover. The structures of TRIS-, DMSP-, and acrylate-bound forms of Fe(III)-DddQ together with Zn(II)-DddQ structures nicely illustrated various states of the enzyme along the reaction pathway (Brummett & Dey, 2016; Li, Wei, et al., 2014).

Based on structural and biochemical studies of DddQ, two mechanisms of DMSP cleavage have been proposed that invoke different tyrosine residues acting as Lewis base to initiate catalysis (Brummett & Dey, 2016; Li, Wei, et al., 2014) (Fig. 6). On the basis of MES-bound Zn(II)-DddQ (4LA2) and DMSP-bound Zn(II)-Y131A DddQ (4LA3) structures, Tyr131 was suggested to be the catalytic base to abstract a proton from the α-carbon of DMSP. However, in the Fe(III)-DddQ structures, Tyr131 existed in different conformations and was shown to be responsible for holding the metal ion at the active site and/or for positioning of the substrate. The DMSP-bound Fe(III)-DddQ structure showed that DMSP was positioned optimally to undergo the β-elimination reaction and is within proximity of Tyr120 that acts as the catalytic base for removing a proton from the C2 carbon, generating acrylate and DMS (Brummett & Dey, 2016). Thus, the structures of Fe(III)-DddQ revealed conformational changes associated with substrate and product binding, and represent the "resting," "catalytic," and "product" states of the enzyme in the catalytic cycle (Fig. 6).

5.2.3.4 DddW Preferentially Uptakes Iron

A vast majority of proteins containing the cupin fold require Fe(II) for their catalytic activity (Chavez et al., 2011; Dunwell et al., 2001; Hajnal et al., 2013; Khuri et al., 2001). The DMSP lyase DddW from *R. pomeroyi* DSS-3 is a 17-kDa enzyme consisting of a C-terminal cupin domain (Fig. 4). Biochemical studies demonstrated that DddW exists as a dimer in solution and requires metal ions for structural integrity and proper folding of the protein (Brummett et al., 2015). The same study showed that *E. coli* cells expressing DddW when grown in minimal media, lacking any metal ion, resulted in insoluble protein. However, addition of selective transition metal ions during protein expression led to the incorporation of the corresponding metal ion resulting in soluble protein. Likewise, when cells were grown with a mixture of metal ions, all the metal ions were incorporated into the enzyme at various ratios, but iron is preferentially up taken (Brummett et al., 2015). Thus, DddW is promiscuous in its metal recognition with iron as its cofactor of choice. Such promiscuity also exists with DddQ and DddP both of which were isolated with different metal ions,

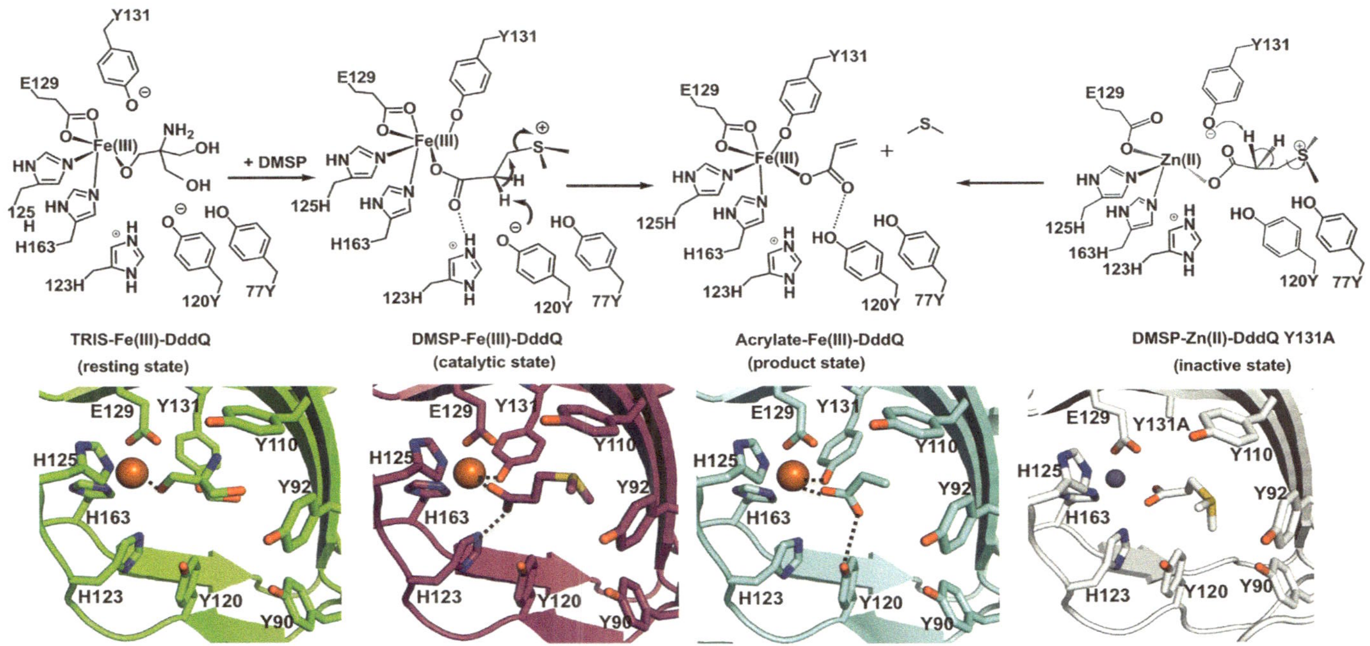

Fig. 6 The proposed mechanism for Fe(III)-DddQ. In the absence of substrate, a TRIS molecule is bound to iron center. When DMSP approaches the active site, it displaces TRIS and coordinates iron in a monodentate fashion via its carboxylate oxygen. His123, Tyr131, and Tyr120 anchor DMSP and position it for catalysis. The proper placement of DMSP allows Tyr120 to be in close proximity to abstract proton from the C2 carbon. Subsequent removal of a proton forms acrylate that can bind to Fe(III), releasing DMS. An alternate mechanism (*right*) based on Zn(II)-DddQ structures has been proposed that invokes Tyr131 as the attacking base. The figure is based on Brummett & Dey, 2016 and generated using the PDBs 5JSO, 5JSP, 5JSR, and 4LA3.

which became evident from their crystal structures as discussed earlier. The Fe(II) cofactor was required for DddW-catalyzed DMSP cleavage as addition of the metal chelator ethylenediaminetetraacetic acid abolishes its enzymatic activity. Measurements of metal-binding affinity and catalytic activity of DddW indicated that Fe(II) was most likely the preferred catalytic metal ion with a nanomolar ($K_d \sim 5\,nM$) binding affinity, and the specific activity of Fe(II)-DddW (65 μmol of acrylate min^{-1} mg of $DddW^{-1}$) was 200-fold higher than that of iron-bound DddP (0.31 μmol of DMS min^{-1} mg of $DddP^{-1}$). Interestingly, DddW showed highest activities in presence of Fe(II) or Mn(II), among the metal ions examined, with catalytic efficiencies of $3.85 \times 10^3\,M^{-1}\,s^{-1}$ and $2.10 \times 10^3\,M^{-1}\,s^{-1}$ in the presence of Fe(II) and Mn(II), respectively (Table 1). In contrast to the Fe(II) binding affinity, DddW had a low affinity for substrate DMSP as reflected in the relatively high K_m of 8.68 mM. However, the K_m is very similar to those known for other DMSP lyases and the demethylase DmdA (5.4 mM) (Table 1).

The secondary structure of DddW observed by circular dichroism exhibited a single shallow minimum between 210 and 220 nm, which suggested that it has a β-sheet structure in solution consistent with proteins of the cupin superfamily, including DddQ (Brummett et al., 2015). Although a crystal structure of DddW has not been determined, an overlay of a homology model of DddW on the DMSP-Fe(III)-DddQ illustrated the probable metal-binding site comprising of 2-His-1-Glu-1-Tyr center (Fig. 5B). Site-directed mutagenesis of three His and one Glu (H81, H83, E87, and H121) resulted in decreased iron content and complete loss of DMSP lyase activity of DddW, thus indicating that all four residues (H81, H83, E87, and H121) were involved in substrate positioning, iron binding, and/or catalysis (Brummett et al., 2015). While tyrosine variants have not been investigated, two tyrosine residues of DddW (Y89 and Y133) align well on one of the catalytically relevant tyrosines of DddQ (Y131) (Fig. 5B). Based on the biochemical studies, a mechanism of DddW-catalyzed DMSP cleavage has been proposed, wherein, a Fe(II) cofactor assisted in catalysis. The proton abstraction from the β-carbon of DMSP has been proposed to be initiated by either a histidine (H81) or a tyrosine (Y89 or Y133) residue. It will be informative to identify the metal-binding residues and the catalytic base by determining the crystal structure of DddW. Further, studies with tyrosine variants of DddW would allow in the identification of their catalytic significance (metal-binding and/or Lewis base). It is puzzling why DddW would use the redox power of Fe(II) to catalyze a simple elimination reaction, which remains to be determined.

5.2.3.5 DddK From the Most Abundant and Ubiquitous Sar11 Bacteria

In 2016, the *dddK* gene encoding a functional DMSP lyase was identified in many sequenced SAR11 species, which are the most abundant and ubiquitous clade of heterotrophic bacteria in the oceans, often comprising 30% of the surface microbial community (Giovannoni, 2017; Morris et al., 2002; Sun et al., 2016). The *dddk* gene existed in ~5% of the bacteria in the GOS. Until this discovery, *P. ubique* HTCC1062 was thought to catabolize DMSP exclusively via the demethylation pathway using DmdA to produce MMPA and MeSH, which are subsequently used as bacterial carbon and sulfur sources (Curson, Todd, et al., 2011; Kiene & Taylor, 1988a, 1988b; Moran et al., 2012). Sun et al. demonstrated that *Pelagibacter* can demethylate and cleave DMSP via two distinct pathways generating MeSH and DMS, respectively (Sun et al., 2016). Based on the differences in their reported K_m for DMSP (K_m of DmdA, 13.2 and 82 mM for DddK), it was proposed that the demethylation pathway would be favored when intracellular levels of DMSP are low, and vice versa when levels are high (Sun et al., 2016). The SAR11 bacterium requires reduced sulfur, such as DMSP, for growth and uses MeSH (a product of DMSP demethylation), as sulfur source but not DMS. In addition, this organism can utilize acrylate (resulting from DMSP cleavage) as a carbon source. Using SAR11 cultures and analysis of substrate (DMSP) consumed and products (acrylate and MeSH) formed with time, indicated that the lyase pathway predominates in SAR11. Interestingly, DMSP cleavage converted nearly 60% of intracellular DMSP to DMS in *P. ubique* (Sun et al., 2016).

Recent biochemical studies with purified *P. ubique* HTCC1062 DddK (MW of 16.5 kDa) showed that the enzyme is promiscuous in its ability to bind various metal ions with a K_d in the range of 28–50 μM. The catalytic rate enhancement in the presence of various metal ions decreases in the following order: Ni(II) > Mn(II) > Fe(II) > Co(II) > Zn(II) > Cu(II) > Fe(III) (see fig. 2B of Schnicker et al., 2017). Interestingly, binding of metal ions to DddK was cooperative, whereas kinetic data in the presence of various DMSP concentrations were not (see figs. 2 and S2 of Schnicker et al., 2017). The dependence of DMSP concentration on DddK activity was shown to follow Michaelis–Menten kinetics at pH 8.5 and room temperature. In the presence of Ni(II), DddK catalyzed the cleavage of DMSP with a K_m for DMSP of 5.1 ± 0.4 mM, which is within the range reported for DMSP lyases (0.4–22 mM) in the literature (Johnston et al., 2016). The higher K_m values reported for the DMSP lyases have been addressed before but are largely believed to be related to the high intracellular concentrations of DMSP present in reported organisms (Johnston et al., 2016). However, the reported K_m of Ni(II)-DddK for DMSP is lower than that determined previously for DddK (82 mM) (Sun et al., 2016), which is believed to be a result of differences in metal speciation, occupancy, or buffer system (Schnicker et al., 2017). The overall catalytic efficiency (k_{cat}/K_m of $608 \pm 0.1\,M^{-1}\,s^{-1}$) of Ni(II)-DddK is comparable with that observed for DddW, but 1000 times higher than that of DddQ. Interestingly, the catalytic efficiency of Ni(II)-DddK was found to be similar to that reported for the DMSP demethylation enzyme DmdA, which is also from the same bacterium as DddK (Table 1).

The amino acid sequence of DddK had exposed that it contained conserved cupin motifs. A multiple sequence alignment of DddK with DddL, DddQ, and DddW showed conserved residues required for metallocofactor binding (Fig. 4). Metal analysis of the violet colored as-isolated homodimeric DddK showed the presence of a mixture of metal ions (iron, zinc, nickel, and manganese per monomer) (Schnicker et al., 2017). Interestingly, as-isolated DddK exhibited an absorption maximum at ~530 nm, due to Fe(III)-tyrosinate ligand-to-metal charge transfer (LMCT) transition, similar to that observed previously for Fe(III)-DddQ (Brummett & Dey, 2016). Mutation of tyrosine residues located in the active site of DddK to alanine (Y64A and Y122A) confirmed that Tyr64 is responsible for the observed LMCT in DddK (Fig. 7A).

DddK is the first cupin-fold containing DMSP lyase for which the crystal structure of a relatively active enzyme has been determined (Schnicker et al., 2017). The structures of DddK with coordinated nickel and iron/zinc established the presence of the DSBH fold and revealed key active site residues involved in metal ion binding and DMSP catabolism (Fig. 7). The structure of inactive FeZn-DddK showed the presence of both iron and zinc at the active site at different occupancies (0.6 Fe and 0.4 Zn). The octahedral Fe(III) center is coordinated to 2-His, a bidentate Glu, Tyr64, and a water (Fig. 7B). Zn(II) is bound by 3-His, 1-Glu, and a water resulting in a 5-coordinate metal center with a trigonal bipyramidal geometry (see fig. 4C of Schnicker et al., 2017). In the Ni(II)-DddK structure, Ni(II) is coordinated by 3-His-1-Glu and a bidendate diacrylate resulting in an octahedral complex. The presence of metal-bound diacrylate in Ni(II)-DddK did not allow Tyr64 to bind to the Ni(II) center (Fig.7C). This structure of the catalytically competent diacrylate-bound Ni(II)-DddK represented the substrate-bound form of the enzyme. On the basis of structural superposition of the two DddK structures (fig.4E of Schnicker et al., 2017) it has been proposed that the iron position possibly mimics the resting state,

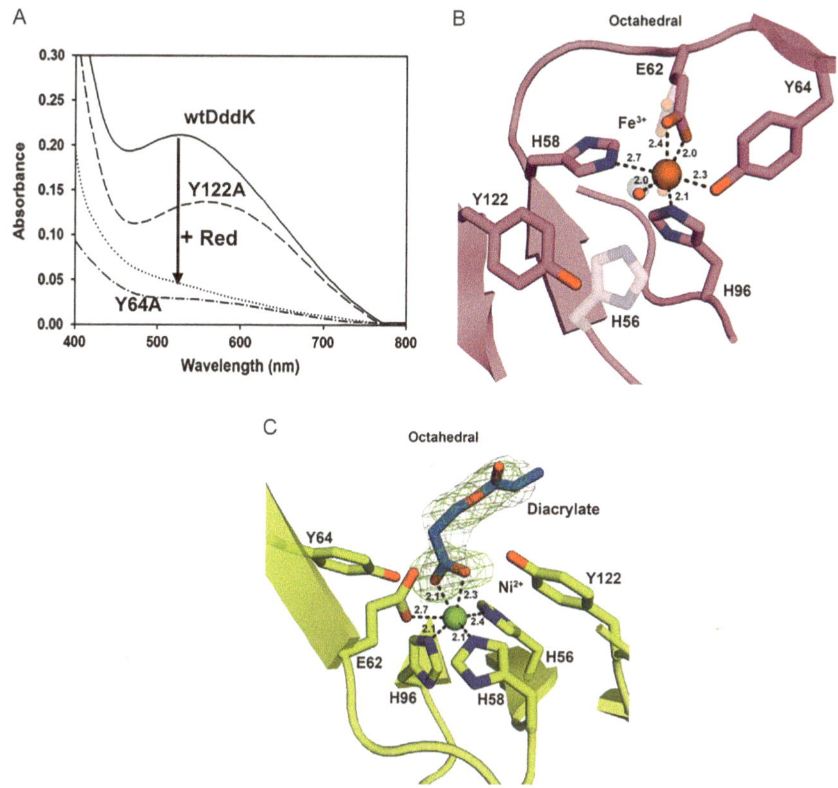

Fig. 7 (A) UV–visible spectral properties of wild-type DddK and tyrosine variants, Y64A and Y122A, together with site-directed mutagenesis confirmed that Tyr64 directly coordinates with the metal center. Activity studies with the DddK variants indicated that these two residues are required for DMSP lyase activity of DddK. Active sites of: (B) FeZn-DddK (*magenta*) showing the Fe(III) coordination center. Ligands not involved in Fe(III) coordination were made transparent. (C) Diacrylate (shown with map, *green mesh*) is bound in a bidentate mode to Ni(II) in Ni-DddK (*green*). Protein residues and diacrylate (*cornflower blue*) are shown as *sticks*: carbon in *backbone color*, oxygen in *red*, and nitrogen in *blue*. Iron (*rust*) and nickel (*green*) are shown as *spheres*. *Adapted from Schnicker, N. J., De Silva, S. M., Todd, J. D., Dey, M., 2017. Structural and biochemical insights into dimethylsulfoniopropionate cleavage by cofactor-bound DddK from the prolific marine bacterium Pelagibacter. http://dx.doi.org/10.1021/acs.biochem.7b00099.*

and the nickel/zinc position represents the active form of the enzyme, with a possible shift in metal ion along the reaction pathway as observed for DddP (Hehemann et al., 2014; Wang et al., 2015). On the basis of these two crystal structures and the biochemical data discussed earlier, a mechanism of DddK-catalyzed β-elimination of DMSP has been proposed where Tyr64 acts as the Lewis base to abstract a proton from the C2 carbon of DMSP (see fig. 7 of Schnicker et al., 2017). Likewise, mutation of Tyr122 to alanine (Y122A) completely abolished the DMSP lyase activity of the variant. Thus, it is reasonable that Tyr122, which is located in the active site, is essential and is likely not involved in proton abstraction but is probably important for substrate positioning and/or transition state stabilization. However, it is conceivable that in the presence of DMSP, the active site of DddK could reorganize positioning Tyr122 to initiate catalysis.

6. SUMMARY AND PERSPECTIVE

Over the past 10 years significant progress has been made in identifying and defining the DMSP catabolic genes and enzymes, which has uncovered a remarkable diversity in the enzymes cleaving DMSP to DMS, in addition to the demethylase. A number of questions remain about the enzymology of DMSP catabolism.

Although it has been proposed that general base catalysis is involved in the DMSP cleavage reaction catalyzed various enzymes in the lyase pathway and in Alma1/DddP, it needs to be firmly established whether there is a generally applicable mechanism for proton abstraction or if these enzymes may be using different mechanisms (tyrosine vs histidine). While various structures of inactive, less active with bound substrate/product, and relatively active lyases clearly defined the metal-binding site within the cupin domain, the catalytically relevant base has not been resolved. Furthermore, no crystal structures have been reported for DddW for which a histidine perhaps acts as the Lewis base.

Though DddY was the first lyase to have been purified, it is unique in that it exists in bacterial periplasm and does not resemble any enzyme with known function, and its mechanism remains to be determined. The expression of DddL is directly regulated by acrylate; thus, biochemical and gene expression studies will be required to characterize the enzyme in vitro and further identify how the lyase activity of DddL is dependent on metallocofactors as it has conserved cupin motifs.

With the exception of Alma1, the catalytic efficiencies of all bacterial DMSP catabolic enzymes are fairly low, raising questions whether or not DMSP is the natural substrate, which has not been resolved. Although there is sufficient evidence for metal promiscuity and preferences of the lyases, it is important to identify the catalytically relevant metal ion for these lyases isolated from corresponding native organism.

Thus, studies of the enzymes involved in the lyase pathway have expanded the repertoire of metalloproteins that catalyze nonredox reactions.

Now that several crystal structures of DddQ, DddP, DddK, and the demethylase DmdA are available to serve as a framework to support various aspects of DMSP catabolic reactions, mechanistic hypotheses can be posed and tested using various biochemical and structural techniques.

ACKNOWLEDGMENTS

M.D. was supported by start-up funds from the University of Iowa College of Liberal Arts and Sciences and funding from the National Science Foundation (CLP 1506181). I am grateful for the contributions of Adam Brummett, Nicholas Schnicker, and Saumya De Silva in my laboratory to some of the work on DMSP lyases described here.

Conflict of Interest: The author declares no conflicts of interest.

REFERENCES

Alcolombri, U., Ben-Dor, S., Feldmesser, E., Levin, Y., Tawfik, D. S., & Vardi, A. (2015). MARINE SULFUR CYCLE. Identification of the algal dimethyl sulfide-releasing enzyme: A missing link in the marine sulfur cycle. *Science*, *348*(6242), 1466–1469. http://dx.doi.org/10.1126/science.aab1586.

Alcolombri, U., Laurino, P., Lara-Astiaso, P., Vardi, A., & Tawfik, D. S. (2014). DddD is a CoA-transferase/lyase producing dimethyl sulfide in the marine environment. *Biochemistry*, *53*(34), 5473–5475. http://dx.doi.org/10.1021/bi500853s.

Brummett, A. E., & Dey, M. (2016). New mechanistic insight from substrate- and product-bound structures of the metal-dependent dimethylsulfoniopropionate lyase DddQ. *Biochemistry*, *55*(44), 6162–6174. http://dx.doi.org/10.1021/acs.biochem.6b00585.

Brummett, A. E., Schnicker, N. J., Crider, A., Todd, J. D., & Dey, M. (2015). Biochemical, kinetic, and spectroscopic characterization of Ruegeria pomeroyi DddW—A mononuclear iron-dependent DMSP lyase. *PLoS One*, *10*(5), e0127288. http://dx.doi.org/10.1371/journal.pone.0127288.

Chavez, F. A., Banerjee, A., & Sljivic, B. (2011). Modeling the metal binding site in cupin proteins. In L. Pramatarova (Ed.), *On biomimetics* (pp. 3–28). Europe and China: InTech. ISBN: 978-953-307-271-5. http://dx.doi.org/10.5772/18262. www.intechopen.com.

Curson, A. R., Rogers, R., Todd, J. D., Brearley, C. A., & Johnston, A. W. B. (2008). Molecular genetic analysis of a dimethylsulfoniopropionate lyase that liberates the climate-changing gas dimethylsulfide in several marine alpha-proteobacteria and Rhodobacter sphaeroides. *Environmental Microbiology*, *10*(3), 757–767. http://dx.doi.org/10.1111/j.1462-2920.2007.01499.x.

Curson, A. R., Sullivan, M. J., Todd, J. D., & Johnston, A. W. (2010). Identification of genes for dimethyl sulfide production in bacteria in the gut of Atlantic Herring (Clupea harengus). *The ISME Journal*, *4*(1), 144–146. http://dx.doi.org/10.1038/ismej.2009.93.

Curson, A. R., Sullivan, M. J., Todd, J. D., & Johnston, A. W. (2011a). DddY, a periplasmic dimethylsulfoniopropionate lyase found in taxonomically diverse species of Proteobacteria. *The ISME Journal*, *5*(7), 1191–1200. http://dx.doi.org/10.1038/ismej.2010.203.

Curson, A. R., Todd, J. D., Sullivan, M. J., & Johnston, A. W. (2011b). Catabolism of dimethylsulphoniopropionate: Microorganisms, enzymes and genes. *Nature Reviews. Microbiology*, *9*(12), 849–859. http://dx.doi.org/10.1038/nrmicro2653.

de Souza, M. P., & Yoch, D. C. (1995). Purification and characterization of dimethylsulfoniopropionate lyase from an alcaligenes-like dimethyl sulfide-producing marine isolate. *Applied and Environmental Microbiology*, *61*, 21–26.

Dickschat, J. S., Rabe, P., & Citron, C. A. (2015). The chemical biology of dimethylsulfoniopropionate. *Organic & Biomolecular Chemistry, 13*(7), 1954–1968. http://dx.doi.org/10.1039/c4ob02407a.

Dunwell, J. M., Culham, A., Carter, C. E., Sosa-Aguirre, C. R., & Goodenough, P. W. (2001). Evolution of functional diversity in the cupin superfamily. *Trends in Biochemical Sciences, 26*(12), 740–746.

Giovannoni, S. J. (2017). SAR11 bacteria: The most abundant plankton in the oceans. *Annual Review of Marine Science, 9*, 231–255. http://dx.doi.org/10.1146/annurev-marine-010814-015934.

Hajnal, I., Lyskowski, A., Hanefeld, U., Gruber, K., Schwab, H., & Steiner, K. (2013). Biochemical and structural characterization of a novel bacterial manganese-dependent hydroxynitrile lyase. *The FEBS Journal, 280*(22), 5815–5828. http://dx.doi.org/10.1111/febs.12501.

Hehemann, J. H., Law, A., Redecke, L., & Boraston, A. B. (2014). The structure of RdDddP from Roseobacter denitrificans reveals that DMSP lyases in the DddP-family are metalloenzymes. *PLoS One, 9*(7), e103128, 1–8. http://dx.doi.org/10.1371/journal.pone.0103128.

Huang, Y., Komoto, J., Konishi, K., Takata, Y., Ogawa, H., Gomi, T., et al. (2000). Mechanisms for auto-inhibition and forced product release in glycine N-methyltransferase: Crystal structures of wild-type, mutant R175K and S-adenosylhomocysteine-bound R175K enzymes. *Journal of Molecular Biology, 298*(1), 149–162. http://dx.doi.org/10.1006/jmbi.2000.3637.

Johnston, A. W., Green, R. T., & Todd, J. D. (2016). Enzymatic breakage of dimethylsulfoniopropionate—A signature molecule for life at sea. *Current Opinion in Chemical Biology, 31*, 58–65. http://dx.doi.org/10.1016/j.cbpa.2016.01.011.

Keller, M. D. (1989). Dimethyl sulfide production and marine phytoplankton the importance of species composition and cell size. *Biological Oceanography, 6*, 375–382. http://dx.doi.org/10.1080/01965581.1988.10749540.

Kettle, A. J., & Andreae, M. O. (2000). Flux of dimethylsulfide from the oceans: A comparison of updated data sets and flux models. *Journal of Geophysical Research: Atmospheres, 105*(D22), 26793–26808. http://dx.doi.org/10.1029/2000jd900252.

Khuri, S., Bakker, F. T., & Dunwell, J. M. (2001). Phylogeny, function, and evolution of the cupins, a structurally conserved, functionally diverse superfamily of proteins. *Molecular Biology and Evolution, 18*(4), 593–605.

Kiene, R. P. B., & Bates, T. S. (1990). Biological removal of dimethyl sulfide from sea water. *Nature, 345*, 702–705.

Kiene, R. P., Linn, L. J., & Bruton, J. A. (2000). New and important roles for DMSP in marine microbial communities. *Journal of Sea Research, 43*, 209–224.

Kiene, R. P., Linn, L. J., Gonzalez, J., Moran, M. A., & Bruton, J. A. (1999). Dimethylsulfoniopropionate and methanethiol are important precursors of methionine and protein-sulfur in marine bacterioplankton. *Applied and Environmental Microbiology, 65*(10), 4549–4558.

Kiene, R. P., & Taylor, B. F. (1988a). Biotransformations of organosulfur compounds in sediments via 3-mercaptopropionate. *Nature, 332*, 148–150.

Kiene, R. P., & Taylor, B. F. (1988b). Demethylation of dimethylsulfoniopropionate and production of thiols in anoxic marine sediments. *Applied and Environmental Microbiology, 54*, 2208–2212.

Kirkwood, M., Le Brun, N. E., Todd, J. D., & Johnston, A. W. B. (2010). The dddP gene of Roseovarius nubinhibens encodes a novel lyase that cleaves dimethylsulfoniopropionate into acrylate plus dimethyl sulfide. *Microbiology, 156*(Pt. 6), 1900–1906. http://dx.doi.org/10.1099/mic.0.038927-0.

Li, C. Y., Chen, X. L., Xie, B. B., Su, H. N., Qin, Q. L., & Zhang, Y. Z. (2014b). Reply to Tawfik et al.: DddQ is a dimethylsulfoniopropionate lyase involved in dimethylsulfoniopropionate catabolism in marine bacterial cells. *Proceedings of the National Academy of Sciences of the United States of America, 111*(20), E2080. http://dx.doi.org/10.1073/pnas.1403460111.

Li, C.-Y., Wei, T., Zhang, S.-H., Chen, X.-L., Gao, X., Wang, P., et al. (2014a). Molecular insight into bacterial cleavage of oceanic dimethylsulfoniopropionate into dimethyl sulfide. *Proceedings of the National Academy of Sciences of the United States of America, 111*, 1026–1031.

Malin, G. (2006). Oceans. New pieces for the marine sulfur cycle jigsaw. *Science, 314*(5799), 607–608. http://dx.doi.org/10.1126/science.1133279.

Moran, M. A., Reisch, C. R., Kiene, R. P., & Whitman, W. B. (2012). Genomic insights into bacterial DMSP transformations. *Annual Review of Marine Science, 4*, 523–542. http://dx.doi.org/10.1146/annurev-marine-120710-100827.

Morris, R. M., Rappe, M. S., Connon, S. A., Vergin, K. L., Siebold, W. A., Carlson, C. A., et al. (2002). SAR11 clade dominates ocean surface bacterioplankton communities. *Nature, 420*(6917), 806–810. http://dx.doi.org/10.1038/nature01240.

Nicolet, Y., Piras, C., Legrand, P., Hatchikian, C. E., & Fontecilla-Camps, J. C. (1999). Desulfovibrio desulfuricans iron hydrogenase: The structure shows unusual coordination to an active site Fe binuclear center. *Structure, 7*(1), 13–23.

Otte, M. L., Wilson, G., Morris, J. T., & Moran, B. M. (2004). Dimethylsulphoniopropionate (DMSP) and related compounds in higher plants. *Journal of Experimental Botany, 55*(404), 1919–1925. http://dx.doi.org/10.1093/jxb/erh178.

Raina, J. B., Dinsdale, E. A., Willis, B. L., & Bourne, D. G. (2010). Do the organic sulfur compounds DMSP and DMS drive coral microbial associations? *Trends in Microbiology, 18*(3), 101–108. http://dx.doi.org/10.1016/j.tim.2009.12.002.

Raina, J. B., Tapiolas, D. M., Foret, S., Lutz, A., Abrego, D., Ceh, J., et al. (2013). DMSP biosynthesis by an animal and its role in coral thermal stress response. *Nature, 502*(7473), 677–680. http://dx.doi.org/10.1038/nature12677.

Raina, J. B., Tapiolas, D., Willis, B. L., & Bourne, D. G. (2009). Coral-associated bacteria and their role in the biogeochemical cycling of sulfur. *Applied and Environmental Microbiology, 75*(11), 3492–3501. http://dx.doi.org/10.1128/AEM.02567-08.

Reisch, C. R., Moran, M. A., & Whitman, W. B. (2008). Dimethylsulfoniopropionate-dependent demethylase (DmdA) from Pelagibacter ubique and Silicibacter pomeroyi. *Journal of Bacteriology, 190*(24), 8018–8024. http://dx.doi.org/10.1128/JB.00770-08.

Reisch, C. R., Moran, M. A., & Whitman, W. B. (2011a). Bacterial catabolism of dimethylsulfoniopropionate (DMSP). *Frontiers in Microbiology, 2*, 172. http://dx.doi.org/10.3389/fmicb.2011.00172.

Reisch, C. R., Stoudemayer, M. J., Varaljay, V. A., Amster, I. J., Moran, M. A., & Whitman, W. B. (2011b). Novel pathway for assimilation of dimethylsulphoniopropionate widespread in marine bacteria. *Nature, 473*(7346), 208–211. http://dx.doi.org/10.1038/nature10078.

Roy, K.-M. (2000). Thiols and organic sulfides. In *Ullmann's encyclopedia of industrial chemistry*. Berlin: Wiley-VCH Verlag GmbH & Co. KGaA. http://dx.doi.org/10.1002/14356007.a26_767.

Rusch, D. B., Halpern, A. L., Sutton, G., Heidelberg, K. B., Williamson, S., Yooseph, S., et al. (2007). The Sorcerer II Global Ocean Sampling expedition: Northwest Atlantic through eastern tropical Pacific. *PLoS Biology, 5*(3), e77. http://dx.doi.org/10.1371/journal.pbio.0050077.

Schnicker, N. J., De Silva, S. M., Todd, J. D., & Dey, M. (2017). Structural and biochemical insights into dimethylsulfoniopropionate cleavage by cofactor-bound DddK from the prolific marine bacterium *Pelagibacter*. http://dx.doi.org/10.1021/acs.biochem.7b00099.

Schubert, H. L., Blumenthal, R. M., & Cheng, X. (2003). Many paths to methyltransfer: A chronicle of convergence. *Trends in Biochemical Sciences, 28*(6), 329–335. http://dx.doi.org/10.1016/S0968-0004(03)00090-2.

Schuller, D. J., Reisch, C. R., Moran, M. A., Whitman, W. B., & Lanzilotta, W. N. (2012). Structures of dimethylsulfoniopropionate-dependent demethylase from the marine organism Pelagabacter ubique. *Protein Science*, *21*, 289–298.

Seymour, J. R., Simo, R., Ahmed, T., & Stocker, R. (2010). Chemoattraction to dimethylsulfoniopropionate throughout the marine microbial food web. *Science*, *329*(5989), 342–345. http://dx.doi.org/10.1126/science.1188418.

Simo, R. (2001). Production of atmospheric sulfur by oceanic plankton: Biogeochemical, ecological and evolutionary links. *Trends in Ecology & Evolution*, *16*(6), 287–294.

Sullivan, M. J., Curson, A. R., Shearer, N., Todd, J. D., Green, R. T., & Johnston, A. W. (2011). Unusual regulation of a leaderless operon involved in the catabolism of dimethylsulfoniopropionate in Rhodobacter sphaeroides. *PLoS One*, *6*(1), e15972. http://dx.doi.org/10.1371/journal.pone.0015972.

Sun, J., Todd, J. D., Thrash, J. C., Qian, Y., Qian, M. C., Temperton, B., et al. (2016). The abundant marine bacterium Pelagibacter simultaneously catabolizes dimethylsulfoniopropionate to the gases dimethyl sulfide and methanethiol. *Nature Microbiology*, *1*(8), 16065. http://dx.doi.org/10.1038/nmicrobiol.2016.65.

Tangerman, A. (2009). Measurement and biological significance of the volatile sulfur compounds hydrogen sulfide, methanethiol and dimethyl sulfide in various biological matrices. *Journal of Chromatography B*, *877*(28), 3366–3377. http://dx.doi.org/10.1016/j.jchromb.2009.05.026.

Tangerman, A., & Winkel, E. G. (2007). Intra- and extra-oral halitosis: Finding of a new form of extra-oral blood-borne halitosis caused by dimethyl sulphide. *Journal of Clinical Periodontology*, *34*(9), 748–755. http://dx.doi.org/10.1111/j.1600-051X.2007. 01116.x.

Todd, J. D., Curson, A. R., Kirkwood, M., Nicholson, P., & Johnston, A. W. B. (2009). The dddP gene, encoding a novel enzyme that converts dimethylsulfoniopropionate into dimethyl sulfide, is widespread in ocean metagenomes and marine bacteria and also occurs in some Ascomycete fungi. *Environmental Microbiology*, *11*(6), 1376–1385. http://dx.doi.org/10.1111/j.1462-2920.2009.01864.x.

Todd, J. D., Curson, A. R., Nikolaidou-Katsaraidou, N., Brearley, C. A., Watmough, N. J., Chan, Y., et al. (2010). Molecular dissection of bacterial acrylate catabolism—Unexpected links with dimethylsulfoniopropionate catabolism and dimethyl sulfide production. *Environmental Microbiology*, *12*(2), 327–343. http://dx.doi.org/10.1111/j.1462-2920.2009.02071.x.

Todd, J. D., Rogers, R., Li, Y. G., Wexler, M., Bond, P. L., Sun, L., et al. (2007). Structural and regulatory genes required to make the gas dimethyl sulfide in bacteria. *Science*, *315*(5812), 666–669. http://dx.doi.org/10.1126/science.1135370.

Wang, P., Chen, X. L., Li, C. Y., Gao, X., Zhu, D. Y., Xie, B. B., et al. (2015). Structural and molecular basis for the novel catalytic mechanism and evolution of DddP, an abundant peptidase-like bacterial dimethylsulfoniopropionate lyase: A new enzyme from an old fold. *Molecular Microbiology*, *98*(2), 289–301. http://dx.doi.org/10.1111/mmi.13119.

Yoch, D. C. (2002). Dimethylsulfoniopropionate: Its sources, role in the marine food web, and biological degradation to dimethylsulfide. *Applied and Environmental Microbiology*, *68*(12), 5804–5815. http://dx.doi.org/10.1128/aem.68.12.5804-5815.2002.

FURTHER READING

Tapiolas, D. M., Motti, C. A., Holloway, P., & Boyle, S. G. (2010). High levels of acrylate in the Great Barrier Reef coral Acropora millepora. *Coral Reefs*, *29*(3), 621–625. http://dx.doi.org/10.1007/s00338-010-0608-3.

http://www.mbari.org/science/upper-ocean-systems/chemical-sensor-group/periodic-table-of-elements-in-the-ocean/.

CHAPTER SEVEN

Sortase Transpeptidases: Structural Biology and Catalytic Mechanism

Alex W. Jacobitz*, Michele D. Kattke*, Jeff Wereszczynski†, Robert T. Clubb*,[1]

*The Molecular Biology Institute and the UCLA-DOE Institute of Genomics and Proteomics, University of California, Los Angeles, CA, United States
†Center for Molecular Study of Condensed Soft Matter, Illinois Institute of Technology, Chicago, IL, United States
[1]Corresponding author: e-mail address: rclubb@mbi.ucla.edu

Contents

1. Introduction 224
2. Structural Biology: Enzyme Structure and Class-Specific Variations 228
   2.1 The Archetypal SaSrtA Enzyme 228
   2.2 Class-Specific Variations 230
3. Structural Biology: Molecular Basis of Substrate Recognition 241
   3.1 Sorting Signal Recognition 241
   3.2 Nucleophile Recognition 246
4. Computational Studies 248
   4.1 The Thr-In Conformation Enables Sortases to Employ a Substrate-Stabilized Active Site 248
   4.2 Sorting Signals Bound to SaSrtA Can Interchange Between Thr-Out and Thr-In Conformers 249
   4.3 Dynamic Sorting Signal Recognition by SaSrtA 250
5. Catalytic Mechanism 251
6. In Vitro Transpeptidation Activity 253
7. Conclusions and Future Directions 258
References 259

Abstract

Gram-positive bacteria use sortase cysteine transpeptidase enzymes to covalently attach proteins to their cell wall and to assemble pili. In pathogenic bacteria sortases are potential drug targets, as many of the proteins that they display on the microbial surface play key roles in the infection process. Moreover, the *Staphylococcus aureus* Sortase A (SaSrtA) enzyme has been developed into a valuable biochemical reagent because of its ability to ligate biomolecules together in vitro via a covalent peptide bond. Here we review what is known about the structures and catalytic mechanism

Advances in Protein Chemistry and Structural Biology, Volume 109
ISSN 1876-1623
http://dx.doi.org/10.1016/bs.apcsb.2017.04.008

of sortase enzymes. Based on their primary sequences, most sortase homologs can be classified into six distinct subfamilies, called class A–F enzymes. Atomic structures reveal unique, class-specific variations that support alternate substrate specificities, while structures of sortase enzymes bound to sorting signal mimics shed light onto the molecular basis of substrate recognition. The results of computational studies are reviewed that provide insight into how key reaction intermediates are stabilized during catalysis, as well as the mechanism and dynamics of substrate recognition. Lastly, the reported in vitro activities of sortases are compared, revealing that the transpeptidation activity of SaSrtA is at least 20-fold faster than other sortases that have thus far been characterized. Together, the results of the structural, computational, and biochemical studies discussed in this review begin to reveal how sortases decorate the microbial surface with proteins and pili, and may facilitate ongoing efforts to discover therapeutically useful small molecule inhibitors.

1. INTRODUCTION

Bacteria display a variety of proteins on their surface that to enable them to effectively interact with their environment. Gram-positive bacteria use sortase cysteine transpeptidase enzymes to covalently attach proteins to their cell wall, and to assemble pili. Sortases in pathogenic bacteria are frequently important virulence factors, as many of the proteins that they display have key roles in the infection process, such as promoting bacterial adhesion, nutrient acquisition, and the evasion and suppression of the immune response (Cascioferro, Totsika, & Schillaci, 2014; Schneewind & Missiakas, 2012, 2014; Siegel, Liu, & Ton-That, 2016; Spirig, Weiner, & Clubb, 2011). As a result, a significant amount of effort has been put forth to elucidate the mechanism of sortase-mediated catalysis and to discover small-molecule sortase inhibitors that could function as potent antiinfective agents (Bradshaw et al., 2015; Cascioferro et al., 2014; Clancy, Melvin, & McCafferty, 2010; Maresso & Schneewind, 2008; Suree, Jung, & Clubb, 2007). Moreover, sortases have been developed into valuable biochemical reagents to ligate distinct biomolecules together via a covalent peptide bond. This in vitro transpeptidation activity has been harnessed for a variety of useful applications, including among others, covalently attaching proteins to cells, attaching fluorophores or drugs to antibodies, cyclizing proteins, and immobilizing peptides on solid surfaces (Antos, Truttmann, & Ploegh, 2016; Popp & Ploegh, 2011; Ritzefeld, 2014; Schmidt, Toplak, Quaedflieg, & Nuijens, 2017; Schmohl & Schwarzer, 2014; Voloshchuk, Liang, & Liang, 2016).

Sortases perform two distinct functions in bacteria: (1) attach proteins directly to the cell wall or (2) assemble pili, long proteinaceous fibers that

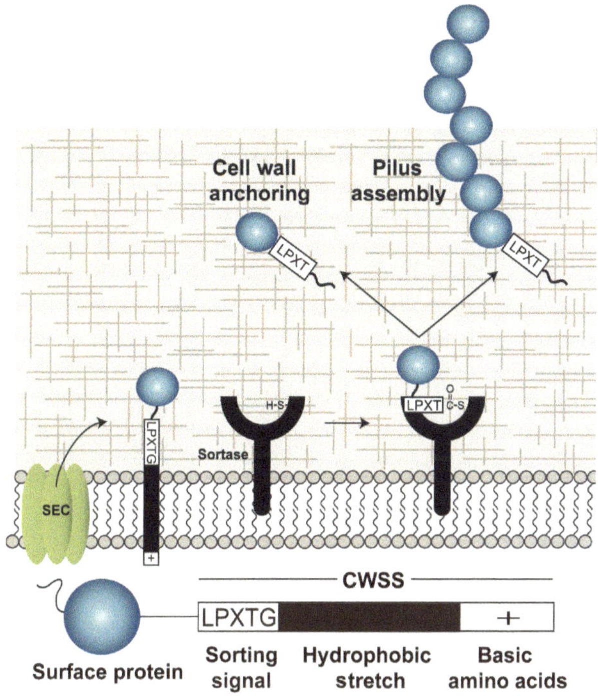

Fig. 1 Sortase enzymes attach proteins to the cell wall and assemble pili. (A) Overview of anchoring and pilus assembly reactions. A protein that is to be displayed (*blue*) contains an N-terminal secretion signal and a C-terminal cell wall sorting signal (CWSS). The CWSS contains an LPXTG-like sorting signal sequence that is processed by the sortase, a nonpolar polypeptide segment (*black*), and a C-terminal segment of positively charged residues (+). After secretion through the Sec translocon, the protein remains embedded in the lipid bilayer via the nonpolar segment within the CWSS. The sortase enzyme then cleaves between the threonine and glycine residues to form a sortase–protein thioacyl intermediate in which the active site cysteine is covalently linked to the carbonyl carbon atom of the threonine. There are two basic types of sortases: (1) cell wall anchoring sortases that attach protein to the crossbridge peptide of the cell wall and (2) pilin polymerase sortases that covalently link pilin subunits together via lysine–isopeptide bonds. In both cases, the enzymes function as transpeptidases. Some sortases are capable of performing both functions, attaching proteins to the cell wall and polymerizing pili.

project from the microbial surface (Fig. 1). Both reactions are mechanistically related and operate on secreted proteins that contain a C-terminal cell wall sorting signal (CWSS). The CWSS contains a five-residue sortase recognition motif, frequently LPXTG, that is followed by a hydrophobic domain and a positively charged cytoplasmic anchor that retains the protein substrate in the membrane (Schneewind, Model, & Fischetti, 1992). The Sortase A enzyme from *Staphylococcus aureus* (SaSrtA) has been studied in detail and is archetypal (Mazmanian, Liu, Ton-That, & Schneewind, 1999; Ton-That, Liu, Mazmanian, Faull, & Schneewind, 1999). SaSrtA attaches surface proteins to the cell wall by recognizing an LPXTG motif within the CWSS of its protein substrate (Fig. 2A). Catalysis begins when SaSrtA's active site cysteine residue nucleophilically attacks the carbonyl carbon in the peptide bond between the Thr and Gly residues in the sorting signal (Fig. 2A, step 1) (Clancy et al., 2010; Connolly & Clubb, 2005). This generates a tetrahedral intermediate that quickly collapses to form a semi-stable thioacyl intermediate in which sortase is covalently linked via its cysteine residue to its protein substrate (Fig. 2A, step 2). SaSrtA then recognizes a second substrate, the cell wall precursor, lipid II (Fig. 2A, step 3) (Perry, Ton-That, Mazmanian, & Schneewind, 2002; Ruzin et al., 2002) and catalyzes a reaction in which the N-terminal primary amine group within the cross-bridge peptide nucleophilically attacks the carbonyl carbon atom within the thioacyl bond. This second transient tetrahedral intermediate resolves into the protein–lipid II product in which the components are joined via a peptide bond (Fig. 2A, step 4) (Perry et al., 2002; Schneewind, Fowler, & Faull, 1995; Ton-That, Faull, & Schneewind, 1997). The lipid II-linked protein is then incorporated into the peptidoglycan via the transpeptidation and glycosylation reactions that synthesize the cell wall. In contrast to attaching proteins to the cell wall, a second type of sortase, frequently called "pilin polymerases," construct bacterial pili by polymerizing pilin protein subunits (Fig. 2B) (Hendrickx, Budzik, Oh, & Schneewind, 2011; Kline, Dodson, Caparon, & Hultgren, 2010; Mandlik, Swierczynski, Das, & Ton-That, 2008; Siegel et al., 2016; Spirig et al., 2011; Ton-That & Schneewind, 2003). These pilin-assembling enzymes employ a similar transpeptidation reaction as SaSrtA, but instead of using lipid II as a nucleophile to attach proteins to the cell wall, a lysine amino group located within a protein pilin subunit is used as a secondary substrate to attack the sortase–protein thioacyl intermediate (Fig. 2B, steps 3 and 4). This reaction constructs pili by covalently linking protein subunits together via lysine–isopeptide bonds. Both types of sortase-catalyzed processes occur on the extracellular membrane, where the enzyme and its substrate are membrane associated (Cozzi et al., 2011; Spirig et al., 2011; Wu et al., 2012).

At present, 3330 gene sequences encoding sortase enzymes have been identified within 1098 species of bacteria (Finn et al., 2016). Sortases are primarily found in Gram-positive bacteria, but are also present to a lesser extent in some species of Gram-negative and archaebacteria. Based on their primary sequences, most sortase homologs can be classified into six distinct subfamilies, called class A–F enzymes (Dramsi, Trieu-Cuot, & Bierne, 2005;

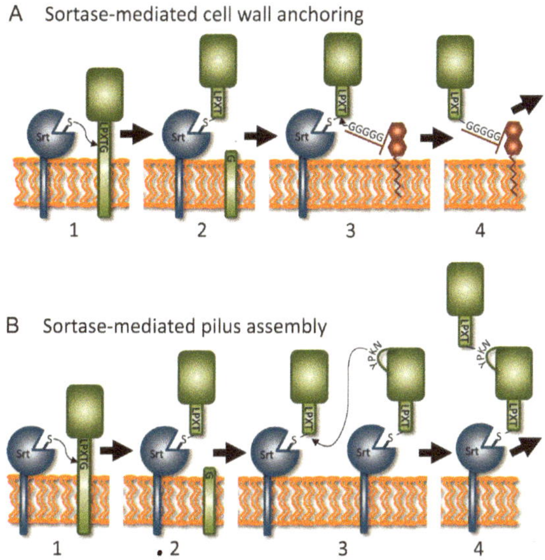

Fig. 2 Mechanism of cell wall protein anchoring and pilus assembly. Sortases perform two basic functions in bacteria: (1) attach proteins to the cell wall and (2) join proteins together to construct pili. (A) In the cell wall anchoring reaction, the sortase and substrate are both membrane bound. The reaction occurs via four distinct steps. Sortase first recognizes a sorting signal motif within the CWSS and nucleophilically attacks the threonine's carbonyl carbon atom via its active site cysteine residue (for demonstration purposes the LPXTG sorting signal recognized by class-A type enzymes is shown, *step 1*). The LPXTG sorting signal is then cleaved to produce a sortase–substrate thioacyl intermediate (*step 2*). Next, the crossbridge peptide from a lipid II molecule nucleophilically attacks the thioacyl intermediate (*step 3*). Lastly, a new peptide bond is formed between the lipid II molecule and surface protein to produce a protein–lipid II intermediate that is incorporated into cell wall by the transglycosylation and transpeptidation reactions that synthesize the peptidoglycan (*step 4*). (B) In the pilus assembly reaction, steps 1–2 produce a sortase–substrate thioacyl intermediate, similar to the cell wall anchoring reaction. In this reaction, the sortase recognizes a pilin protein that contains a CWSS. However, a lysine residue within the pilin motif from an adjacent pilin protein performs the nucleophilic attack on the thioacyl intermediate (*step 3*). A new protein–protein isopeptide bond is formed that covalently links the pilin proteins (*step 4*). This assembly process is repeated to build an isopeptide-linked pilus shaft that contains multiple pilin proteins. Depending on the type of pilus, distinct tip and base pilin proteins can be located at the ends of the pilus shaft, which are incorporated through a similar mechanism and involve covalent linkages via lysine-derived isopeptide bonds. Finally, the intact pilus is attached to the cell wall via sortase-catalyzed attachment of the pilus to lipid II, similar to cell wall protein display. Some sortases are capable of performing both functions, attaching proteins to the cell wall and functioning as pilin polymerases.

Spirig et al., 2011). Class A, B, and D enzymes are prevalent in bacteria within the Firmicutes phylum, while class E and F enzymes predominate in Actinobacteria. Class C enzymes are found in both Firmicutes and Actinobacteria. Similar to SaSrtA (a class A enzyme), all sortases contain a

His–Cys–Arg catalytic triad (Ilangovan, Ton-That, Iwahara, Schneewind, & Clubb, 2001; Marraffini, Ton-That, Zong, Narayana, & Schneewind, 2004; Ton-That, Mazmanian, Alksne, & Schneewind, 2002), and their primary sequences harbor a highly conserved TLXTC motif that contains the catalytically essential cysteine residue (Clancy et al., 2010). All sortases characterized to date catalyze a transpeptidation reaction that joins an LPXTG-like sorting signal within the CWSS of their protein substrate to an amino nucleophile. However, their sorting signal and nucleophile substrate specificities can vary substantially. These distinct specificities enable microbes to utilize more than one type of sortase to elaborate their surface, with each sortase operating nonredundantly to display or assemble distinct proteins on the cell surface (Comfort & Clubb, 2004; Pallen, Lam, Antonio, & Dunbar, 2001).

A number of excellent reviews have been written that describe the overall function of sortases in bacteria, their development as biochemical reagents, and efforts to discover therapeutically useful sortase inhibitors (Antos et al., 2016; Bradshaw et al., 2015; Cascioferro et al., 2014; Clancy et al., 2010; Maresso & Schneewind, 2008; Popp & Ploegh, 2011; Ritzefeld, 2014; Schmidt et al., 2017; Schmohl & Schwarzer, 2014; Schneewind & Missiakas, 2012, 2014; Siegel et al., 2016; Spirig et al., 2011; Suree et al., 2007; Voloshchuk et al., 2016). In this chapter, we review what is known about their atomic structures and the molecular basis of substrate recognition and catalysis.

2. STRUCTURAL BIOLOGY: ENZYME STRUCTURE AND CLASS-SPECIFIC VARIATIONS

2.1 The Archetypal SaSrtA Enzyme

The NMR structure of SaSrtA determined by the Clubb and Schneewind groups was the first reported structure of a sortase enzyme (Ilangovan et al., 2001) (Fig. 3A). The primary sequence of this class A enzyme exhibits features that are generally conserved in other sortases. It has three components: (i) an N-terminal signal sequence that enables it to be transported across the membrane through the Sec translocon, (ii) a nonpolar segment of amino acids that embed the enzyme in the bilayer, (iii) and a conserved, water-soluble C-terminal catalytic domain that contains the His–Cys–Arg triad (Marraffini et al., 2004; Ton-That et al., 2002). The structure of the catalytic domain was determined, residues 60–206 of SaSrtA ($SaSrtA_{\Delta 59}$). This structure revealed the now canonical "sortase fold" that contains a closed eight-stranded β-barrel architecture (Fig. 3A). The atomic coordinates of $SaSrtA_{\Delta 59}$ were precisely defined, with the exception of a 19 amino acid

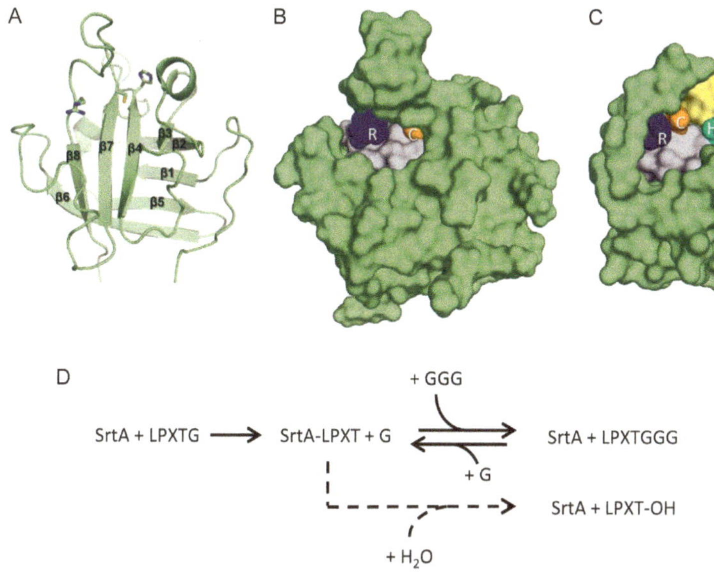

Fig. 3 Structure and transpeptidation reaction of representative class A sortases. (A) *S. aureus* SrtA (SaSrtA) NMR structure (*cartoon*), showcasing an eight-stranded β-barrel with active site His120, Cys184, and Arg197 residues (*sticks*). (B) SaSrtA NMR structure (*green surface*) with active residues Arg (*blue*) and Cys (*orange*). The active site His is occluded by a closed $\beta 7/\beta 8$ loop, and there is no obvious groove for a full-length peptide to exit the active site. (C) *S. pyogenes* SrtA (SpySrtA) structure (*green surface*) with active site Arg (*blue*), Cys (*orange*), and His (*cyan*) residues. An open $\beta 7/\beta 8$ loop creates a clear channel that can be seen running between active Cys and His residues, indicating the potential exit channel (*yellow*) for the full-length peptide substrate. (D) In vitro, SaSrtA catalyzes a reversible transpeptidation reaction (*top*, indicated by *solid arrows*) in which it joins LPXTG and $(Gly)_3$ peptides. In the absence of glycine oligopeptide, SaSrtA acts a protease and cleaves the LPXTG peptide between its threonine and glycine residues (*bottom*, indicated by a *dashed arrow*). In this spurious pathway, a water molecule, instead of lipid II, performs the second nucleophilic attack to cleave the thioacyl bond between sortase and substrate, thereby hydrolyzing the peptide. On the cell surface, hydrolysis is presumably undesirable, as proteolysis separates the protein from its membrane anchor, releasing it from the microbe. Transpeptidation occurs faster than the rate of proteolysis in vitro, making SaSrtA a valuable bioconjugation reagent ($k_{cat} = 0.28 \pm 0.02$ and $0.086 \pm 0.015\,s^{-1}$, respectively). Although all sortases are thought to catalyze transpeptidation reactions on the cell surface, this activity has only been reconstituted in vitro for a few sortases in addition to SaSrtA (listed in Table 2).

flexible loop that connects strands $\beta 6$ to $\beta 7$ (the $\beta 6/\beta 7$ loop). Subsequently, a crystal structure of $SaSrtA_{\Delta 59}$ was determined at 2.0 Å resolution that is very similar to the solution structure; the Cα coordinates in structurally ordered parts of the crystal and NMR structures have an RMSD of

1.97 Å (Zong, Bice, Ton-That, Schneewind, & Narayana, 2004). In both structures, residues in the His–Cys–Arg triad are positioned adjacent to one another within the active site (His120, Cys184, and Arg197 in SaSrtA). As described in detail later, subsequent studies revealed that the LPXTG sorting signal substrate of SaSrtA binds to a pocket that is positioned adjacent to the active site cysteine (Fig. 3B) (Suree et al., 2009). The base of the pocket in SaSrtA is formed by residues in strands $\beta 4$ and $\beta 7$, while the walls are formed by surface loops that connect strand $\beta 6$ to $\beta 7$ ($\beta 6/\beta 7$ loop), strand $\beta 3$ to $\beta 4$ ($\beta 3/\beta 4$ loop), and strand $\beta 2$ to helix H2 ($\beta 2$/H2 loop). All sortases are thought to utilize similarly positioned sorting signal binding grooves. Interestingly, SaSrtA requires Ca^{2+} for efficient catalysis, as the removal of this ion reduces activity fivefold (Ilangovan et al., 2001; Naik et al., 2006); other divalent cations also affected enzyme activity, but to a lesser extent. NMR chemical shift mapping and biochemical studies revealed that Ca^{2+} binds to a pocket located between the $\beta 3/\beta 4$ and $\beta 6/\beta 7$ loops, where it is likely coordinated by the side chains of Glu105, Glu108, Asp112, Glu171, and a backbone carbonyl from Asn114 (Ilangovan et al., 2001; Naik et al., 2006). NMR dynamics experiments indicate that Ca^{2+} binding alters the mobility and structure of the $\beta 6/\beta 7$ active site loop, thereby allosterically regulating the enzyme's affinity for the sorting signal. Interestingly, this mechanism of metal-dependent regulation appears to be unique to SaSrtA, as no other structurally characterized member of the sortase-superfamily contains a structurally similar Ca^{2+} binding pocket.

2.2 Class-Specific Variations

Structures of class A, B, C, D, and E sortases have been reported, revealing unique class-specific variations that likely impact function and modulate substrate specificity. Table 1 lists the structures of sortases that have thus far been determined, and Fig. 4 displays representative class A–E structures for comparison. As expected, all types of sortases contain a conserved catalytic domain that adopts a sortase fold (Fig. 4, blue). However, there are significant class-specific variations that are localized to four distinct structural foci: (i) the N-terminal segment that precedes the catalytic domain (red), (ii) the loop between strands $\beta 6$ and $\beta 7$ (the $\beta 6/\beta 7$ loop) (green), (iii) the loop between strands $\beta 7$ and $\beta 8$ (the $\beta 7/\beta 8$ loop) (yellow), and (iv) the C-terminal polypeptide segment that follows the catalytic domain (Fig. 4). Below, we discuss these differences and highlight their functional implications if they are known.

Table 1 Structurally Characterized Sortase Enzymes

Organism and Sortase	**PDB**	**Bound Ligands or Substrates**	**Method**
Class A			
S. aureus SrtA (SaSrtA)	1IJA	N/A	NMR
S. aureus SrtA-C184A (SaSrtA)	1T2O	N/A	X-ray
S. aureus SrtA (SaSrtA)	1T2P	N/A	X-ray
S. aureus SrtA (SaSrtA)	1T2W	LPETG	X-ray
S. aureus SrtA (SaSrtA)	2KID	Cbz–LPAT*, Ca^{2+}	NMR
S. aureus SrtA (SaSrtA)	2MLM	Benzo[*d*]isothiazol-3-one based inhibitor	NMR
B. anthracis SrtA (BaSrtA)	2KW8	N/A	NMR
B. anthracis SrtA (BaSrtA)	2RUI	Boc–LPAT*	NMR
S. pyogenes SrtA	3FN5	N/A	X-ray
S. pyogenes SrtA	3FN6	Cys in suflphenic acid form	X-ray
S. pyogenes SrtA	3FN7	N/A	X-ray
S. agalactiae SrtA	3RCC	Zn^{2+}	X-ray
S. pneumoniae SrtA	4O8L	N/A	X-ray
S. pneumoniae SrtA-C207A	4O8T	N/A	X-ray
S. mutans SrtA	4TQX	Chalcone	X-ray
Class B			
S. aureus SrtB (SaSrtB)	1NG5	N/A	X-ray
S. aureus SrtB (SaSrtB)	1QWZ	MTSET	X-ray
S. aureus SrtB (SaSrtB)	1QX6	E-64	X-ray
S. aureus SrtB (SaSrtB)	1QXA	Gly_3	X-ray
S. aureus SrtB (SaSrtB)	4FLD	Cbz–NPQT*	X-ray
B. anthracis SrtB	1RZ2	N/A	X-ray
B. anthracis SrtB	2OQW	AAEK1	X-ray
B. anthracis SrtB	2OQZ	AAEK2	X-ray
S. pyogenes SrtB	3PSQ	Zn^{2+}, Cl^{-}	X-ray
C. difficile SrtB	4UX7	N/A	X-ray

Continued

Table 1 Structurally Characterized Sortase Enzymes—cont'd

Organism and Sortase	PDB	Bound Ligands or Substrates	Method
Class C			
A. oris SrtC-1	2XWG	Ca^{2+}	X-ray
S. pneumoniae SrtC-1	2W1J	Glycerol	X-ray
S. *pneumoniae* SrtC-3	2W1K	N/A	X-ray
S. pneumoniae SrtC-1-H131D	2WTS	Alanine	X-ray
S. pneumoniae SrtC-2	3G66	N/A	X-ray
S. pneumoniae SrtC-2	3G69	SO_4^{2-}	X-ray
S. agalactiae SrtC-1Pilus Island-2a	3O0P	N/A	X-ray
S. agalactiae SrtC-1 Pilus Island-1 "Type III"	3RBI	N/A	X-ray
S. agalactiae SrtC1- Pilus Island-1 C184A; KDPYS to IPNTG	3RBJ	N/A	X-ray
S. agalactiae SrtC-1 Pilus Island-1 "Type II"	3RBK	N/A	X-ray
S. agalactiae SrtC1 Pilus Island-1 "Type I"—open lid	3TB7	N/A	X-ray
S. agalactiae SrtC-1 Pilus Island-1	3TBE	MTSET	X-ray
S. agalactiae SrtC-2 Pilus Island-1	4G1H	Ca^{2+}	X-ray
S. agalactiae SrtC-1 Pilus Island-1	4G1J	N/A	X-ray
S. suis SrtC-1	3RE9	N/A	X-ray
Class D			
B. anthracis SrtD	2LN7	N/A	NMR
C. perfringens SrtD	4D70	None	X-ray
Class E			
S. coelicolor SrtE-1	5CUW	N/A	X-ray

2.2.1 Class A Housekeeping Enzymes: Variable Active Site Loops and N-Termini May Modulate Substrate Recognition

These enzymes are typified by the aforementioned SaSrtA sortase (Fig. 4A). They have been proposed to perform a housekeeping role in the cell by anchoring a large number of functionally distinct proteins to the cell wall

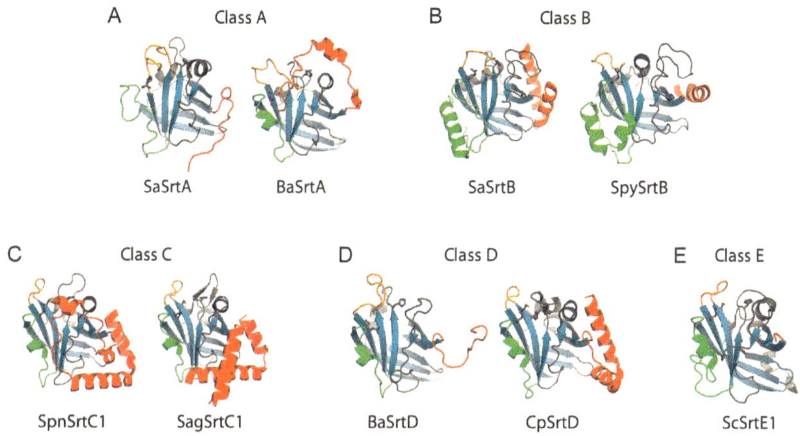

Fig. 4 Structural variation by class of sortase. Sortases representative of the major themes seen for each class are displayed (*cartoon*) with active site residues (*sticks*). The hallmark sortase β-barrel (*blue*) and major sources of structural variability are highlighted, including N-terminus (*red*), $\beta6/\beta7$ loop (*green*), and $\beta7/\beta8$ loop (*orange*). Panels A–E show representative class A–E enzymes, respectively.

(Comfort & Clubb, 2004). Bioinformatics and biochemical analyses indicate that they recognize sorting signals that contain an LPXTG consensus, where X is any amino acid. In addition to SaSrtA, class A enzyme structures have been reported for *Streptococcus pyogenes* (Race et al., 2009), *Bacillus anthracis* (Weiner et al., 2010), *Streptococcus agalactiae* (Khare, Krishnan, et al., 2011), and *Streptococcus mutans* (Wallock-Richards et al., 2015). A common feature is the presence of a short helix within the $\beta6/\beta7$ loop, which has been shown to contact the sorting signal upon substrate binding (Chan et al., 2015; Suree et al., 2009). However, the dynamics of the loop can vary substantially. In the apo-form of SaSrtA, the loop is structurally disordered and mobile, therefore, lacking the $\beta6/\beta7$ helix (Ilangovan et al., 2001; Naik et al., 2006). However, sorting signal binding triggers a disordered-to-ordered transition, resulting in helix formation and helix–substrate interactions in the holo-enzyme (described below) (Suree et al., 2009). In contrast, all other class A enzyme structures contain a preformed binding pocket for the sorting signal in which the $\beta6/\beta7$ loop adopts an ordered state that contains the short helix.

Interestingly, class A enzymes exhibit structural variations near their catalytic histidine residue, causing some structures to contain a second groove that leads into the active site. This structural variation was first highlighted in the crystal structure of *S. pyogenes* SrtA (SpySrtA) (Race et al., 2009). Its catalytic domain adopts the same canonical eight-stranded β-barrel sortase fold typified by SaSrtA, but the positioning of its Cys-S sulfhydryl group differs. In SpySrtA, this sulfhydryl group points toward the active site His-δN and is separated by a distance of 5.4 Å, whereas in SaSrtA it points away from His-δN such that they are separated by 6.5 Å. These differences are illustrated by Fig. 3B and C, which show the structures of SaSrtA and SpySrtA, respectively. The side chain positioning in the SpySrtA structure results in the formation of a unique groove that leads into the active site; the walls are defined by residues in helix H1 and the $\beta 7/\beta 8$ loop, and the base is defined by residues in the $\beta 4/\beta 5$ loop (Fig. 3C, yellow). This groove is positioned adjacent to the active site, opposite the sorting signal binding groove, providing a potential binding site for amino acids located C-terminal to the LPXTG motif in the protein substrate, or for the lipid II substrate. In contrast, this groove is partially masked in the structure of apo-SaSrtA because of interactions between residues in the $\beta 7/\beta 8$ loop and helix H1. The structure of the *S. agalactiae* SrtA (SagSrtA) has also been determined, and like SpySrtA, contains the same secondary groove adjacent to the active site. The SagSrtA structure is unique for the conformation of its $\beta 6/\beta 7$ loop; however, it seems likely that this difference may be a byproduct of the buffer used to crystallize the protein, as residues in the loop coordinate a Zn^{2+} ion in conjunction with an adjacent protein in the asymmetric unit (Khare, Krishnan, et al., 2011).

Some class A enzymes contain a flexible, N-terminal appendage that may modulate substrate binding. In the NMR structure of the *B. anthracis* SrtA (BaSrtA) sortase, the appendage, formed by residues Asp^{57} to Val^{79}, wraps around the body of the protein to contact the enzyme's active site (Fig. 4A, colored red) (Weiner et al., 2010). The first eight residues, Asp^{57} to Pro^{64}, adopt an extended conformation and partially shield the active site His^{126} residue from the solvent, while the remainder of the appendage contains a short alpha helix and wraps around the surface of the catalytic domain to contact helix H2 and the $\beta 2$/H2 loop. Recent studies suggest that the N-terminal appendage modulates substrate access to the enzyme, possibly increasing the efficiency of protein display by reducing unproductive hydrolytic cleavage of enzyme–protein covalent intermediates that form during the cell wall-anchoring reaction (Chan et al., 2015). Conformational plasticity in a related N-terminal segment has also recently been observed in the structure of the class A sortase from *S. mutans* SrtA (SmSrtA) (Wallock-Richards et al., 2015). The SmSrtA crystal structure reveals a dimer in which an extended N-terminal helix preceding the catalytic domain interacts with residues in the active site of a symmetry-related molecule. Although these intermolecular interactions may be an artifact of crystallization, they highlight the proclivity of amino acids preceding the catalytic domain to interact with the enzyme's active site, which has now been observed in BaSrtA and several class C enzymes (described below). NMR studies also revealed that the SaSrtA and BaSrtA enzymes exhibit distinct active site conformational dynamics even though they recognize sorting signals that contain an LPXTG motif (Ilangovan et al., 2001; Naik et al., 2006; Weiner et al., 2010). In BaSrtA, the $\beta6/\beta7$ and $\beta7/\beta8$ loops adopt rigid and mobile states prior to engaging the sorting signal, respectively; however, these loop dynamics are reversed in SaSrtA (the $\beta6/\beta7$ and $\beta7/\beta8$ loops are mobile and structured, respectively). The coordinates for the class A enzyme from *Streptococcus pneumoniae* (SpnSrtA) have also been deposited in the PDB, but its biological significance is unclear as the protein adopts an unusual β-strand swapped dimer and a paper describing this structure has not been published.

2.2.2 Mixed Function Class B Enzymes: An Extended β6/β7 Loop Is Used to Recognize Noncanonical Sorting Signals

Class B enzymes have diverse functions, with members of this subfamily either anchoring proteins to the cell wall or acting as pilin polymerases that assemble pili (Kang, Coulibaly, Proft, & Baker, 2011; Lazzarin et al., 2015; Mazmanian, Ton-That, Su, & Schneewind, 2002). The sorting signals recognized by class B enzymes vary, but are predicted to have a NPX[T/S][N/G/S] consensus instead of LPXTG (Comfort & Clubb, 2004). Five crystal structures of class B sortases have been reported. When compared to the canonical SaSrtA structure, two major differences are apparent as class B enzymes contain: (i) additional helices located N-terminal to the catalytic domain (Fig. 4B, colored red) and (ii) a much longer $\beta6/\beta7$ loop that contains an additional α-helix (Fig. 4B, colored green). Structures of class B enzymes that attach proteins to the cell wall were determined first. In 2004, structures of the *S. aureus* SrtB (SaSrtB) enzyme bound to several nonspecific sulfhydryl modifiers were elucidated (Zong, Mazmanian, Schneewind, & Narayana, 2004). As described below, subsequent studies of SaSrtB bound to its signal peptide revealed that the extended $\beta6/\beta7$ loop is involved in recognizing its distinct NPQTN sorting signal substrate (Jacobitz et al., 2014). The role of the additional N-terminal helices remains unknown, but they may be important for dictating this class B enzyme's preference for anchoring substrates to buried, uncrosslinked portions of the cell wall (Marraffini & Schneewind, 2005; Mazmanian et al., 2003). The apo-structures of SaSrtB and *B. anthracis* SrtB (BaSrtB) were also reported in 2004 and are structurally similar (C_α coordinate RMSDs 3.2 Å) (Zhang et al., 2004). These structures differ in that a portion of the $\beta7/\beta8$ loop in BaSrtB appears to be dynamic, as electron density for this region is missing. Minor conformational variations also occur in the $\beta6/\beta7$ loop and the short loop that connects helices H1 and H2 in these enzymes. A structure of BaSrtB bound to a aryl (beta-amino) ethyl ketone inhibitor has also been determined, revealing only small structural differences with the apo-form of the enzyme (Maresso et al., 2007). Very recently, the atomic structure of the SrtB enzyme from *Clostridium difficile* (CdSrtB) was determined by X-ray crystallography (Chambers, Roberts, Shone, & Acharya, 2015) and is nearly identical in structure to SaSrtB (RMSD = 1.93 Å for all C_α atoms). Interestingly, unlike the SaSrtB and BaSrtB proteins that attach heme-binding proteins to the cell wall, CdSrtB appears to play a more generalized function, as a genetic analysis has predicted that it attaches seven proteins to the cell wall, none of which are thought to be involved in iron acquisition (Chambers et al., 2015).

Some members of the class B subfamily function as pilin polymerases, instead of attaching proteins to the cell wall. Structures of two polymerizing class B sortases have been determined, *S. pyogenes* (SpySrtB) and *S. pneumoniae* (SpnSrtB) (also referred to as SrtG-1) (Kang et al., 2011; Shaik et al., 2015). This work revealed that class B enzymes that attach proteins to the cell wall or assemble pili adopt generally similar tertiary structures. However, the polymerizing enzymes are unique because they contain an additional short helix and β-strand within the extended $\beta6/\beta7$ loop (Fig. 4B, SpySrtB). The latter alteration adds an uncommon ninth β-strand to the protein that is not inserted into the conserved β-barrel core, but instead paired with a portion of strand $\beta6$ on the protein's surface. Understanding how class B sortases can have similar structures yet distinct functions is a major unresolved question.

2.2.3 Class C Pilin Polymerases: An N-Terminal "Lid" Regulates Sorting Signal Substrate Access

Class C sortases can function as polymerases that link pilin proteins together via lysine–isopeptide bonds to construct pili (Fig. 2) (Ton-That & Schneewind, 2003). In some instances, a single class C enzyme can also perform double duty, acting as both a pilin polymerase and a cell wall anchoring sortase that attaches proteins to the peptidoglycan. Members of this class recognize proteins with sorting signals that contain the consensus [L/I]PXTG (Comfort & Clubb, 2004). Because of their unique polymerizing function, class C enzymes have been actively studied with a total of 15 structures being reported to date. Interestingly, nearly all of these enzymes contain a "lid" structure, an elongated N-terminal region that occludes the active site (Fig. 4C, red) (Mandlik et al., 2008; Manzano et al., 2008; Manzano, Izoré, Job, Di Guilmi, & Dessen, 2009; Spirig et al., 2011). Members of this group also harbor a unique C-terminal nonpolar helix that is important for function and likely embedded in the membrane (Cozzi et al., 2011; Wu et al., 2012).

The first class C enzymes to be structurally characterized were SrtC-1 and SrtC-3 from *S. pneumoniae* (SpnSrtC1 and SpnSrtC3) by Manzano et al. (2008). This work revealed the presence of an N-terminal lid that contains a conserved DP(F/W/Y) motif. The aspartic acid residue in the lid motif favorably contacts the conserved active site arginine residue in the His–Cys–Arg triad. The lid is also bound to the active site via sulfur–aromatic interactions between the active site cysteine and aromatic residues in the lid (Phe, Trp, or Tyr depending upon the enzyme). Owing to these key interactions, the Asp and aromatic residues in the lid motif are referred to as "anchors" (Manzano et al., 2008, 2009). Interestingly, the B-factors for residues comprising the lid are elevated, suggesting that they are mobile in solution and transiently detach from the enzyme's active site. Subsequent structures of class C sortases from *S. agalactiae* (Cozzi et al., 2011, 2012; Khare, Fu, Huang, Ton-That, & Narayana, 2011), *A. oris* (Persson, 2011), and *S. pneumoniae* (Neiers et al., 2009) supported this idea, revealing similar patterns of elevated B factors or stretches of missing electron density in regions flanking the lid anchor residues. Several other class C structures have been determined, which reveal only small shifts in lid positioning or subtle differences in the amount of electron density that define their lids.

It has been proposed that the lid regulates enzyme activity (Manzano et al., 2008, 2009) (add references Mazmanian et al., 2002; Zong, Mazmanian, et al., 2004). When it adopts the closed state observed in nearly all crystal structures, the lid occludes the binding site for the sorting signal and holds the enzyme in an inactive state. The enzyme can then become activated by partial dislodgement of the lid, enabling binding of the sorting signal, and formation of the enzyme–substrate thioacyl intermediate. This notion is compatible with modeling studies of the sortase–signal complex, as well as in vitro data that have demonstrated that mutants harboring alterations in the lid exhibit increased rates of sorting signal hydrolysis (Cozzi et al., 2011) and, in some instances, reduced stability (Manzano et al., 2009). However, the role of the lid in catalysis is not fully understood, since cellular studies of pilin polymerases containing mutations that should presumably dislodge the lid exhibit wild-type transpeptidation activity in vivo (Cozzi et al., 2011; Wu et al., 2012). Moreover, structural data have shown that the lid does not completely block access to the enzyme active site, since the crystal structure of *S. agalactiae* SrtC-1 (SagSrtC1) shows the enzyme bound to the nonspecific sulfhydryl modifier (MTSET) despite adopting a closed-lid state (Khare, Fu, et al., 2011).

Recent NMR and computational studies of SpnSrtC1 suggest that the lid in class C enzymes prefers to adopt a closed and rigid state (Jacobitz et al., 2016). This work revealed that the lid in SpnSrtC1 adopts a rigid conformation in solution that is devoid of large magnitude conformational excursions that occur on mechanistically relevant timescales. Additionally, point mutations in the lid were shown to induce dynamic behavior that correlates with increased hydrolytic activity and sorting signal substrate access to the active site cysteine. These results support the notion that the lid in this class C enzyme has a negative regulatory function, and imply that a significant energetic barrier must be surmounted to dislodge it from the active site and initiate pilus biogenesis. Presumably, an as of yet unidentified factor(s) must pry the lid open to hold the enzyme in a catalytically active state that can assemble pili.

One structure of the SagSrtC1 enzyme appears to have captured the lid in an "open" conformation, providing insight into the mechanism of lid-opening that is expected to precede binding of the sorting signal (Fig. 4C, compare left to right) (Khare, Fu, et al., 2011). The structure of SagSrtC1 was determined from multiple crystalline forms, and one of these structures, solved in space group C2, showcases the lid in an "open" conformation. The enzyme maintains the typical sortase fold and, excluding the N-terminal extension preceding the β-barrel core, is extremely similar to SagSrtC1 structures previously solved in space groups $P2_12_12_1$ and $P3_12$ with an average backbone RMSD of 0.72 Å (Khare, Fu, et al., 2011). However, in space group C2, residues A38–E71, which typically form the lid structure, instead form an extended helical structure with the aromatic lid anchor residue of the conserved DP(F/W/Y) motif (Y51) displaced from the active site by over 30 Å to a position where it stacks against the backbone of helix H2. An additional crystal structure of *Streptococcus suis* SrtC-1 (SsSrtC1) also maintains a similar "open" conformation of the lid, with the same extended helix replacing what was expected to be a closed lid (Lu et al., 2011). As nearly all structures of class C enzymes possess a closed lid, which NMR and computational studies suggest is immobile, it is tempting to speculate that the unique open structure observed for SagSrtC1 and SsSrtC1 arose from the solvent conditions used to obtain this crystal form of the protein. Finally, when regions outside of the lid are compared in other class C structures, perhaps the most significant deviation from the norm is the addition of a short C-terminal α-helix opposite the active site in the structure of SpnSrtC3; however, its functional importance has not been determined (Manzano et al., 2008).

2.2.4 Class D Enzymes: Specialized Sortases That Attach Cell Wall Proteins That Contain an LPXTA Sorting Signal

Class D sortases predominate in *Bacilli* species and recognize an unusual LPXTA motif consensus, in which an alanine (underlined) replaces the canonical glycine residue (Comfort & Clubb, 2004). Currently, only two structures of class D enzymes have been reported. In 2012, the NMR structure of the *B. anthracis* SrtD enzyme was published (BaSrtD, and also previously referred to as SrtC) (Robson et al., 2012). BaSrtD anchors proteins required for efficient sporulation to the surface of the cell wall. The catalytic domain of BaSrtD adopts the conserved eight-stranded β-barrel sortase fold (Fig. 4D). Structurally, it is most similar to members of the class A subfamily as it contains an ordered 3_{10} helix within the $\beta6/\beta7$ loop and lacks the elongated $\beta6/\beta7$ loop and lid that are found in class B and C enzymes, respectively (Fig. 4). Interestingly, ultracentrifugation studies indicate that isolated BaSrtD forms a dimeric structure with a K_D of 89 μM (Robson et al., 2012). Based on resonance line broadening effects observed in its NMR spectrum, BaSrtD dimerization is potentially mediated by residues in the structurally disordered $\beta2/\beta3$ and $\beta4$/H1 surface loops that are positioned adjacent to the active site histidine residue. It has been suggested that this disordered surface may mediate interactions with lipid II or other factors on the cell surface, but this has not been demonstrated experimentally. Recently, the structure of the *Clostridium perfringens* SrtD (CpSrtD) sortase was determined at 1.99 Å (Suryadinata et al., 2015). Similar to BaSrtD, CpSrtD adopts a β-barrel sortase fold that contains a short helix within the $\beta6/\beta7$ loop. However, CpSrtD also contains two alpha helices at its N-terminus, and unlike BaSrtA, its $\beta2/\beta3$ and $\beta4$/H1 loops are structurally ordered (the $\beta2/\beta3$ loop also contains a two-turn alpha helix). Furthermore, CpSrtD exists as a monomer according to crystallographic and dynamic light scattering studies. Interestingly, CpSrtD exhibits catalytic activity in vitro that is enhanced in a magnesium-dependent manner, making it one of only two known sortases (the other being SaSrtA) whose activity is modulated by metal ions.

The origin of this stimulatory effect is not known in CpSrtD, but in SaSrtA, metal binding increases enzyme activity by modulating the structure and dynamics of the $\beta6/\beta7$ loop (described below).

2.2.5 Class E Enzymes: A Novel LAXTG Sorting Signal for Anchoring of Surface Proteins in Actinobacteria

Class E enzymes predominate in soil and freshwater-dwelling Actinobacteria and have not been studied extensively. Bioinformatic predictions suggest that members of this group recognize an unusual LAXTG sorting signal motif in which the highly conserved proline residue is replaced with alanine (underlined) (Comfort & Clubb, 2004). Two class E sortases from *Streptomyces coelicolor*, ScSrtE1 and ScSrtE2, have been shown to display chaplin proteins to promote aerial hyphae development (Duong et al., 2012). In vitro studies indicate that ScSrtE1 and ScSrtE2 can hydrolyze LAETG- and LAHTG-containing peptides, cleaving the peptide bond after the threonine residue. The enzymes exhibit promiscuous activity, as they also cleave an LAETG peptide at a secondary site following the alanine (Duong et al., 2012). Very recently, we reported the first crystal structure of a class E sortase, the 1.93 Å resolution structure of ScSrtE1 (Fig. 4E) (Kattke et al., 2016). The structure is similar to class A enzymes, as its $\beta6/\beta7$ loop contains a single 3_{10} helix. However, variations in the conformation of its $\beta3/\beta4$ and $\beta6/\beta7$ loops are evident. ScSrtE1 contains a 21 amino acid insertion immediately following the 3_{10} helix in the $\beta6/\beta7$ loop. This long insertion is similar in length to that observed in class B sortases, but is distinctly devoid of secondary structure, whereas class B sortases contain an additional alpha helix. In the ScSrtE1 structure, the active site is bound to a tripeptide that is presumably a proteolytic protein fragment that was present in the crystallization buffer. The coordinates of the tripeptide and computational modeling with substrate mimics suggest that ScSrtE1 and other members of this group may use a class *E*-specific tyrosine residue present within their $\beta3/\beta4$ loops to recognize the alanine backbone within the LAXTG substrate. It is possible that the tyrosine participates in a hydrogen bond with the amide nitrogen of the alanine residue within the signal, an interaction that is not possible in substrates containing a proline at this position. However, the role of the conserved tyrosine in dictating substrate specificity was not experimentally determined because single amino acid mutants of ScSrtE1 that altered this position in the protein were unstable. Class F enzymes are also prevalent in Actinobacteria, but they have not been structurally or biochemically characterized.

3. STRUCTURAL BIOLOGY: MOLECULAR BASIS OF SUBSTRATE RECOGNITION

All sortases characterized to date catalyze a transpeptidation reaction that joins an LPXTG-like sorting signal within the CWSS of their protein substrate to an amino nucleophile (Fig. 2). However, depending upon the type of sortase, the chemical structure of these substrates can vary substantially. For example, biochemical and bioinformatics analyses suggest that class A, B, C, D, and E enzymes have evolved specificities for distinct LPXTG, NPX[T/S][N/G/S], [L/I]PXTG, LPXTA, and LAXTG sorting signals, respectively (unambiguous differences from LPXTG are underlined) (Comfort & Clubb, 2004). In addition, they can either recognize nucleophiles that originate from lipid II (sortases that anchor proteins to the cell wall), or a lysine residue located within another protein (sortases that function as pilin polymerases) (Fig. 2). Typically, microbes encode genes for more than one sortase enzyme (Pallen et al., 2001). Their distinct substrate specificities enable multiple sortases to nonredundantly operate, with different types of sortases "sorting" distinct proteins to the cell surface or assembling pili. Below, we summarize what is currently known about the molecular basis of substrate recognition.

3.1 Sorting Signal Recognition

The sorting signal recognized by sortases is located within the CWSS of the protein substrate, which in turn is embedded in the bilayer via its hydrophobic domain (Fig. 1) (Schneewind et al., 1992). Our structural studies have shed light onto how class A and B enzymes recognize LPATG and NPQTN sorting signals, respectively (Chan et al., 2015; Jacobitz et al., 2014; Suree et al., 2009). Since sortases can hydrolyze their cognate sorting signals and bind to them weakly in vitro, these structural studies made use of a substrate analog developed by the Jung group in which the threonine moiety contains a sulfhydryl group in place of its carbonyl atom (Jung et al., 2005). The analog (hereafter called T*), mimics the threonine residue in the native sorting signal substrate, but forms a disulfide bond with the active site cysteine residue. This leads to the production of a stable enzyme–substrate complex that is suitable for structural studies.

Using T*-containing sorting signal analogs the atomic structures of three sortase–substrate complexes have been determined by the Clubb group: the class A SaSrtA–LPXT* (Fig. 5A), class B SaSrtB–NPQT* (Fig. 5B), and

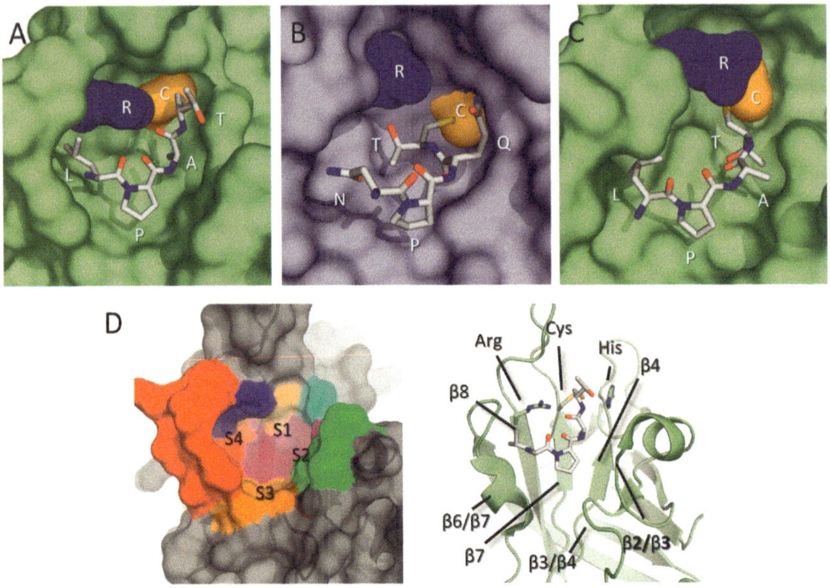

Fig. 5 Sorting signal recognition. (A) The SaSrtA–LPAT* complex. (B) The SaSrtB–NPQT* complex. (C) The BaSrtA–LPAT* complex, shown with N-terminal appendage removed from view for clarity. Enzymes are shown as surface representations with SrtA types in *light green* and SaSrtB in *light blue*, substrate mimics are shown as *gray sticks*. Active site Cys and Arg residues are shown as *gold* and *blue* surfaces, respectively. (D) Conserved recognition sites for sortase enzymes. Left, SaSrtA shown as a transparent surface representation with recognition subsites determined from the combination of sortase structures color coded as follows: S4 is shown in *red*, S3 in *orange*, S2 in *green*, and S1 in *magenta*, and active site Arg in *blue*, Cys in *gold*, and His in *cyan*. Right, Cartoon diagram of SaSrtA with secondary structure elements that contribute to substrate binding labeled for clarity.

class A BaSrtA–LPAT* (Fig. 5C) complexes (Chan et al., 2015; Jacobitz et al., 2014; Suree et al., 2009). To facilitate a discussion of the binding interactions that govern signal recognition, we henceforth refer to residues in each sorting signal in relation to their positioning relative to the scissile bond. Residues in an L–P–X–T–G sorting signal are referred to as P4–P3–P2–P1–P1′ and their corresponding binding sites on the enzyme as subsites S4–S3–S2–S1–S1′, respectively.

3.1.1 *Sorting Signal Recognition by Class A Enzymes*

The NMR structure of the SaSrtA–LPAT* complex provided the first insight into the molecular basis of sorting signal recognition (Suree et al.,

2009) (Fig. 5A). In the structure of the complex, the peptide substrate binds to a pocket adjacent to the active site cysteine whose base is formed by residues located in strands $\beta 4$ and $\beta 7$, and whose walls are formed by surface loops that connect strand $\beta 6$ to $\beta 7$ ($\beta 6/\beta 7$ loop), strand $\beta 3$ to $\beta 4$ ($\beta 3/\beta 4$ loop), and strand $\beta 2$ to helix H2 ($\beta 2$/H2 loop). The location of this binding site is consistent with chemical shift mapping studies (Liew et al., 2004). Binding of the signal to SaSrtA causes a major reorganization of the active site, including a disordered-to-ordered transition of the $\beta 6/\beta 7$ loop to create a short 3_{10} helix that contacts the signal, as well as displacement of the $\beta 7/\beta 8$ loop. The latter change may have a regulatory role, exposing the active site histidine to the solvent, and possibly facilitating the binding of the secondary lipid II substrate only after the sorting signal has first bound to the enzyme. Additional ^{15}N relaxation analyses revealed that the $\beta 6/\beta 7$ loop, which is highly dynamic in the unbound state, rigidifies upon peptide binding (Naik et al., 2006; Suree et al., 2009). The peptide binding mode in the SaSrtA–LPAT* complex differs substantially from a previously reported crystal structure of SaSrtA noncovalently bound to a LPETG peptide (Zong, Bice, et al., 2004). In the crystal structure, the peptide is presumably nonspecifically bound, which is not surprising as the sorting signal substrate binds to SaSrtA with very weak affinity ($K_m = 7.33\,\text{mM}$) (Frankel et al., 2005).

The structure of the SaSrtA–LPAT* complex reveals how class A enzymes recognize the P4 and P3 residues within the LPXTG sorting signal, Leu and Pro, respectively (Suree et al., 2009). The P4 leucine side chain is positioned within a large hydrophobic S4 pocket that is formed by residues V161, V166, V168, and L169 in the reordered $\beta 6/\beta 7$ loop, and I199 on strand $\beta 8$. The P3 Pro residue plays a key architectural role, forming a kink in the sorting signal that enables it to adopt an "L-shaped" structure in which the C-terminal end is positioned toward the active site cysteine. Numerous hydrophobic contacts to the proline are formed by the S3 site. Surprisingly, in the structure of the SaSrtA–LPAT* complex, the positioning of the P1 and P2 side chains is incompatible with biochemical and bioinformatics data (Comfort & Clubb, 2004; Fischetti, Pancholi, & Schneewind, 1990; Kruger, Otvos, et al., 2004). Subsequent structural and computational studies revealed the origin of this discrepancy and are described in Section 3.1.3.

The recently determined structure of the class A BaSrtA–LPAT* complex revealed a generally similar binding mode for the LPXTG sorting signal, but also indicated fundamental differences in the conformational dynamics and structure of the active site (Chan et al., 2015). Similar to SaSrtA, the peptide adopts an L-shaped conformation in the BaSrtA–LPAT* complex by virtue of a kink at position P3, and the $\beta6/\beta7$ loop is instrumental in building the S4 site that dictates specificity. However, four significant differences in the binding mechanism are apparent. First, in contrast to SaSrtA, the $\beta6/\beta7$ loop in BaSrtA is structured in the absence of substrate (Weiner et al., 2010) and only experiences modest changes upon signal binding (the coordinates of the C_α backbone atoms in the loop in the apo- and signal-bound forms of BaSrtA have an RMSD of 0.76 Å). Thus, the $\beta6/\beta7$ loop forms a preformed binding pocket for the P4 residue in BaSrtA, whereas in SaSrtA, the loop is flexible prior to signal binding (Naik et al., 2006; Suree et al., 2009). Second, the $\beta7/\beta8$ loop, which was disordered in the unmodified BaSrtA structure, undergoes a disordered-to-ordered transition as a result of binding to the substrate. A comparison of the apo- and bound-forms of BaSrtA reveals that the active site cysteine residue is displaced ~7 Å upon peptide binding (Chan et al., 2015; Weiner et al., 2010). This movement allows the $\beta7/\beta8$ loop to form new intraprotein interactions with residues within the $\beta4/\beta5$ loop, which presumably stabilize the $\beta7/\beta8$ loop and cause it to become ordered. In contrast, the $\beta7/\beta8$ loop of SaSrtA is structurally ordered in the apo-state and is displaced when the signal binds. Third, unlike SaSrtA, the BaSrtA enzyme contains an N-terminal appendage that partially encapsulates the sorting signal (Chan et al., 2015; Suree et al., 2009). The N-terminal appendage precedes the catalytic domain in the primary sequence and wraps around the body of the protein to contact the active site. The side chain of Ile61 in the appendage forms extensive contacts with the active site histidine, while the hydroxyl group of Ser59 is positioned to form a hydrogen bond with the backbone carbonyl oxygen of the proline residue in the bound peptide. Interestingly, NMR and in vitro kinetics data suggest that the appendage transiently detaches from the isolated enzyme when the sorting signal is bound. A model of the thioacyl intermediate constructed from the coordinates of the complex suggests that the N-terminal appendage may obstruct nucleophile access to the active site, potentially increasing the efficiency of protein display by reducing the unproductive hydrolytic cleavage of enzyme–protein covalent intermediates. A fourth difference between the BaSrtA–LPAT* and SaSrtA–LPAT* complexes is the positioning of the P1 and P2 side chains, which are described in Section 3.1.3.

3.1.2 Sorting Signal Recognition by Class B Enzymes

The structure of the SaSrtB–NPQT* complex provided the first-ever insight into signal recognition by class B enzymes (Jacobitz et al., 2014), which binds an NPQTN sorting signal in the case of SaSrtB (Fig. 5B) (Mazmanian et al., 2002). SaSrtB uses a rigid pocket to bind to the "L"-shaped peptide. Specifically, the P4 asparagine residue is primarily recognized by contacts to the $\beta6/\beta7$ loop. However, as compared to class A enzymes, the P4 residue is more solvent exposed, concordant with the more hydrophilic nature of the asparagine side chain. The binding mode of the sorting signal in class A and B enzymes is compatible with biochemical data from the McCafferty group. This data demonstrated that replacement of the $\beta6/\beta7$ loop in SaSrtA with the corresponding loop from SaSrtB resulted in a chimeric protein that could recognize the SaSrtB-specific NPQTN sorting signal motif (Bentley et al., 2007). The P3 proline residue also appears to have the same function as in the BaSrtA–LPAT* and SaSrtA–LPAT* structures, altering the trajectory of the peptide so that its C-terminal end points toward the active site. As with BaSrtA, the $\beta6/\beta7$ loop in SaSrtB forms a rigid, preformed binding pocket for the sorting signal; the structures of BaSrtA in the free- and bound-states are nearly identical (the C_α coordinates have an RMSD $= 0.44$ Å). Moreover, within the BaSrtA–LPAT* and SaSrtB–NPQT* complexes, the P1 and P2 side chains in the bound substrates adopt similar positions relative to the enzyme active site that are distinct from their positioning observed in the structure of the SaSrtA–LPAT* complex (see Section 3.1.3).

3.1.3 Sorting Signal Conformational Heterogeneity: Thr-In vs Thr-Out

A comparison of the structures of the SaSrtA–LPAT* and SaSrtB–NPQT* complexes revealed that there were major differences in the conformation of the P1 and P2 residues of the bound sorting signals (Jacobitz et al., 2014). In particular, the P2 glutamine residue in the SaSrtB–NPQT* complex rests along the wall of the pocket and points out toward the solvent, while in the SaSrtA–LPAT* structure, the analogous P2 residue (alanine) points toward the base of the active site. The P1 threonine residues in these complexes also adopt fundamentally distinct conformations. In the SaSrtB–NPQT* structure, the threonine is buried in the active site (Thr-in position) where it forms two hydrogen bonds with the active site arginine residue, whereas in the SaSrtA–LPAT* structure, it is projected toward solvent and hydrogen bonds with the active site histidine residue (Thr-out position). Interestingly, an inspection of the recently determined structure of the BaSrtA–LPAT* complex reveals a P1 and P2 positioning that is similar to what is seen in the SaSrtB–NPQT* substrate complex; the P1 threonine side chain adopts a Thr-in conformation in which it is buried within the active site. Thus, the Thr-in conformation of the peptide can be adopted by signal peptides bound to both class A and B enzymes.

Several lines of evidence suggest that the Thr-in conformer observed in the structures of the BaSrtA–LPAT* and SaSrtB–NPQT* complexes represents the catalytically active form of the bound substrate. First, in these structures, the side chain of the P2 residue projects into the solvent and is not recognized. This makes sense as the P2 residue within the sorting signals, also called the "X" position, is not recognized by sortases according to bioinformatics (Comfort & Clubb, 2004) and biochemical data (Kruger, Otvos, et al., 2004). This is in contrast to the SaSrtA–LPAT* complex in which the peptide adopts the Thr-out conformer and the side chain at the P2 "X" position projects into the active site where it contacts A118 and I182. Second, the high degree of sequence conservation at site P1, which contains a threonine residue in ~95% of predicted sorting signals, suggests functional relevance of the Thr-in sorting signal conformer observed in the BaSrtA–LPAT* and SaSrtB–NPQT* complexes. This is in marked contrast to the Thr-out conformer observed in the SaSrtA–LPAT* complex that projects into the solvent and is not recognized by the enzyme. Finally, as described in Section 4, computational modeling of sortase reaction intermediates indicates that the Thr-in conformer likely facilitates the creation of an oxyanion hole for substrate stabilization during thioacyl intermediate formation.

3.2 Nucleophile Recognition

Unlike the mechanism of sorting signal recognition, sortase binding to their secondary substrate, the amino nucleophile to which the sorting signal is joined, remains poorly understood. Depending upon the type of sortase, cell wall-anchoring vs pilin polymerizing, the nucleophile can either originate from a lipid II molecule or a lysine residue within a pilin protein substrate, respectively. Thus far, only lipid II recognition by sortases has been explored, but the mechanism of binding still remains enigmatic. Experimental studies have investigated how the SaSrtA enzyme interacts with polyglycine peptides that mimic the pentaglycine crossbridge peptide moiety within the intact *S. aureus* lipid II molecule, as large quantities of the intact, water-soluble portion of lipid II are difficult to obtain. NMR chemical shift mapping of the SaSrtA–LPAT* complex using a triglycine peptide titrant revealed a low affinity, continuous interaction surface on the enzyme that contains portions of the $\beta7/\beta8$ loop, $\beta4$/H2 loop, and an N-terminal segment of helix H1 (Suree et al., 2009). The mapping data coarsely define the interaction surface, but do not provide specific details about triglycine–enzyme interactions. Interestingly, significant chemical shift changes are not observed when apo-SaSrtA is titrated with the triglycine peptide, but only when added to the SaSrtA–LPAT* complex. This observation is compatible with the proposed bi–bi mechanism of catalysis (Frankel et al., 2005; Huang et al., 2003) and suggests that LPXTG sorting signal binding may direct catalysis forward by causing alterations in the $\beta7/\beta8$ loop that unmask the binding surface for lipid II. Crystals of SaSrtB modified with a small molecule sulfhydryl modifier were soaked with a triglycine peptide, and the structure was determined by molecular replacement. In this complex, the N-terminal amine of the triglycine peptide is 6.4 Å from the active site histidine, which is occluded from solvent by a closed $\beta7/\beta8$ loop (Zong, Mazmanian, et al., 2004). While the binding mode of the triglycine peptide to SaSrtB is generally compatible with the NMR chemical shift data from the SaSrtA–LPAT* complex, it does not fit the accepted view of the mechanism that would have the incoming nucleophile deprotonated by the active site histidine residue.

Interestingly, other class A sortases appear to contain a preformed binding site for the crossbridge peptide in lipid II. In the crystal structures of the SpySrtA, SagSrtA, and SmSrtA sortases, their $\beta7/\beta8$ loops are displaced in the absence of the sorting signal, creating a surface-exposed groove adjacent to the active site histidine and cysteine residues (Fig. 3C, yellow) (Khare, Krishnan, et al., 2011; Race et al., 2009; Wallock-Richards et al., 2015). Recent docking studies of BaSrtA highlight a potential binding site for the amino component of lipid II in *B. anthracis* (diaminopimelic acid) (Chan et al., 2015). This site is positioned adjacent to the active site histidine, between the $\beta7/\beta8$ and $\beta4$/H3 loops, and would position the nucleophile near the electrophilic carbonyl carbon atom in the thioacyl bond. Interestingly, exposure of this site requires partial displacement of the N-terminal appendage, which has been postulated to mask the acyl-linked reaction intermediate from hydrolysis. Conclusive determination of the molecular basis of amino nucleophile selectivity will require structure determination of sortases bound to their secondary substrates. The mechanism of pilin protein substrate recognition by pilin polymerases also remains to be determined.

4. COMPUTATIONAL STUDIES

4.1 The Thr-In Conformation Enables Sortases to Employ a Substrate-Stabilized Active Site

Computational studies have leveraged the experimentally determined structures of sortase–substrate complexes to gain insight into the mechanism of catalysis. Jacobitz et al. generated an energy minimized model of the thioacyl intermediate using the coordinates of the SaSrtB–NPQT* complex (Jacobitz et al., 2014). The model of the thioacyl intermediate is shown in Fig. 6 and reveals that the Thr-in conformation adopted by the highly conserved P1 threonine residue (see Section 3.1.3) enables it to hydrogen bond to the active site arginine residue. This interaction positions the active site arginine proximal to the thioacyl bond, where it presumably can stabilize the two high-energy tetrahedral oxyanion intermediates that form during catalysis (Jacobitz et al., 2014). In this capacity, threonine–arginine interactions facilitate the formation of an oxyanion hole in which the arginine neutralizes the negative charge of the oxyanion. Further stabilization of the oxyanion is achieved by interactions from the backbone amide

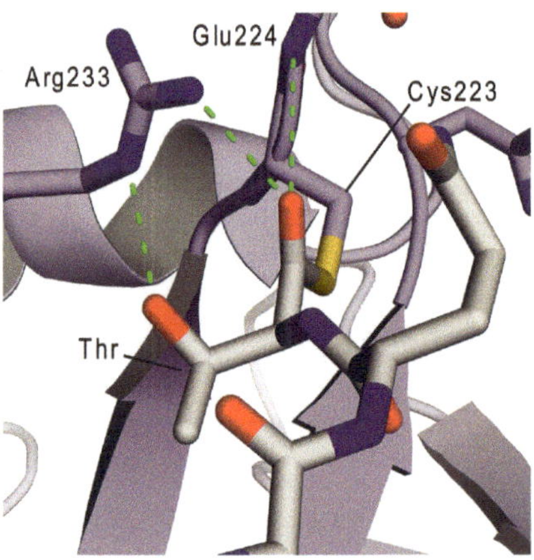

Fig. 6 The substrate-stabilized oxyanion hole. The energy minimized model of the SaSrtB-NPQT thioacyl intermediate displayed with SaSrtB (*light blue cartoon*), residues in the active site and oxyanion hole (*sticks*), and NPQT substrate (*gray sticks*). The side chain hydroxyl of the substrate's P1 Thr residue and backbone carbonyl participate in a hydrogen bonding network with the active site Arg, and the backbone amide of Glu224 that together build an oxyanion hole to stabilize the high energy tetrahedral reaction intermediates. *Reproduced from Jacobitz, A. W., Wereszczynski, J., Yi, S. W., Amer, B. R., Huang, G. L., Nguyen, A. V., et al. (2014). Structural and computational studies of the Staphylococcus aureus Sortase B-substrate complex reveal a substrate-stabilized oxyanion hole,* The Journal of Biological Chemistry, 289, *8891–8902, p. jbc.M113.509273.*

of E224 that is located immediately C-terminal to the active site cysteine. This novel "substrate-stabilized oxyanion hole" presumably increases substrate specificity for a threonine residue at position P1 and is compatible with biochemical studies that have shown that the enzyme is unable to utilize sorting signals that contain conservative mutations at the P1 position (e.g., Ser or Val instead of Thr) (Jacobitz et al., 2014). A similar computational strategy was used to model the structure of the BaSrtA thioacyl intermediate using the experimentally determined coordinates of the BaSrtA–LPAT* complex (Chan et al., 2015). This model displayed an analogous hydrogen bonding interaction between the P1 threonine residue and the active site arginine. As in the SaSrtB model, the P1 threonine carbonyl atom in the thioacyl bond is in close proximity to the active site arginine guanidino group, suggesting that both enzymes employ-related substrate-stabilized oxyanion holes to facilitate catalysis (Chan et al., 2015). However, it should be noted that the active site conformations of the two models differ slightly, since in the model of the BaSrtA thioacyl intermediate, the active site arginine adopts a more extended structure that allows it to form an additional hydrogen bond to the P3 proline residue.

4.2 Sorting Signals Bound to SaSrtA Can Interchange Between Thr-Out and Thr-In Conformers

As described in Section 3.1.3, a comparison of the experimentally determined structures of the SaSrtA–LPAT*, SaSrtB–NPQT*, and BaSrtA–LPAT* revealed fundamental differences in the positioning of the P1 and P2 residues in the bound sorting signal (Chan et al., 2015; Jacobitz et al., 2014; Suree et al., 2009).

To rectify the discrepancies between the SaSrtA–LPAT* Thr-out and SaSrtB–NPQT* Thr-in conformations, Jacobitz et al. performed molecular dynamics (MD) simulations for three thioacyl intermediate systems: SaSrtA–LPAT, SaSrtB–NPQT, and SaSrtB–NPAT (Jacobitz et al., 2014). Through the use of umbrella sampling calculations, the free energy landscape of transitions between the Thr-out and Thr-in states was mapped for each complex. These studies indicated that SaSrtA–LPAT could adopt both the Thr-in and Thr-out states with equal probability, while for both NPQT and NPAT substrates, SaSrtB could only sample the Thr-in state. Based on these results, it appears as if the Thr-in conformation is likely the more evolutionarily conserved state, and that the inherent flexibility of SaSrtA allows for the Thr-out conformation with an LPAT substrate that was captured by Suree et al. (2009).

4.3 Dynamic Sorting Signal Recognition by SaSrtA

MD studies of SaSrtA have provided additional insight into the sorting signal recognition process. By performing conventional and accelerated MD simulations of both the sorting signal free and bound states, Kappel et al. proposed that sorting signal binding is a mixture of conformational selection and induced fit mechanisms (Kappel, Wereszczynski, Clubb, & McCammon, 2012). For example, the $\beta 6/\beta 7$ loop appears to follow the conformational selection paradigm: it sampled a range of stable conformations in the apo-state, some of which were relatively close to the bound configurations. In contrast, the $\beta 7/\beta 8$ "open" state from the NMR structure was only stable in the presence of a bound sorting signal, suggesting an induced fit mechanism. In addition, analysis of the sorting signal-bound conformations showed that an allosteric network runs throughout the protein, linking the calcium ion, sorting signal, and proposed lipid-II binding regions to one another. In a complementary study, Moritsugu et al. used the multiscale enhanced sampling method to probe the allosteric effects of the calcium ion and sorting signal (Moritsugu, Terada, & Kidera, 2012). Simulations of each combination of bound states showed that binding of both molecules is required to stabilize the $\beta 6/\beta 7$ and $\beta 7/\beta 8$ loops in conformations observed in the NMR–LPAT* structure. Overall, these simulations point toward a mechanism in which calcium, sorting signal, and potentially lipid II binding are modulated by a dynamic network that includes the $\beta 6/\beta 7$ loop region in SaSrtA.

Other aspects of the SaSrtA recognition and catalytic processes have also been explored by computational studies. Biswas et al. used a hybrid MD and biochemical experimental approach to probe the roles of the conserved sorting signal Leu and Pro residues in substrate binding (Biswas, Pawale, Choudhury, & Roy, 2014). Comparative simulations with LPAT, APAT, and LAAT substrates demonstrated that contacts between the leucine side chain and SaSrtA contribute to stabilize the $\beta 6/\beta 7$ loop, whereas the kink that is induced by the proline appears to be essential for recognition. In another study, Tian and Eriksson performed simulations in which His120 and Cys184 were in their zwitterionic and neutral forms (Tian & Eriksson, 2011). Their results showed that Arg197 adopts distinct conformations based upon the charged state of the protein, which helps to stabilize the catalytically active form. It should be noted that each of these studies was performed with the sorting signal in the Thr-out state. Although the global effects of the Thr-in and Thr-out states on the induced fit/conformational selection process, allosteric networks, and recognition processes are likely similar, subtle differences may exist that influence some of the fine details that resulted from these simulations.

5. CATALYTIC MECHANISM

The current model of the molecular mechanism of the SaSrtA enzyme is presented in Fig. 7. Kinetic studies indicate that catalysis occurs through a ping-pong mechanism that begins when the sortase recognizes the CWSS of a membrane anchored protein (Frankel et al., 2005; Huang et al., 2003). The LPXTG-type sorting signal within the CWSS binds to a groove on the sortase whose base is formed by residues in the $\beta 6/\beta 7$ loop, strands $\beta 4$, $\beta 7$, $\beta 8$, and whose walls are formed by residues in the $\beta 2/\beta 3$ and $\beta 3/\beta 4$ loops (Fig. 4D). Here the sorting signal's L-shaped structure dictated by the highly conserved proline residue at P3 (>90% conserved) orients the residue P4 for recognition in subsite S4 on the $\beta 6/\beta 7$ loop, as the C-terminal end of the sorting signal is directed toward the active site (Suree et al., 2009). In order for catalysis to proceed, the enzyme must contain a properly charged active site in which the cysteine and histidine residues are in their thiolate and imidazolium forms, respectively. Based on pK_a measurements of their active site residues (Connolly et al., 2003; Frankel et al., 2005; Weiner et al., 2010), when removed from the cell surface, in isolation less than 1% of the SaSrtA and BaSrtA sortases possess an appropriately charged active site that can perform catalysis. This explains why sortases in isolation catalyze reactions very slowly, as presumably only a small fraction of these enzymes are active (Section 6). The notion that the active site is predominantly dormant is substantiated by structural analyses of SaSrtA, which revealed that its active site cysteine and histidine residues are not close enough to form a thiolate-imidazolium ion-pair (the Cys-S-His-δN distance is 6.5 and 7.6 Å in the NMR and crystal structures, respectively) (Ilangovan et al., 2001; Zong, Bice, et al., 2004). In the rare instance that the cysteine and histidine residues are appropriately ionized, the cysteine thiolate nucleophilically attacks the carbonyl carbon on the P1 residue (Fig. 7B). This leads to the formation of the first tetrahedral intermediate, which is likely stabilized by an oxyanion hole that is formed by the active site arginine residue and a backbone amide in the $\beta 7/\beta 8$ loop (Jacobitz et al., 2014). In nearly all sorting signals a threonine occurs at the P1 position, which based on structural and MD simulations helps to stabilize the oxyanion hole by forming a hydrogen

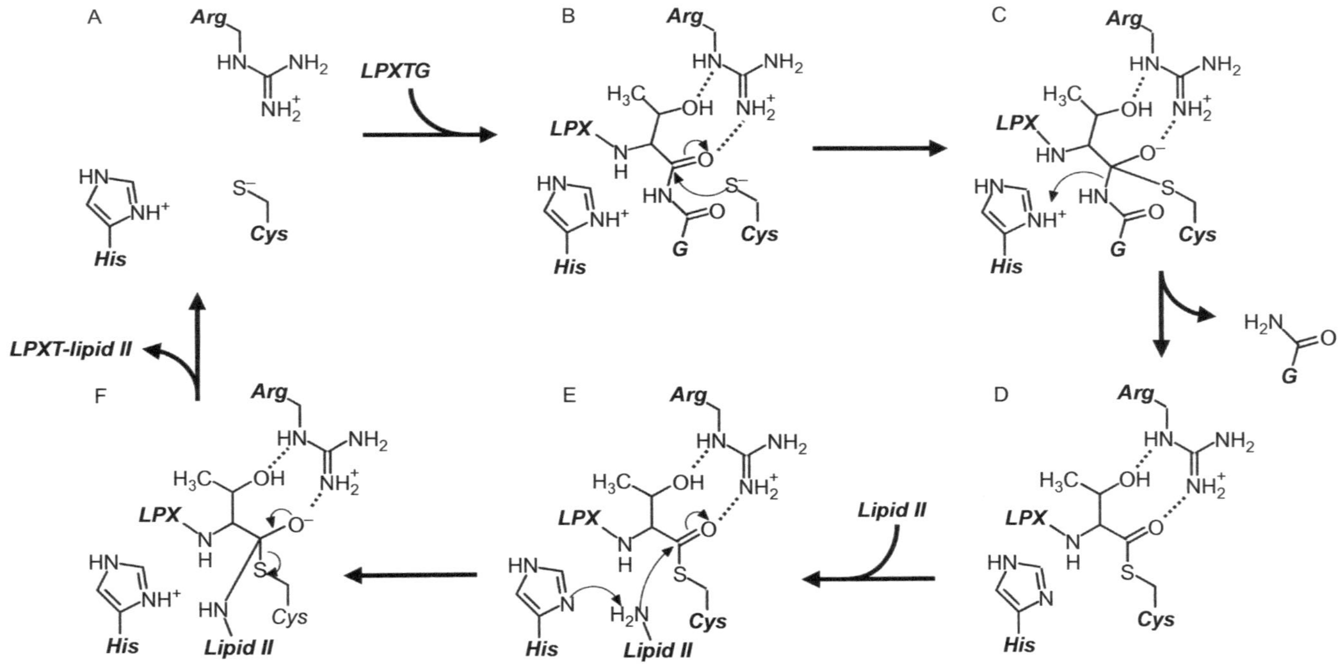

Fig. 7 Molecular mechanism of sortase enzymes. The active site of sortase consists of a His–Cys–Arg triad, and in its active form, the His and Cys residues form a thiolate–imidazolium ion-pair (A). The reaction begins with recognition of an appropriate sorting signal (here, the LPXTG sorting signal for SrtA types is shown), and the active site Cys residue performs nucleophilic attack on the carbonyl carbon at the substrate's P1 position (B). An oxyanion tetrahedral intermediate is stabilized by the nearby Arg residue that is likely oriented by interactions with the side chain of the substrate's P1 residue, which is a threonine in over 95% of all substrates (C). The active His residue concomitantly donates a proton to the leaving group, and the tetrahedral transition state then collapses to form a semistable, thioacyl intermediate between the substrate's P1 residue and the active site Cys (D). Next, the secondary substrate (here shown as lipid II used by cell wall anchoring sortases) enters the active site, where its terminal amine is deprotonated by the active His residue before performing nucleophilic attack on the carbonyl carbon in the thioacyl bond (E); this second tetrahedral intermediate (F) collapses to form a peptide bond between the two substrates, and the product is finally released to leave the regenerated active site (A).

bond to the active site arginine (Jacobitz et al., 2014) (Fig. 7C). The first oxyanion is a transient intermediate, and quickly collapses to form a semi-stable thioacyl intermediate as the scissile peptide bond is broken. In the thioacyl intermediate, the substrate's P1-Thr and the active site cysteine are joined via a thioacyl bond (Fig. 7D) (Aulabaugh et al., 2007). This process is presumably assisted by the conserved active site histidine residue, which may act as a general acid to protonate the amino leaving group. Beyond stabilizing the oxyanion intermediate, the side chain of the active site arginine residue may orient the substrate in the active site by forming a hydrogen bond to the backbone carbonyl atom in the P2 residue (Chan et al., 2015; Suree et al., 2009; Tian & Eriksson, 2011). In the next step of the reaction, a secondary substrate bearing an amine group enters the active site and is presumably deprotonated by the active site histidine residue to facilitate its nucleophilic attack on the thioacyl bond (Fig. 7E). A second tetrahedral intermediate then forms which may also be stabilized by an oxyanion hole that is constructed with the assistance of the sorting signal's P1 residue (Fig. 7F). The second tetrahedral intermediate then quickly collapses to form the peptide bond linked product. Sortases that anchor proteins to the cell wall join the protein substrate to the cell wall precursor lipid II, whereas pilin polymerases join the protein to a lysine amine located within another pilin protein. These enzymes are believed to use a similar mechanism to catalyze transpeptidation.

6. IN VITRO TRANSPEPTIDATION ACTIVITY

Many sortase enzymes exhibit in vitro transpeptidation and/or proteolytic activity. In vitro activity was first demonstrated for the prototypical SaSrtA enzyme by Ton-That et al. (1999). Their assay utilized FRET-based detection of activity using a reporter LPXTG sorting signal peptide that contained donor and quencher fluorophores at each end. Cleavage of this peptide by SaSrtA, liberates the donor from the quencher enabling enzyme activity to be detected as an increase in fluorescence. Initial studies demonstrated that SaSrtA catalyzes an in vitro transpeptidation that joins two peptides, one that contains an LPXTG sorting signal motif and a second peptide that contains N-terminal glycine residues (Huang et al., 2003; Ton-That, Mazmanian, Alksne, & Schneewind, 2000). Studies using this assay also demonstrated that the active site histidine, cysteine, and arginine residues are important for catalysis (Marraffini et al., 2004; Ton-That et al., 2002). Although the FRET-based assay is easy to employ, inner filter effects can occur at high substrate concentrations leading to inaccurate measurements of the enzyme's kinetic parameters (Kruger, Dostal, et al., 2004). Subsequent development of a medium-throughput HPLC enzyme assay enabled more accurate measurement of the kinetic parameters and revealed that transpeptidation occurs with a $k_{cat} = 0.28 \pm 0.02\,s^{-1}$, and K_m values for its LPXTG and secondary Gly_5 peptide substrates of $7.33 \pm 1.01\,mM$ and $196 \pm 64\,\mu M$, respectively (Frankel et al., 2005). In the absence of the Gly_5 peptide SaSrtA acts as a protease, hydrolytically cleaving the sorting signal between the threonine and glycine residues with a $k_{cat} = 0.086 \pm 0.015\,s^{-1}$. As hydrolysis occurs much slower than transpeptidation, it can be largely avoided when the Gly_5 peptide is in excess. As described in Section 5, when sortase is purified and removed from the cell surface only a small fraction (~0.06%) contains a properly ionized active site that can catalyze transpeptidation. This small subpopulation is much more enzymatically active with an estimated k_{cat}/K_m greater than $10^5\,M^{-1}$ (Frankel et al., 2005). On the cell surface, the transpeptidation reaction catalyzed by membrane-associated SaSrtA may occur at a rate that is faster than the rate of the reconstituted in vitro transpeptidation reaction that employs short peptide substrates. This is because pulse-chase labeling experiments using intact cells indicate that the reaction is complete in <3 min, and it is likely that individual sortase enzymes attach several proteins to the cell wall during this time (Schneewind et al., 1992). The increased in vivo rate may result from the fact that the sortase and both of its substrates are embedded in the membrane. It is also possible that there exists yet to be identified factors on the cell surface that facilitate sortase association with its substrates and/or the conversion of its active site into the properly ionized state.

Since the original work on SaSrtA, the in vitro enzymatic activities of several other sortases have been characterized (summarized in Table 2). Interestingly, none of these enzymes are as active as SaSrtA and in many instances only their proteolytic activity has been demonstrated. Generally, their activities have not been rigorously characterized as only the amount of product generated by the sortase has been measured after a specific incubation time. At present, only the SaSrtA, SpySrtA, and SaSrtB sortases have been shown to catalyze a transpeptidation reaction that joins two peptides together via a peptide bond. The native SaSrtA enzyme catalyzes transpeptidation ~20–500-fold faster than the other enzymes making it a useful bioconjugation reagent. Moreover, the activity of SaSrtA has been improved using directed evolution approaches, resulting in tetramutant enzyme that is ~140-fold more active than the native SaSrtA (Chen

Table 2 Activity of Sortase Enzymes In Vitro

Sortase	Primary Substrate (K_m mM)	Secondary Substrate (K_m μM)	Cleavage k_{cat} (s^{-1})	Transpeptidation k_{cat} (s^{-1})
Class A				
SaSrtA$_{\Delta 24}$ (Frankel, Kruger, Robinson, Kelleher, & McCafferty, 2005)	Abz-**LPETG**-Dap(Dnp) (7.33 ± 1.01)	Gly_5 (196 ± 64)	0.086 ± 0.015	0.28 ± 0.02
SaSrtA$_{\Delta 24}$ (Kruger, Dostal, & McCafferty, 2004)	Abz-**LPETG**-Dap(Dnp) (5.5 mM)	Gly_5 (140)	NR	0.27
SaSrtA$_{\Delta 24}$ (Bentley, Lamb, & McCafferty, 2008)	Abz-**LPETG**G-Dap(Dnp) (8.76 ± 0.78)[a]	Gly_5 (NR)	NR	1.10 ± 0.06[a]
SaSrtA$_{\Delta 59}$ (Chen, Dorr, & Liu, 2011)	Abz-**LPETG**K-(Dnp) (7.6 ± 0.5)	Gly_3 (140 ± 30)	NR	1.5 ± 0.2
SaSrtA$_{\Delta 59}$ Evolved tetramutant (Chen et al., 2011)	Abz-**LPETG**K-(Dnp) (0.17 ± 0.03)	Gly_3 (4800 ± 700)	NR	4.8 ± 0.8
SpySrtA$_{\Delta 81}$ (Race et al., 2009)	Abz-**LPETG**G-Dap(Dnp) (0.83 ± 0.11)	Ala_2 (NR)	NR	0.0136 ± 0.0011
BaSrtA$_{\Delta 56}$ (Weiner, Robson, Marohn, & Clubb, 2010)	Abz-**LPET**G-Dap(Dnp) (0.038 ± 4)[b]	m-DAP[c]	0.0004 ± 0.0001[b]	[c]
BaSrtA$_{\Delta 56}$ (Chan et al., 2015)	Abz-**LPET**G-Dap(Dnp) (0.306 ± 0.023)	m-DAP[c]	$3.6 \pm 0.2 \times 10^{-5}$	[c]
BaSrtA$_{\Delta 64}$ (Chan et al., 2015)	Abz-**LPET**G-Dap(Dnp) (0.173 ± 0.011)	m-DAP[c]	$5.7 \pm 0.2 \times 10^{-5}$	[c]
SmutSrtA$_{\Delta 40}$ (Wallock-Richards et al., 2015)	Dabcyl-QA**LPETG**EE-Edans (0.0904 ± 0.0047)[b]	NR	Yes	NR

Continued

Table 2 Activity of Sortase Enzymes In Vitro—cont'd

Sortase	Primary Substrate (K_m mM)	Secondary Substrate (K_m μM)	Cleavage k_{cat} (s^{-1})	Transpeptidation k_{cat} (s^{-1})
Class B				
SaSrtB$_{\Delta 21}$ (Bentley, Gaweska, Kielec, & McCafferty, 2007)	Abz-KVE**NPQTN**AGT-Dap(DNP) (7.8 ± 2)	Gly_5 (NR)	NR	$5.4 \pm 0.5 \times 10^{-4}$
SaSrtB$_{\Delta 31}$ (Jacobitz et al., 2014)	SNKDKVEN**PQTN**AGT (1.8)	Gly_5 (NR)	NR	1.010^{-4}
BaSrtB$_{\Delta 37}$ (Maresso, Chapa, & Schneewind, 2006)	Abz -KTD**NPKTG**DEA-Dap(DNP)	NR	Yes	NR
CdSrtB$_{\Delta 26}$ (van Leeuwen et al., 2014)	KIVK**SPKTG**DETQLMK KPPV**PPKTG**DSTTIGK	$Gly_{4/5}$ or Ala-D-Glu-DAP	NR	NR
Class C				
SpnSrtC1$_{17–228}$ (Manzano et al., 2008)	**IPQTG** in $RrgB_{30–633}$	YPKN in $RrgB_{30–633}$	No	Yes
SagSrtC1$_{43–254}$ from PI-2a (Cozzi et al., 2011)	Dabcyl-KKVT**IPQTG**GIGT-Edans (0.0138)	NR	Yes	NR
SagSrtC1$_{42–305}$ from PI-1 (Cozzi et al., 2012)	Dabcyl-RPPGV**FPKTG**GIG-Edans (0.01358 ± 0.00063)	NR	$1.16 \pm 0.044 \times 10^{-3}$	NR
	Dabcyl-RPS**IPNTG**GIG-Edans (0.03100 ± 0.00462)	NR	$1.77 \pm 0.101 \times 10^{-3}$	NR
	Dabcyl-RGGL**IPKTG**EQQ-Edans (0.01639 ± 0.00250)	NR	$0.77 \pm 0.038 \times 10^{-3}$	NR

SagSrtC2$_{42–283}$ from PI-1 (Cozzi et al., 2012)	Dabcyl-RPPGV**FPKTG**GIG-Edans (0.006385 ± 0.00142)	NR	$1.04 \pm 0.058 \times 10^{-3}$	NR
	Dabcyl-RGGL**IPKTG**EQQ-Edans (0.02733 ± 0.00435)	NR	$4.36 \pm 0.256 \times 10^{-4}$	NR
	Dabcyl-RPS**IPNTG**GIG-Edans (0.05715 ± 0.00354)	NR	$5.56 \pm 0.174 \times 10^{-3}$	NR
Class D				
BaSrtD$_{\Delta55}$ (Robson, Jacobitz, Phillips, & Clubb, 2012)	VQGEK**LPNTA**SNN	m-DAP (NR)	Yes	NR
BaSrtD$_{\Delta55}$ (Marraffini & Schneewind, 2006)	Abz-GEK**LPNTA**SNN-Dnp	m-DAP (NR)	Yes	NR
CpSrtD$_{\Delta23–187}$ (Suryadinata, Seabrook, Adams, Nuttall, & Peat, 2015)	A$\beta_{1–16}$-**LPQTG**S	NR	Yes	NR

[a]Values reported from fluorescence assay and subject to inner filter effect and are likely underestimates of true parameters.
[b]These values calculated assuming a hydrolytic shunt mechanism.
[c]The enzyme reportedly does not perform this reaction in vitro.
Sorting signals for all substrates are highlighted in bold.
Errors are reported where published.
"Yes" indicates the reaction was performed in vitro but kinetics parameters were not reported.
NR, not reported.
m-DAP, mesodiaminopimelic acid.

et al., 2011). Additional rate enhancements have been achieved by altering the reaction conditions and by fusing the nucleophile substrate to SaSrtA (Amer, Macdonald, Jacobitz, Liauw, & Clubb, 2016). The reader is referred to a number of excellent reviews describing the use of sortase as a bioconjugation reagent (Proft, 2010; Tsukiji & Nagamune, 2009; Wu & Guo, 2012).

7. CONCLUSIONS AND FUTURE DIRECTIONS

Sortase enzymes are ubiquitous in Gram-positive bacteria where they attach proteins to the cell wall and construct pili. Their important role in displaying virulence factors makes them promising drug targets (Bradshaw et al., 2015; Cascioferro et al., 2014; Clancy et al., 2010; Maresso & Schneewind, 2008; Suree et al., 2007), while their ability catalyze in vitro transpeptidation has made them useful bioconjugation reagents (Antos et al., 2016; Bentley et al., 2008; Maresso et al., 2006; Popp & Ploegh, 2011; Ritzefeld, 2014; Schmidt et al., 2017; Schmohl & Schwarzer, 2014; Tsukiji & Nagamune, 2009; Voloshchuk et al., 2016). Considerable effort has been put forth to elucidate the molecular mechanism of catalysis, class-specific structural features that dictate function, and the molecular basis of substrate recognition. Structural and computational studies of three sortase–peptide complexes have provided insight into the initial steps of catalysis—binding of the sorting signal to the active site and formation of the first tetrahedral and thioacyl intermediates (Chan et al., 2015; Jacobitz et al., 2014; Suree et al., 2009). This work has shown that class A and B sortases recognize their sorting signal substrates in a similar manner. The side chains of the P1, P3, and P4 residues are recognized. The bound signal adopts an L-shaped conformation as a result of kink introduced at the proline P3 residue, positioning the side chain of the P4 residue within a pocket formed by the $\beta6/\beta7$ loop, and directing the P1 and P2 residues toward the active site cysteine. The positioning of P1 and P2 exhibit conformational heterogeneity, with substrate assisted catalysis occurring when the threonine P1 side chain contacts the active arginine residue to stabilize the oxyanion hole (Jacobitz et al., 2014). Other classes of sortases may bind their sorting signals in a generally similar manner, with class C pilin polymerases requiring unlatching of a lid structure to enable signal access (Manzano et al., 2008), and class E enzymes using a unique surface to recognize alanine at position P3 instead of proline (Kattke et al., 2016). The second-half of the transpeptidation reaction remains poorly understood—nucleophilic attack of the thioacyl enzyme–substrate intermediate by an amine group and peptide bond formation. No conclusive evidence has emerged to pinpoint the location of the secondary substrate binding site for lipid II or pilin proteins in cell wall anchoring or pilin assembling sortases, respectively. Deciphering how some sortases function as polymerases, while others attach proteins to the cell wall will require the development of robust biochemical assays to monitor pilus assembly and novel substrate analogs to visualize nucleophile recognition. Finally, although many small molecule sortase inhibitors have been identified, they have yet to be developed into a drug to treat bacterial infections. Given the rising prevalence of antibiotic-resistant bacteria, pressure to develop viable sortase inhibitors as therapeutics is growing and will undoubtedly lead to the discovery and characterization of more potent and specific compounds.

REFERENCES

Amer, B. R., Macdonald, R., Jacobitz, A. W., Liauw, B., & Clubb, R. T. (2016). Rapid addition of unlabeled silent solubility tags to proteins using a new substrate-fused sortase reagent. *Journal of Biomolecular NMR*, *64*(3), 197–205.

Antos, J. M., Truttmann, M. C., & Ploegh, H. L. (2016). Recent advances in sortase-catalyzed ligation methodology. *Current Opinion in Structural Biology*, *38*, 111–118.

Aulabaugh, A., Ding, W., Kapoor, B., Tabei, K., Alksne, L., Dushin, R., et al. (2007). Development of an HPLC assay for *Staphylococcus aureus* sortase: Evidence for the formation of the kinetically competent acyl enzyme intermediate. *Analytical Biochemistry*, *360*(1), 14–22.

Bentley, M. L., Gaweska, H., Kielec, J. M., & McCafferty, D. G. (2007). Engineering the substrate specificity of *Staphylococcus aureus* Sortase A. The beta6/beta7 loop from SrtB confers NPQTN recognition to SrtA. *The Journal of Biological Chemistry*, *282*(9), 6571–6581.

Bentley, M. L., Lamb, E. C., & McCafferty, D. G. (2008). Mutagenesis studies of substrate recognition and catalysis in the Sortase A transpeptidase from *Staphylococcus aureus*. *The Journal of Biological Chemistry*, *283*(21), 14762–14771.

Biswas, T., Pawale, V. S., Choudhury, D., & Roy, R. P. (2014). Sorting of LPXTG peptides by archetypal Sortase A: Role of invariant substrate residues in modulating the enzyme dynamics and conformational signature of a productive substrate. *Biochemistry*, *53*(15), 2515–2524.

Bradshaw, W. J., Davies, A. H., Chambers, C. J., Roberts, A. K., Shone, C. C., & Acharya, K. R. (2015). Molecular features of the sortase enzyme family. *The FEBS Journal*, *282*(11), 2097–2114.

Cascioferro, S., Totsika, M., & Schillaci, D. (2014). Sortase A: An ideal target for anti-virulence drug development. *Microbial Pathogenesis*, 77, 105–112.

Chambers, C. J., Roberts, A. K., Shone, C. C., & Acharya, K. R. (2015). Structure and function of a Clostridium difficile sortase enzyme. *Scientific Reports*, *5*, 9449.

Chan, A. H., Yi, S. W., Terwilliger, A. L., Maresso, A. W., Jung, M. E., & Clubb, R. T. (2015). Structure of the Bacillus anthracis Sortase A enzyme bound to its sorting signal: A flexible amino-terminal appendage modulates substrate access. *The Journal of Biological Chemistry*, *290*(42), 25461–25474.

Chen, I., Dorr, B. M., & Liu, D. R. (2011). A general strategy for the evolution of bond-forming enzymes using yeast display. *Proceedings of the National Academy of Sciences of the United States of America*, *108*(28), 11399–11404.

Clancy, K. W., Melvin, J. A., & McCafferty, D. G. (2010). Sortase transpeptidases: Insights into mechanism, substrate specificity, and inhibition. *Biopolymers*, *94*(4), 385–396.

Comfort, D., & Clubb, R. T. (2004). A comparative genome analysis identifies distinct sorting pathways in Gram-positive bacteria. *Infection and Immunity*, *72*(5), 2710–2722.

Connolly, K., & Clubb, R. (2005). No title. In G. Waksman, M. Caparon, & C. Hultgren (Eds.), *Structural biology of bacterial pathogenesis* (pp. 101–127). Washington, DC: ASM Press.

Connolly, K. M., Smith, B. T., Pilpa, R., Ilangovan, U., Jung, M. E., & Clubb, R. T. (2003). Sortase from *Staphylococcus aureus* does not contain a thiolate-imidazolium ion pair in its active site. *The Journal of Biological Chemistry*, *278*(36), 34061–34065.

Cozzi, R., Malito, E., Nuccitelli, A., D'Onofrio, M., Martinelli, M., Ferlenghi, I., et al. (2011). Structure analysis and site-directed mutagenesis of defined key residues and motives for pilus-related sortase C1 in group B Streptococcus. *The FASEB Journal*, *25*(6), 1874–1886.

Cozzi, R., Prigozhin, D., Rosini, R., Abate, F., Bottomley, M. J., Grandi, G., et al. (2012). Structural basis for group B Streptococcus pilus 1 Sortases C regulation and specificity. *PLoS One*, *7*(11). e49048.

Dramsi, S., Trieu-Cuot, P., & Bierne, H. (2005). Sorting sortases: A nomenclature proposal for the various sortases of Gram-positive bacteria. *Research in Microbiology*, *156*(3), 289–297.

Duong, A., Capstick, D. S., Di Berardo, C., Findlay, K. C., Hesketh, A., Hong, H.-J., et al. (2012). Aerial development in Streptomyces coelicolor requires sortase activity. *Molecular Microbiology*, *83*(5), 992–1005.

Finn, R. D., Coggill, P., Eberhardt, R. Y., Eddy, S. R., Mistry, J., Mitchell, A. L., et al. (2016). The Pfam protein families database: Towards a more sustainable future. *Nucleic Acids Research*, *44*(D1), D279–D285.

Fischetti, V., Pancholi, V., & Schneewind, O. (1990). Conservation of a hexapeptide sequence in the anchor region of surface proteins from Gram-positive cocci. *Molecular Microbiology*, *4*(9), 1603–1605.

Frankel, B. A., Kruger, R. G., Robinson, D. E., Kelleher, N. L., & McCafferty, D. G. (2005). *Staphylococcus aureus* sortase transpeptidase SrtA: Insight into the kinetic mechanism and evidence for a reverse protonation catalytic mechanism. *Biochemistry*, *44*(33), 11188–11200.

Hendrickx, A. P. A., Budzik, J. M., Oh, S.-Y., & Schneewind, O. (2011). Architects at the bacterial surface—Sortases and the assembly of pili with isopeptide bonds. *Nature Reviews. Microbiology*, *9*(3), 166–176.

Huang, X., Aulabaugh, A., Ding, W., Kapoor, B., Alksne, L., Tabei, K., et al. (2003). Kinetic mechanism of *Staphylococcus aureus* sortase SrtA. *Biochemistry*, *42*(38), 11307–11315.

Ilangovan, U., Ton-That, H., Iwahara, J., Schneewind, O., & Clubb, R. T. (2001). Structure of sortase, the transpeptidase that anchors proteins to the cell wall of *Staphylococcus aureus*. *Proceedings of the National Academy of Sciences*, *98*(11), 6056–6061.

Jacobitz, A. W., Wereszczynski, J., Yi, S. W., Amer, B. R., Huang, G. L., Nguyen, A. V., et al. (2014). Structural and computational studies of the *Staphylococcus aureus* Sortase B-substrate complex reveal a substrate-stabilized oxyanion hole. *The Journal of Biological Chemistry*, *289*, 8891–8902. p. jbc.M113.509273.

Jacobitz, A. W., Naziga, E. B., Yi, S. W., McConnell, S. A., Peterson, R., Jung, M. E., et al. (2016). The 'lid' in the *Streptococcus pneumoniae* SrtC1 sortase adopts a rigid structure that regulates substrate access to the active site. *The Journal of Physical Chemistry B*, *120*(33), 8302–8312.

Jung, M. E., Clemens, J. J., Suree, N., Liew, C. K., Pilpa, R., Campbell, D. O., et al. (2005). Synthesis of (2R,3S) 3-amino-4-mercapto-2-butanol, a threonine analogue for covalent inhibition of sortases. *Bioorganic & Medicinal Chemistry Letters*, *15*(22), 5076–5079.

Kang, H. J., Coulibaly, F., Proft, T., & Baker, E. N. (2011). Crystal structure of Spy0129, a Streptococcus pyogenes Class B Sortase involved in pilus assembly. *PLoS One*, *6*(1). e15969.

Kappel, K., Wereszczynski, J., Clubb, R. T., & McCammon, J. A. (2012). The binding mechanism, multiple binding modes, and allosteric regulation of *Staphylococcus aureus* Sortase A probed by molecular dynamics simulations. *Protein Science*, *21*(12), 1858–1871.

Kattke, M. D., Chan, A. H., Duong, A., Sexton, D. L., Sawaya, M. R., Cascio, D., et al. (2016). Crystal structure of the Streptomyces coelicolor Sortase E1 transpeptidase provides insight into the binding mode of the novel class E sorting signal. *PLoS One*, *11*(12). e0167763.

Khare, B., Krishnan, V., Rajashankar, K. R., I-Hsiu, H., Xin, M., Ton-That, H., et al. (2011a). Structural differences between the *Streptococcus agalactiae* housekeeping and pilus-specific sortases: SrtA and SrtC1. *PloS One*, *6*(8), e22995.

Khare, B., Fu, Z.-Q., Huang, I.-H., Ton-That, H., & Narayana, S. V. L. (2011b). The crystal structure analysis of group B Streptococcus sortase C1: A model for the 'lid' movement upon substrate binding. *Journal of Molecular Biology*, *414*(4), 563–577.

Kline, K. A., Dodson, K. W., Caparon, M. G., & Hultgren, S. J. (2010). A tale of two pili: Assembly and function of pili in bacteria. *Trends in Microbiology*, *18*(5), 224–232.

Kruger, R. G., Otvos, B., Frankel, B. A., Bentley, M., Dostal, P., & McCafferty, D. G. (2004a). Analysis of the substrate specificity of the *Staphylococcus aureus* Sortase transpeptidase SrtA†. *Biochemistry*, *43*(6), 1541–1551.

Kruger, R. G., Dostal, P., & McCafferty, D. G. (2004b). Development of a high-performance liquid chromatography assay and revision of kinetic parameters for the *Staphylococcus aureus* sortase transpeptidase SrtA. *Analytical Biochemistry*, *326*(1), 42–48.

Lazzarin, M., Cozzi, R., Malito, E., Martinelli, M., D'Onofrio, M., Maione, D., et al. (2015). Noncanonical sortase-mediated assembly of pilus type 2b in group B Streptococcus. *The FASEB Journal*, *29*(11), 4629–4640.

van Leeuwen, H. C., Klychnikov, O. I., Menks, M. A. C., Kuijper, E. J., Drijfhout, J. W., & Hensbergen, P. J. (2014). *Clostridium difficile* sortase recognizes a (S/P)PXTG sequence motif and can accommodate diaminopimelic acid as a substrate for transpeptidation. *FEBS Letters*, *588*(23), 4325–4333.

Liew, C. K., Smith, B. T., Pilpa, R., Suree, N., Ilangovan, U., Connolly, K. M., et al. (2004). Localization and mutagenesis of the sorting signal binding site on Sortase A from *Staphylococcus aureus*. *FEBS Letters*, *571*(1–3), 221–226.

Lu, G., Qi, J., Gao, F., Yan, J., Tang, J., & Gao, G. F. (2011). A novel 'open-form' structure of sortase C from Streptococcus suis. *Proteins*, *79*(9), 2764–2769.

Mandlik, A., Swierczynski, A., Das, A., & Ton-That, H. (2008). Pili in Gram-positive bacteria: Assembly, involvement in colonization and biofilm development. *Trends in Microbiology*, *16*(1), 33–40.

Manzano, C., Contreras-Martel, C., El Mortaji, L., Izoré, T., Fenel, D., Vernet, T., et al. (2008). Sortase-mediated pilus fiber biogenesis in Streptococcus pneumoniae. *Structure (London, England: 1993)*, *16*(12), 1838–1848.

Manzano, C., Izoré, T., Job, V., Di Guilmi, A. M., & Dessen, A. (2009). Sortase activity is controlled by a flexible lid in the pilus biogenesis mechanism of Gram-positive pathogens. *Biochemistry*, *48*(44), 10549–10557.

Maresso, A. W., & Schneewind, O. (2008). Sortase as a target of anti-infective therapy. *Pharmacological Reviews*, *60*(1), 128–141.

Maresso, A. W., Chapa, T. J., & Schneewind, O. (2006). Surface protein IsdC and sortase B are required for heme-iron scavenging of Bacillus anthracis. *Journal of Bacteriology, 188*(23), 8145–8152.

Maresso, A. W., Wu, R., Kern, J. W., Zhang, R., Janik, D., Missiakas, D. M., et al. (2007). Activation of inhibitors by sortase triggers irreversible modification of the active site. *The Journal of Biological Chemistry, 282*(32), 23129–23139.

Marraffini, L. A., & Schneewind, O. (2005). Anchor structure of staphylococcal surface proteins V. Anchor structure of the sortase B substrate IsdC. *The Journal of Biological Chemistry, 280*(16), 16263–16271.

Marraffini, L. A., & Schneewind, O. (2006). Targeting proteins to the cell wall of sporulating Bacillus anthracis. *Molecular Microbiology, 62*(5), 1402–1417.

Marraffini, L. A., Ton-That, H., Zong, Y., Narayana, S. V. L., & Schneewind, O. (2004). Anchoring of surface proteins to the cell wall of *Staphylococcus aureus*. A conserved arginine residue is required for efficient catalysis of Sortase A. *The Journal of Biological Chemistry, 279*(36), 37763–37770.

Mazmanian, S. K., Liu, G., Ton-That, H., & Schneewind, O. (1999). *Staphylococcus aureus* sortase, an enzyme that anchors surface proteins to the cell wall. *Science, 285*(5428), 760–763.

Mazmanian, S. K., Ton-That, H., Su, K., & Schneewind, O. (2002). An iron-regulated sortase anchors a class of surface protein during *Staphylococcus aureus* pathogenesis. *Proceedings of the National Academy of Sciences of the United States of America, 99*(4), 2293–2298.

Mazmanian, S. K., Skaar, E. P., Gaspar, A. H., Humayun, M., Gornicki, P., Jelenska, J., et al. (2003). Passage of heme-iron across the envelope of *Staphylococcus aureus*. *Science (New York, NY)., 299*(5608), 906–909.

Moritsugu, K., Terada, T., & Kidera, A. (2012). Disorder-to-order transition of an intrinsically disordered region of sortase revealed by multiscale enhanced sampling. *Journal of the American Chemical Society, 134*(16), 7094–7101.

Naik, M. T., Suree, N., Ilangovan, U., Liew, C. K., Thieu, W., Campbell, D. O., et al. (2006). *Staphylococcus aureus* Sortase A transpeptidase. Calcium promotes sorting signal binding by altering the mobility and structure of an active site loop. *The Journal of Biological Chemistry, 281*(3), 1817–1826.

Neiers, F., Madhurantakam, C., Fälker, S., Manzano, C., Dessen, A., Normark, S., et al. (2009). Two crystal structures of pneumococcal pilus sortase C provide novel insights into catalysis and substrate specificity. *Journal of Molecular Biology, 393*(3), 704–716.

Pallen, M. J., Lam, A. C., Antonio, M., & Dunbar, K. (2001). An embarrassment of sortases—A richness of substrates? *Trends in Microbiology, 9*(3), 97–101.

Perry, A. M., Ton-That, H., Mazmanian, S. K., & Schneewind, O. (2002). Anchoring of surface proteins to the cell wall of *Staphylococcus aureus*. III. Lipid II is an in vivo peptidoglycan substrate for sortase-catalyzed surface protein anchoring. *The Journal of Biological Chemistry, 277*(18), 16241–16248.

Persson, K. (2011). Structure of the Sortase AcSrtC-1 from Actinomyces oris. *Acta Crystallographica. Section D, Biological Crystallography, 67*(Pt. 3), 212–217.

Popp, M. W.-L., & Ploegh, H. L. (2011). Making and breaking peptide bonds: Protein engineering using Sortase. *Angewandte Chemie International Edition, 50*(22), 5024–5032.

Proft, T. (2010). Sortase-mediated protein ligation: An emerging biotechnology tool for protein modification and immobilisation. *Biotechnology Letters, 32*(1), 1–10.

Race, P. R., Bentley, M. L., Melvin, J. A., Crow, A., Hughes, R. K., Smith, W. D., et al. (2009). Crystal structure of Streptococcus pyogenes Sortase A. *The Journal of Biological Chemistry, 284*(11), 6924–6933.

Ritzefeld, M. (2014). Sortagging: A robust and efficient chemoenzymatic ligation strategy. *Chemistry, 20*(28), 8516–8529.

Robson, S. A., Jacobitz, A. W., Phillips, M. L., & Clubb, R. T. (2012). Solution structure of the Sortase required for efficient production of infectious Bacillus anthracis spores. *Biochemistry, 51*(40), 7953–7963.
Ruzin, A., Severin, A., Ritacco, F., Tabei, K., Singh, G., Bradford, P. A., et al. (2002). Further evidence that a cell wall precursor [C(55)-MurNAc-(peptide)-GlcNAc] serves as an acceptor in a sorting reaction. *Journal of Bacteriology, 184*(8), 2141–2147.
Schmidt, M., Toplak, A., Quaedflieg, P. J., & Nuijens, T. (2017). Enzyme-mediated ligation technologies for peptides and proteins. *Current Opinion in Chemical Biology, 38*, 1–7.
Schmohl, L., & Schwarzer, D. (2014). Sortase-mediated ligations for the site-specific modification of proteins. *Current Opinion in Chemical Biology, 22*, 122–128.
Schneewind, O., & Missiakas, D. M. (2012). Protein secretion and surface display in Gram-positive bacteria. *Philosophical Transactions of the Royal Society of London. Series B, Biological Sciences, 367*(1592).
Schneewind, O., & Missiakas, D. (2014). Sec-secretion and sortase-mediated anchoring of proteins in Gram-positive bacteria. *Biochimica et Biophysica Acta, 1843*(8), 1687–1697.
Schneewind, O., Model, P., & Fischetti, V. A. (1992). Sorting of protein A to the staphylococcal cell wall. *Cell, 70*(2), 267–281.
Schneewind, O., Fowler, A., & Faull, K. F. (1995). Structure of the cell wall anchor of surface proteins in *Staphylococcus aureus*. *Science, 268*(5207), 103–106.
Shaik, M. M., Lombardi, C., Maragno Trindade, D., Fenel, D., Schoehn, G., Di Guilmi, A. M., et al. (2015). A structural snapshot of type II pilus formation in Streptococcus pneumoniae. *Journal of Biological Chemistry, 290*, 22581–22592. p. M115.647834.
Siegel, S. D., Liu, J., & Ton-That, H. (2016). Biogenesis of the Gram-positive bacterial cell envelope. *Current Opinion in Microbiology, 34*, 31–37.
Spirig, T., Weiner, E. M., & Clubb, R. T. (2011). Sortase enzymes in Gram-positive bacteria. *Molecular Microbiology, 82*(5), 1044–1059.
Suree, N., Jung, M. E., & Clubb, R. T. (2007). Recent advances towards new anti-infective agents that inhibit cell surface protein anchoring in *Staphylococcus aureus* and other Gram-positive pathogens. *Mini Reviews in Medicinal Chemistry*, 7(10), 991–1000.
Suree, N., Liew, C. K., Villareal, V. A., Thieu, W., Fadeev, E. A., Clemens, J. J., et al. (2009). The structure of the *Staphylococcus aureus* Sortase-substrate complex reveals how the universally conserved LPXTG sorting signal is recognized. *The Journal of Biological Chemistry, 284*(36), 24465–24477.
Suryadinata, R., Seabrook, S. A., Adams, T. E., Nuttall, S. D., & Peat, T. S. (2015). Structural and biochemical analyses of a *Clostridium perfringens* sortase D transpeptidase. *Acta Crystallographica. Section D, Biological Crystallography, 71*(7), 1505–1513.
Tian, B.-X., & Eriksson, L. A. (2011). Catalytic mechanism and roles of Arg197 and Thr183 in the *Staphylococcus aureus* Sortase A enzyme. *The Journal of Physical Chemistry. B, 115*(44), 13003–13011.
Ton-That, H., & Schneewind, O. (2003). Assembly of pili on the surface of Corynebacterium diphtheriae. *Molecular Microbiology, 50*(4), 1429–1438.
Ton-That, H., Faull, K. F., & Schneewind, O. (1997). Anchor structure of staphylococcal surface proteins. A branched peptide that links the carboxyl terminus of proteins to the cell wall. *The Journal of Biological Chemistry, 272*(35), 22285–22292.
Ton-That, H., Liu, G., Mazmanian, S. K., Faull, K. F., & Schneewind, O. (1999). Purification and characterization of sortase, the transpeptidase that cleaves surface proteins of *Staphylococcus aureus* at the LPXTG motif. *Proceedings of the National Academy of Sciences of the United States of America, 96*(22), 12424–12429.
Ton-That, H., Mazmanian, S. K., Alksne, L., & Schneewind, O. (2000). Anchoring of surface proteins to the cell wall of *Staphylococcus aureus*. Sortase catalyzed in vitro transpeptidation reaction using LPXTG peptide and NH2-GLY3 substrates. *The Journal of Biological Chemistry, 275*(13), 9876–9881.

Ton-That, H., Mazmanian, S. K., Alksne, L., & Schneewind, O. (2002). Anchoring of surface proteins to the cell wall of *Staphylococcus aureus*. Cysteine 184 and histidine 120 of sortase form a thiolate-imidazolium ion pair for catalysis. *The Journal of Biological Chemistry*, *277*(9), 7447–7452.

Tsukiji, S., & Nagamune, T. (2009). Sortase-mediated ligation: A gift from Gram-positive bacteria to protein engineering. *Chembiochem*, *10*(5), 787–798.

Voloshchuk, N., Liang, D., & Liang, J. (2016). Sortase A mediated protein modifications and peptide conjugations. *Current Drug Discovery Technologies*, *12*(4), 205–213.

Wallock-Richards, D. J., Marles-Wright, J., Clarke, D., Maitra, A., Dodds, M., Hanley, B., et al. (2015). Molecular basis of Streptococcus mutans Sortase A inhibition by the flavonoid natural product trans-chalcone. *Chemical Communications*, *51*, 10483–10485.

Weiner, E. M., Robson, S., Marohn, M., & Clubb, R. T. (2010). The Sortase A enzyme that attaches proteins to the cell wall of Bacillus anthracis contains an unusual active site architecture. *The Journal of Biological Chemistry*, *285*(30), 23433–23443.

Wu, Z., & Guo, Z. (2012). Sortase-mediated transpeptidation for site-specific modification of peptides, glycopeptides, and proteins. *Journal of Carbohydrate Chemistry*, *31*(1), 48–66.

Wu, C., Mishra, A., Reardon, M. E., Huang, I.-H., Counts, S. C., Das, A., et al. (2012). Structural determinants of Actinomyces sortase SrtC2 required for membrane localization and assembly of type 2 fimbriae for interbacterial coaggregation and oral biofilm formation. *Journal of Bacteriology*, *194*(10), 2531–2539.

Zhang, R., Wu, R., Joachimiak, G., Mazmanian, S. K., Missiakas, D. M., Gornicki, P., et al. (2004). Structures of sortase B from *Staphylococcus aureus* and Bacillus anthracis reveal catalytic amino acid triad in the active site. *Structure*, *12*(7), 1147–1156.

Zong, Y., Bice, T. W., Ton-That, H., Schneewind, O., & Narayana, S. V. L. (2004a). Crystal structures of *Staphylococcus aureus* Sortase A and its substrate complex. *The Journal of Biological Chemistry*, *279*(30), 31383–31389.

Zong, Y., Mazmanian, S. K., Schneewind, O., & Narayana, S. V. (2004b). The structure of Sortase B, a cysteine transpeptidase that tethers surface protein to the *Staphylococcus aureus* cell wall. *Structure*, *12*(1), 105–112.

CHAPTER EIGHT

Biology, Mechanism, and Structure of Enzymes in the α-D-Phosphohexomutase Superfamily

Kyle M. Stiers, Andrew G. Muenks, Lesa J. Beamer[1]
University of Missouri, Columbia, MO, United States
[1]Corresponding author: e-mail address: beamerl@missouri.edu

Contents

1. Introduction 266
2. Biology 267
  2.1 Subgroups of the Superfamily 267
  2.2 Roles in Bacterial Pathogenesis 274
  2.3 Human Paralogs and Inherited Disease 278
3. Catalytic Mechanism 280
  3.1 Overview 280
  3.2 The "Flip" 281
  3.3 Acid–Base Catalysis 283
4. Three-Dimensional Structure 284
  4.1 Structures in the PDB 284
  4.2 Overall Architecture 287
  4.3 Key Regions of Active Site 289
  4.4 Quaternary Structure and Conformational Mobility of Domain 4 293
5. Conclusions and Future Directions 295
Acknowledgment 296
References 296

Abstract

Enzymes in the α-D-phosphohexomutases superfamily catalyze the reversible conversion of phosphosugars, such as glucose 1-phosphate and glucose 6-phosphate. These reactions are fundamental to primary metabolism across the kingdoms of life and are required for a myriad of cellular processes, ranging from exopolysaccharide production to protein glycosylation. The subject of extensive mechanistic characterization during the latter half of the 20th century, these enzymes have recently benefitted from biophysical characterization, including X-ray crystallography, NMR, and hydrogen–deuterium exchange studies. This work has provided new insights into the unique

Advances in Protein Chemistry and Structural Biology, Volume 109
ISSN 1876-1623
http://dx.doi.org/10.1016/bs.apcsb.2017.04.005

catalytic mechanism of the superfamily, shed light on the molecular determinants of ligand recognition, and revealed the evolutionary conservation of conformational flexibility. Novel associations with inherited metabolic disease and the pathogenesis of bacterial infections have emerged, spurring renewed interest in the long-appreciated functional roles of these enzymes.

1. INTRODUCTION

The α-D-phosphohexomutases (PHMs) comprise an ancient and ubiquitous enzyme superfamily, and are found in all three kingdoms of life. These proteins play key roles in many cellular processes, including carbohydrate metabolism, sugar nucleotide biosynthesis, as well as glycolipid, exopolysaccharide, and peptidoglycan synthesis. All organisms, even simple bacteria, appear to encode at least one member of the superfamily in their genomes. Enzymes in the superfamily share highly conserved catalytic residues and a four-domain structural organization. Nevertheless, overall sequence diversity within the superfamily is high, reflecting its ancient evolutionary origins, and consistent with adaptation of these enzymes for widely varied biological roles (Whitehouse, Tomkins, Lovegrove, Hopkinson, & McMillian, 1998).

Enzymes in the PHM superfamily are typically cytoplasmic and catalyze a reversible, intramolecular phosphoryl transfer on their phosphosugar substrates (Shackelford, Regni, & Beamer, 2004). Often this interconverts the 1- and 6-phospho forms of hexoses

$$\text{sugar 1} - \text{phosphate} \overset{Mg^{2+}}{\rightleftharpoons} \text{sugar 6} - \text{phosphate},$$

although one branch of the family prefers pentose phosphosugars as substrates (Maliekal, Sokolova, Vertommen, Veiga-da-Cunha, & Van Schaftingen, 2007). Active enzyme is phosphorylated on a conserved serine residue and binds a divalent metal ion, usually Mg^{2+}. As the name implies, enzymes in the superfamily are highly specific for the conversion of the α-anomers of D-phosphosugars and are structurally and mechanistically unrelated to the β-D-phosphoglucomutases, which are part of the haloacid dehalogenase superfamily (Burroughs, Allen, Dunaway-Mariano, & Aravind, 2006).

Unlike some other enzyme superfamilies, where the same structural scaffold has been adapted for differing catalytic mechanisms (Rauwerdink & Kazlauskas, 2015), the sequence diversity of the PHMs is primarily associated with different preferences/specificities for their phosphosugar substrates. Four major subgroups are distinguished that utilize the related phosphohexoses of glucose, mannose, glucosamine, and *N*-acetylglucosamine (Shackelford et al., 2004). The diversity of these phosphosugar substrates/products allows these enzymes to participate in a multitude of metabolic pathways in microorganisms, plants, and animals.

Here, we provide an overview of the biology, mechanism, and structure of this widespread enzyme family, with an emphasis on recent studies. In addition to intriguing mechanistic features, several proteins in the superfamily have emerged as model systems for understanding fundamental relationships between protein structure and function, including the role of conformational change and protein flexibility in catalysis. Moreover, recent biological and clinical studies have shed new light on the importance of PHMs to human health and disease, in bacterial pathogenesis, and for efforts such as metabolic engineering.

2. BIOLOGY

The biological roles of the PHMs are as varied as the organisms in which they are found, and too numerous to describe comprehensively. Specific functional roles are determined by their substrate preferences, which in turn define the various subgroups of the superfamily. Section 2.1 presents an overview of the subgroups and is followed by highlights of two topics: known roles for these enzymes in bacterial pathogenesis (Section 2.2) and the five PHM paralogs in the human genome and their relevance to inherited disease (Section 2.3).

2.1 Subgroups of the Superfamily

The four main subgroups of the PHM superfamily are traditionally distinguished according to their substrate preferences (Table 1). Four related phosphosugars are utilized as substrates: glucose, mannose, glucosamine, and *N*-acetylglucosamine (Shackelford et al., 2004). These sugars are structurally similar, differing only in the substituent at the 2′-position of the sugar ring (Fig. 1A). Enzymes in the phosphoglucomutase (PGM) subgroup have high specificity for glucose-based phosphosugar substrates, while those in the phosphomannomutase/phosphoglucomutase (PMM/PGM) subgroup have dual substrate specificity for glucose and mannose phosphosugars. The phosphoglucosamine mutase (PNGM) subgroup uses glucosamine-based phosphosugars, while the *N*-acetylglucosamine phosphomutases (PAGM) prefer *N*-acetylglucosamine phosphosugars. While each subgroup has

Table 1 Summary of Known Subgroups in the PHM Superfamily

Subgroup (EC No.)	Name	Lineage	Preferred Substrate(s)	Functional Roles	Typical Length (No. Residues)	Sugar-Binding Loop Sequence Motif[a]
PGM (5.4.2.2)	Phosphoglucomutase	Bacteria, eukaryotes	G1P, G6P	Glucose homeostasis, protein N-glycosylation	550	-GEESF-
PMM/PGM (5.4.2.8)	Phosphomannomutase/ phosphoglucomutase	Bacteria	G1P, G6P, M1P, M6P	Polysaccharide/EPS biosynthesis	450	-GEMSG-
PNGM (5.4.2.10)	Phosphoglucosamine mutase	Bacteria	GlcN-1P, GlcN-6P	UDP-GlcNAc pathway, peptidoglycan	450	-GE(n/q)SG-
PAGM (5.4.2.3)	Phosphoacetylglucosamine mutase	Eukaryotes	GlcNAc-6P, GlcNAc-1P	Synthesis of UDP-GlcNAc; nucleotide sugar biosynthesis	550	-FEANG-
PGM2 (EC 5.4.2.7)	Phosphopentomutase	Animals, fungi	R1P, R5P	Pentose phosphate synthesis/ breakdown	600+	-FEEAIG-
PGM2L1 (2.7.1.106)	Glucose 1,6-bisphosphate synthase	Vertebrates	G1P, G6P, 1, 3-bisphosphoglycerate	Synthesis of glucose 1,6-bisphosphate	600+	NA

[a]Residues known or predicted to be involved in direct contacts with the sugar (typically substituents at the 3′- and 4′- position) are *underlined*.

Abbreviations: EPS, extracellular polymeric substances; G1P, glucose 1-phosphate; G6P, glucose 6-phosphate; GlcN-1P, glucosamine 1-phosphate; GlcN-6P, glucosamine 6-phosphate; GlcNAc-1P, *N*-acetylglucosamine 1-phosphate; GlcNAc-6P, *N*-acetylglucosamine 6-phosphate; M1P, mannose 1-phosphate; M6P, mannose 6-phosphate; NA, not available; R1P, ribose 1-phosphate; R5P, ribose 5-phosphate.

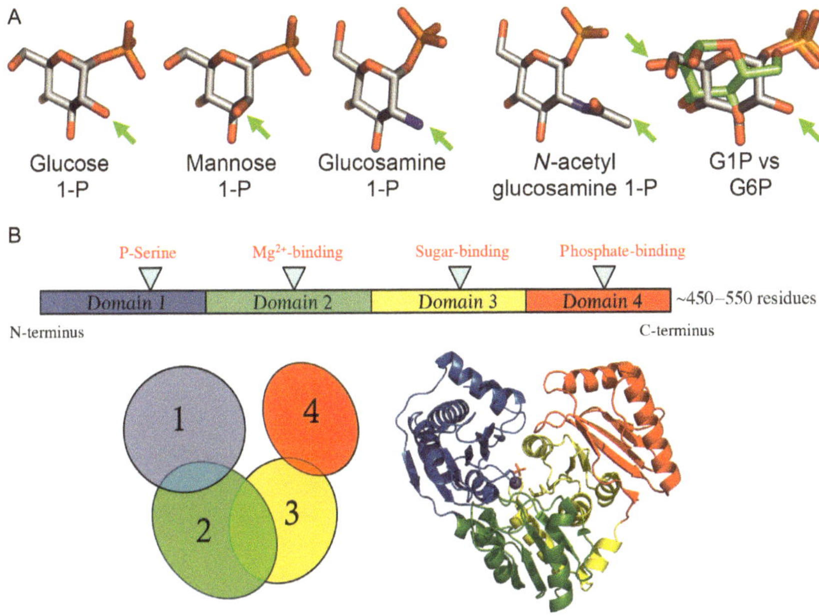

Fig. 1 Overview of phosphosugar substrates and structural architecture of the PHMs. (A) *Stick figures* of the 1-phospho forms of various substrates of the α-D-phosphohexomutases, showing how structural differences are all localized at the 2-position of the sugar (*green arrow*). Also shown is a superposition of glucose 1-phosphate (*white*) and glucose 6-phosphate (*green*) illustrating the two different binding orientations (related by ~180 degree rotation of the sugar ring) for 1- vs 6-phosphosugars. Superposition based on crystal structures of enzyme–ligand complexes of *P. aeruginosa* PMM/PGM (Regni et al., 2004). (B) Schematics of the 4-domain architecture of the PHMs, highlighting the locations of key catalytic and ligand-binding residues. The enzymes are typically 450–550 amino acids in length.

distinct substrate preferences, overlap in substrate usage between the subgroups is also apparent. For example, in addition to the PGMs, it appears that enzymes in the other three subgroups are also able to utilize glucose-based phosphosugars as substrates, at least to some extent (Boles, Liebetrau, Hofmann, & Zimmermann, 1994; Jolly et al., 1999; Kato, Mueller, Wessely, Lan, & Christensen, 2005) (Section 4.3.6).

Categorization of proteins in the superfamily into one of the four subgroups can be challenging, especially without supporting biochemical data. However, as more sequences for the superfamily have become available and more proteins have been studied, some sequence and structural features have begun to emerge that can assist with categorization. One of these is the sugar-binding loop, a key region in the active site of the PHMs, along with the catalytic phosphoserine and metal-binding loop (Figs. 1B and 2). Unlike

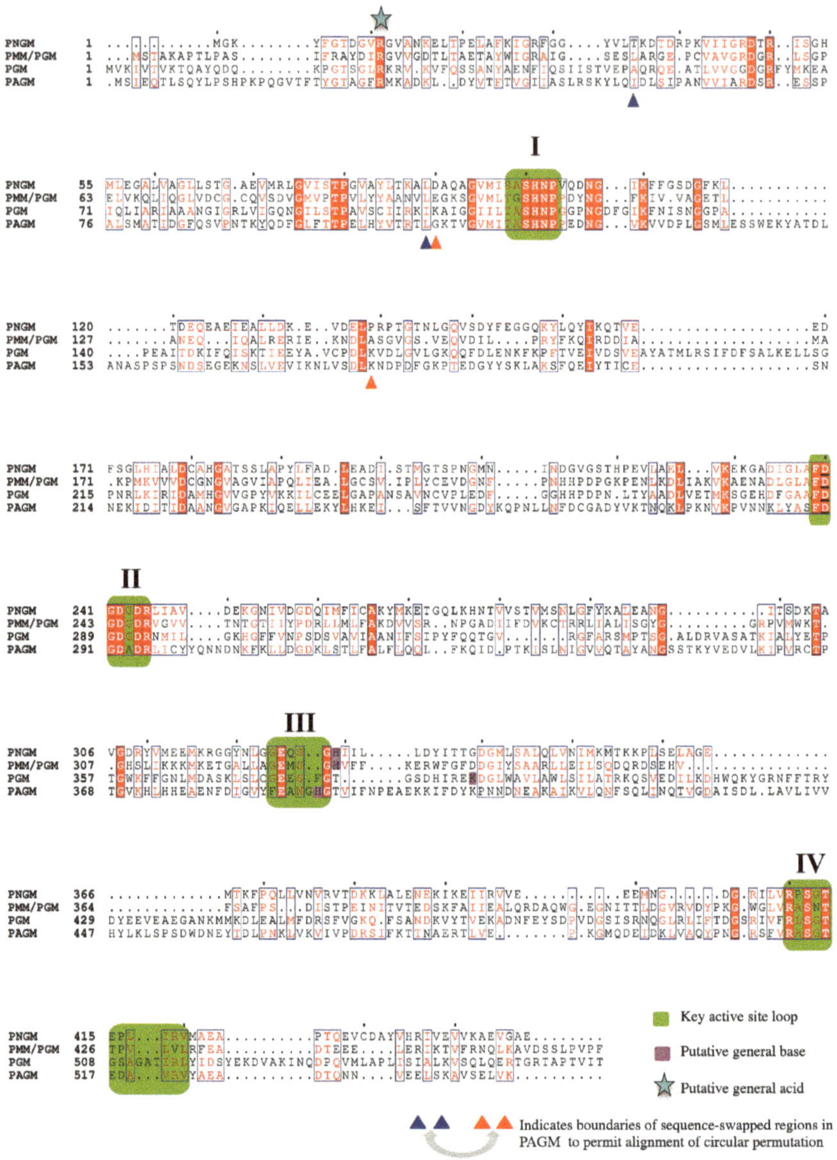

Fig. 2 A multiple sequence alignment of one representative from each of the four major PHM subgroups. Sequences shown are from *B. anthracis* PNGM, *P. aeruginosa* PMM/PGM, human PGM1, and *C. albicans* PAGM. The four (I–IV) highly conserved active site loops are highlighted (Section 4.3). Due to the circular permutation of the PAGM proteins relative to the others, the corresponding regions of sequence were swapped to permit the best sequence alignment (sequence between *orange triangles* was swapped with sequence between *blue triangles*). See figure key for other symbols.

the latter two regions, which are nearly identical across the superfamily, the sugar-binding loop tends to vary in sequence according to subgroup (Table 1) due to contacts between residues in this loop and the sugar ring of the substrates (Section 4.3.3). Phylogenetic distribution and sequence length can also help differentiate among the subgroups. A discussion of these distinguishing features is included in Sections 2.1.1–2.1.4.

Several other less well-characterized subgroups in the PHM family are also known, including the eukaryotic phosphopentomutases (not to be confused with the phosphopentomutases related to alkaline phosphatase; Galperin, Koonin, & Bairoch, 1998) and the glucose 1,6-bisphosphate synthases. More information on these two subgroups is found in Section 2.3. A number of other superfamily members that either lack catalytic activity (Moiseeva, Belkin, Spurr, Koteliansky, & Critchley, 2004; Müller et al., 2002) or are highly divergent in sequence have been reported (Mitra, Cui, Robbins, & Samuelson, 2010), but are not discussed further herein.

2.1.1 Phosphoglucomutases

The PGM subgroup is the most widely distributed evolutionarily, with representatives easily identified in bacteria and eukaryotes, including plants and animals. PGMs appear to be quite specific for glucose, even relative to the closely related mannose, which differs only in the stereochemistry of its 2′-hydroxyl group (Fig. 1A). Rabbit PGM, for example, shows a 6000-fold preference for glucose phosphosugars over their mannose counterparts (Lowry & Passonneau, 1969). Along with a preference for glucose-based substrates, other features common to PGM enzymes are a sequence length of ~550 residues and the -GEESF- sequence motif of the sugar-binding loop (Table 1; Fig. 2).

The role of PGM in glucose metabolism has been appreciated for decades. In humans, the enzyme PGM1 is the key regulator of cellular glucose homeostasis (Fig. 3A; Section 2.3). In higher plants, such as *Arabidopsis*, two (or more) isoforms of PGM exist. The plastidial isoform is essential for starch synthesis, while cytosolic isoform(s) contributes to the syntheses of sucrose and cell wall components (Malinova et al., 2014). In bacteria, PGM is often also a contributor to cell wall and polysaccharide capsular production (Section 2.2).

Our mechanistic understanding of PGM is derived from studies of the enzyme from rabbit muscle (97% identical to human PGM1), which was extensively characterized by Ray and coworkers over several decades (see Ray, Burgner, & Post, 1990 and references therein). These studies remain

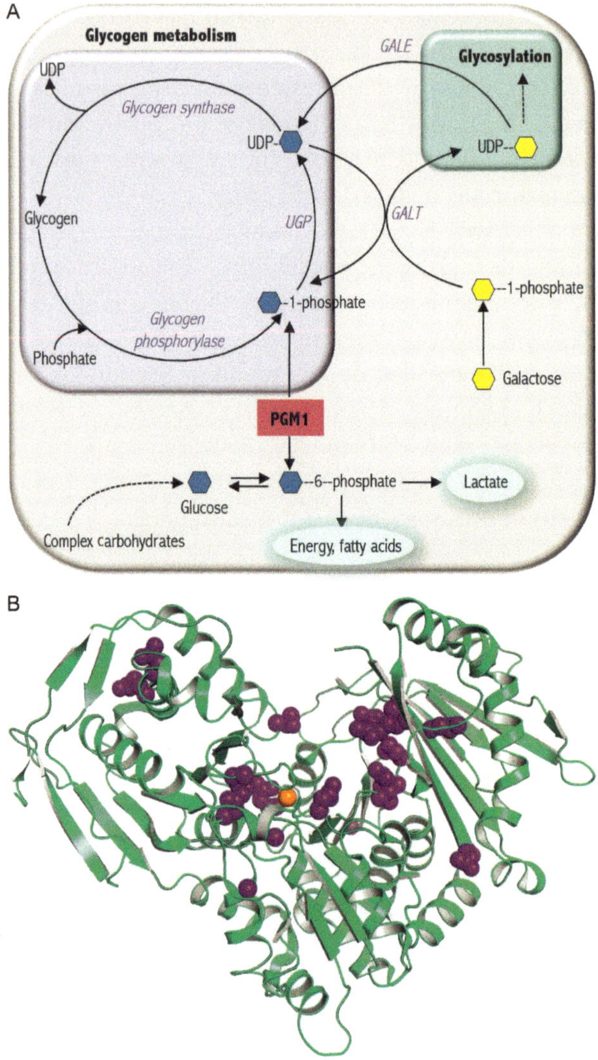

Fig. 3 Metabolic roles and disease-associated variants of human PGM1. (A) The pivotal position of PGM1 determining the storage of glucose as glycogen or its utilization for energy. Also shown is the connection of PGM1 to protein glycosylation, through galactose metabolism. *GALE*, UDP-galactose-4-epimerase; *GALT*, galactose 1-phosphate uridylyltransferase. (B) *Ribbon diagram* of human PGM1 illustrating residues affected by known missense variants (*purple*) associated with PGM1 deficiency. The bound metal ion in the active site is shown as an *orange sphere*.

the foundation for understanding enzyme mechanism of the entire superfamily (Section 3). More limited kinetic studies of PGMs from other sources have also been conducted, including from *Escherichia coli* (Lu & Kleckner, 1994), *Acetobacter xylinum* (Brautaset, Petersen, & Valla, 2000), *Entamoeba histolytica*

(Ortner, Binder, Scheiner, Wiedermann, & Duchêne, 1997), wheat (Davies, Tetlow, Bowsher, & Emes, 2003), *Arabidopsis* (Periappuram et al., 2000), and maize (Manjunath, Lee, VanWinkle, & Bailey-Serres, 1998).

2.1.2 Phosphomannomutase/Phosphoglucomutase

PMM/PGMs are primarily found in bacteria, although archaeal representatives have also been identified (Akutsu et al., 2005; Ray et al., 2005). These proteins can utilize both glucose- and mannose-based substrates with approximately equal efficiency (Naught & Tipton, 2001) and are usually ~450 amino acids in length. The sugar-binding loop of PMM/PGMs often contains a -GEMSG- sequence motif (Table 1; Fig. 2).

Due to their dual substrate specificity, PMM/PGMs can serve multiple metabolic roles. This is perhaps best exemplified by the case of PMM/PGM from the opportunistic human pathogen, *Pseudomonas aeruginosa*. The product of the AlgC gene, PMM/PGM from this organism plays multiple roles in rhamnolipid, lipopolysaccharide (LPS), and alginate biosynthesis, as well as in the production of Psl and Pel, two critical biofilm matrix exopolysaccharides (Govan & Deretic, 1996; Ma et al., 2012; Rocchetta, Burrows, & Lam, 1999; Ye, Zielinski, & Chakrabarty, 1994). *P. aeruginosa* PMM/PGM provides the common precursor mannose 1-phosphate for alginate and B-band LPS, while providing glucose 1-phosphate for the synthesis of rhamnolipid and the LPS-core. PMM/PGMs from other organisms are similarly involved in the synthesis of polysaccharides and other extracellular polymeric substances (Akutsu et al., 2005; Kim et al., 2003).

Curiously, although PMM/PGMs can effectively utilize both glucose and mannose-based substrates, many bacteria have both a PGM and a PMM/PGM (e.g., for *E. coli* the pgm and manB genes), suggesting either a benefit from this functional redundancy or that these enzymes occupy different biosynthetic niches within the organism.

2.1.3 Phosphoglucosamine Mutase

Enzymes in the PNGM subgroup are generally found in bacteria, although some archaeal proteins also appear to belong to this subgroup (Namboori & Graham, 2008). These proteins are also typically ~450 residues in length and have a -GE(N/Q)SG- sequence motif in their sugar-binding loop (Table 1; Fig. 2) (Mehra-Chaudhary, Mick, & Beamer, 2011).

PNGMs participate in the cytoplasmic portion of the UDP-*N*-acetylglucosamine biosynthetic pathway, which provides critical precursors for peptidoglycan biosynthesis (Barreteau et al., 2008). Although eukaryotes also synthesize UDP-*N*-acetylglucosamine, the bacterial pathway differs from the eukaryotic one, making these proteins attractive targets for antimicrobial design (Barreteau et al., 2008; Gautam, Vyas, & Tewari, 2011).

Overall, the PNGMs are less well studied than the other subgroups of the superfamily. A few have been biochemically characterized, including proteins from *E. coli* (Jolly et al., 1999), *Helicobacter pylori* (De Reuse, Labigne, & Mengin-Lecreulx, 1997), *Staphylococcus aureus* (Jolly et al., 1997), *Streptococcus gordonii* (Shimazu et al., 2008), *Methanococcus maripaludis* (Namboori & Graham, 2008), and *Mycobacterium tuberculosis* (Li et al., 2012).

2.1.4 Phospho-N-Acetylglucosamine Mutase

The PAGM enzymes are exclusively eukaryotic and typically 550 residues in length. They share the highly conserved catalytic and ligand-binding residues of the PHM superfamily but have a subgroup specific circular permutation within domain 1 of the protein (Nishitani et al., 2006). This can make them more difficult to identify as a member of the superfamily based on sequence searches. The sequence motif of the sugar-binding loop in the PAGMs is distinct: -FEANG- (Bandini et al., 2012; Kato et al., 2005) (Table 1; Fig. 2), with the asparagine reflecting a role in recognizing the *N*-acetyl group of the 1-phospho form of the sugar (Nishitani et al., 2006).

Like the PNGMs, these enzymes are also involved in the synthesis of UDP-*N*-acetylglucosamine, via the distinct eukaryotic pathway (Barreteau et al., 2008). Essential in organisms from fungi to mice (Fang et al., 2013; Greig et al., 2007; Hofmann, Boles, & Zimmermann, 1994), PAGM has been suggested as a target for design of antifungal agents. In humans, PAGM is known as PGM3, where it functions in the hexosamine biosynthetic pathway and is associated with an inherited enzyme deficiency (Section 2.3).

Only a few PAGM enzymes have been biochemically characterized, including the proteins from *Saccharomyces cerevisiae* (Boles et al., 1994; Hofmann et al., 1994), *Aedes aegypti* (Kato et al., 2005), mice (Greig et al., 2007), and humans (Mio, Yamada-Okabe, Arisawa, & Yamada-Okabe, 2000).

2.2 Roles in Bacterial Pathogenesis

Bacterial proteins in the PHM superfamily are often involved in the biosynthesis of exopolysaccharides and other exoproducts, such as glycolipids. Because of the importance of these molecules to cell wall properties, capsule production, and biofilm formation, many members of the PHM superfamily are implicated in bacterial pathogenesis (Table 2). The organisms listed

Table 2 Roles of PHM Superfamily Members in Bacterial Pathogenesis

	Organism	Enzyme	Role	References
Gram positive	*N. gonorrhoeae*	PMM/PGM	Lipooligosaccharide biosynthesis, pathogenesis	Sandlin and Stein (1994) and Zhou et al. (1994)
	N. meningitidis	PMM/PGM	Lipooligosaccharide biosynthesis, vaccine target	Monteiro, Fortuna-Nevin, Farley, and Pavliak (2003) and Zhou et al. (1994)
	S. aureus	PGM	Cell wall envelope, antibiotic resistance	Gründling and Schneewind (2007)
	S. aureus	PNGM	Antibiotic resistance	Jolly et al. (1997)
	S. gordonii	PGM	Fitness, cell wall properties	Bizzini et al. (2007)
	S. gordonii	PNGM	Growth, biofilm formation, antibiotic sensitivity	Shimazu et al. (2008), Shimazu, Takahashi, Karibe, Mitsuhashi, and Konishi (2012), and Yajima et al. (2009)
	S. iniae	PGM	Virulence, vaccine target	Buchanan et al. (2005)
	S. mutans	PNGM	Reduced growth rate, biofilm attenuation, increased autolysis	Liu, Duan, and Guo (2009)
	S. pneumoniae	PGM	Virulence, capsule production	Hardy, Caimano, and Yother (2000) and Hardy, Magee, Ventura, Caimano, and Yother (2001)
	S. coelicolor	PMM/PGM	Chloramphenicol sensitivity; reduced exp. drug efflux pumps	Rajesh et al. (2012)
	M. tuberculosis	PMM/PGM	Host–pathogen interactions	McCarthy et al. (2005)

Continued

Table 2 Roles of PHM Superfamily Members in Bacterial Pathogenesis—cont'd

	Organism	Enzyme	Role	References
	M. smegmatis	PNGM	Essential	Kang et al. (2013) and Li et al. (2012)
Gram negative	*A. baumannii*	PMM/PGM	Biofilm formation	Sahu, Iyer, Barage, Sonawane, and Chopade (2014)
	B. bronchiseptica	PMM/PGM	Virulence, sensitivity to oxidative stress, and antimicrobials	West et al. (2000)
	B. abortus	PGM	Virulence, resistance to complement, cytotoxicity	Pei and Ficht (2004), Ugalde, Comerci, Leguizamón, and Ugalde (2003), and Ugalde, Czibener, Feldman, and Ugalde (2000)
	B. melitensis	PMM/PGM	Establishment of chronic infections	Paixão et al. (2009)
	B. suis	PGM	Virulence, vaccine candidate	Czibener, Del Giudice, Spera, Fulgenzi, and Ugalde (2016)
	E. coli	PGM	Cell morphology, microbicide sensitivity	Lu and Kleckner (1994)
	E. coli	PNGM	Essential for viability	Mengin-Lecreulx and Heijenoort (1996)
	H. influenzae	PGM	Glycosylation, bacterial adherence to host	Grass et al. (2003) and Swords et al. (2000)
	H. pylori	PNGM	Essential for viability	De Reuse et al. (1997)
	P. aeruginosa	PMM/PGM	Virulence, growth, clearance from host	Goldberg, Coyne, Neely, and Holder (1995) and Tang et al. (1996)
	P. aeruginosa	PNGM	Essential	Tavares, Leitão, and Sá-Correia (2003)
	S. typhimurium	PGM	Fitness, antibiotic resistance	Paterson, Cone, Peters, and Maskell (2009)
	S. typhimurium	PNGM	Viability	Rathi, Sarangi, and Trivedi (2009)
	S. maltophilia	PMM/PGM	Virulence, antibiotic resistance	Liaw, Lee, and Hsueh (2010) and McKay, Woods, MacDonald, and Poole (2003)
	V. furnissii	PMM/PGM	Virulence, O-antigen polysaccharide synthesis	Kim et al. (2003)
	V. cholerae	PMM/PGM	Colonization	Chiang and Mekalanos (1999)
	Y. pestis	PGM	Decreased autoaggregation, reduced resistance to polymyxin B	Felek et al. (2010)

include many notorious human pathogens, such as *M. tuberculosis*, *Neisseria gonorrhoeae*, and *S. aureus*. Effects of these enzymes on the organism were assessed through gene deletion or disruption studies, and demonstrate roles in pathogenesis for three subgroups of the PHM superfamily (PGM, PNGM, and PMM/PGM). In addition, several bacterial strains with deletions of PGM or PMM/PGM are of interest as vaccine candidates due to their attenuated infectivity (Buchanan et al., 2005; Czibener et al., 2016; Ugalde et al., 2003, 2000).

Changes to the bacteria caused by lack of PHM activity often manifest as reduced virulence, slowed growth, increased susceptibility to antibiotics, or differences in cell morphology (Table 2). These important roles in infectivity and pathogenesis suggest that enzymes in the family could be targets for antimicrobial inhibitor design, a possibility that has not yet been realized. Effective enzyme inhibitors would be expected to mimic the effects seen in Table 2. Enzymes in the PNGM subgroup are often essential for viability in Gram-negative organisms (e.g., *E. coli*, *H. pylori*, *P. aeruginosa*, etc.), making them particularly attractive targets for inhibitor design (Barreteau et al., 2008; Gautam et al., 2011). In the case of antibiotic-resistant bacterial strains, inhibitors of the PHMs might also be helpful as adjuvants to traditional antibiotic therapy (Wright, 2016).

Clinically useful inhibitors would need to avoid compromising the activity of human PHMs (Section 2.3). This challenge might be overcome by exploiting the substrate preferences of the subgroups. For example, while inhibitors of bacterial PGM enzymes would likely also affect human PGM, the possibility of designing inhibitors specific to PMM/PGM or PNGMs is attractive, given the absence of these subgroups in humans. In such efforts, the structural differences between glucose and mannose, or between glucose and glucosamine, might be coopted for inhibitor design to provide specificity for enzymes in these two subgroups.

It is worth noting that a full appreciation for the roles of these bacterial enzymes, even from gene deletion studies as in Table 2, is likely clouded by the existence of multiple members of the PHM superfamily in bacterial genomes. As noted earlier, *E. coli* has genes for a PGM and PMM/PGM; it also encodes for a PNGM (*glmM*). A similar situation exists for the human pathogen *P. aeruginosa*, which has genes for PGM (*pgm*), PMM/PGM (*AlgC*), and PNGM (*GlmM*). The possibility of partial functional substitution (e.g., catalysis with nonpreferred substrates) between subgroups in the superfamily has not been systematically addressed.

2.3 Human Paralogs and Inherited Disease

2.3.1 *PGM1*

The human genome encodes five members of the PHM superfamily. PGM1 is the major source of PGM activity in most human tissues, particularly in liver and muscles where the metabolic demand is greatest. This enzyme is well known for its role in regulating glucose homeostasis in cells. It acts as the metabolic pivot point between glucose storage in the form of glycogen and glucose utilization for energy (Fig. 3A). New insights into the importance of human PGM1 in metabolism have been gained due to its recent association with an inherited metabolic disease, PGM1 deficiency. In 2014, a landmark study (Tegtmeyer et al., 2014) showed that patients with mutations of PGM1 have characteristics of both glycogen storage disease and a congenital disorder of glycosylation (CDG) of types I and II (affecting both sugar processing and assembly). Patient phenotypes are variable, even when comparing siblings with the same genotype, but can include dilated cardiomyopathy, hepatopathy, muscle weakness and exercise intolerance, hypoglycemia, delayed puberty, and developmental abnormalities of the head such as cleft palate (Tegtmeyer et al., 2014). These pleiotropic effects are consistent with the critical roles of human PGM1 in metabolism and its widespread expression in tissues. Given this, it is perhaps surprising that some patients with PGM1 deficiency have only mild symptoms (Stojkovic et al., 2009). This may be due, at least in part, to multiple paralogs of PGM1 in the human genome, as discussed later.

Mutations of various types have been associated with PGM1 deficiency, which is autosomal recessive in inheritance. These include nonsense, splicing, and missense variants. In the latter category, ~20 residues out of 562 in the protein are known to have disease-associated mutations (Fig. 3B), but this number is expected to increase as more patients are screened for the condition. Biochemical and structural analyses of PGM1 missense variants have revealed varying origins for enzyme dysfunction, including effects on both enzyme activity and protein folding (Beamer, 2015; Lee, Stiers, Kain, & Beamer, 2014). X-ray crystallographic studies of several disease-related mutations of PGM1 have provided new insights into their effects on enzyme mechanism and structure, and may aid in understanding the complicated disease phenotype (Stiers, Graham, Kain, & Beamer, 2017; Stiers, Kain, Graham, & Beamer, 2016).

2.3.2 *PGM2, -2L1, -3, and -5*

In addition to PGM1, four other proteins in the PHM superfamily are found in humans, known as PGM2, PGM2L1, PGM3, and PGM5. Based on sequence alignments, each of these paralogs is easily categorized as a member of the superfamily but has unique features relative to PGM1. These include tissue-dependent differences in expression, as well as differences in substrate preferences and, in one case, a distinct catalytic reaction. Compared to PGM1, all of these proteins are understudied, but recent efforts have begun to unravel their distinct functionalities and cellular roles.

PGM2 and PGM2L1 (for PGM2-like 1) have been identified as a phosphopentomutase and glucose 1,6-bisphosphate synthase, respectively (Maliekal et al., 2007). These two proteins are found primarily in mammals and are closely related in sequence to each other—60% sequence identity—yet share only 20% sequence identity with PGM1. PGM2 has a preference for pentose phosphosugars, including ribose 1-phosphate and deoxyribose 1-phosphate, and its highest expression levels are found in lung, spleen, and thymus. It contains a conserved -FEEAIG- motif in its sugar-binding loop (Whitehouse et al., 1998). PGM2L1 utilizes 1,3-bisphoglycerate and glucose 1- or 6-phosphate to form glucose 1,6-bisphosphate. This reaction is essentially one-half of the two-step phosphoryl transfer reaction catalyzed by other members of the superfamily. PGM2L1 is highly expressed in the brain, where glucose 1,6-bisphosphate may play a regulatory role. Both PGM2 and PGM2L1 also have PGM activity, although at a lower level than with their preferred substrates.

Human PGM3 belongs to the PAGM subgroup of the superfamily and thus has a preference for *N*-acetylglucosamine phosphosugars as substrates. It shares ~20% sequence identity with PGM1 and has relatively high expression in pancreas, heart, liver, and placenta (Li, Rodriguez, & Banerjee, 2000). PGM3 is an essential enzyme in the hexosamine biosynthetic pathway required for the production of UDP-GlcNAc (Willems, van Engelen, & Lefeber, 2016). Mutations in PGM3 were recently identified as causing an autosomal recessive immunodeficiency syndrome, combined with CDG phenotypic features (Lundin et al., 2015; Sassi et al., 2014; Stray-Pedersen et al., 2014; Zhang et al., 2014). As is the case with PGM1 deficiency, patient phenotypes in PGM3 deficiency are variable and complex.

Of the four PGM1 paralogs, PGM5 is mostly closely related in sequence (65% identity) to PGM1. It is reported to lack enzyme activity (Moiseeva et al., 2004) and has been most studied for its role as a component of the peripheral membrane cytoskeleton in muscle and some nonmuscle cells (Belkin et al., 1994). Its expression is enriched in cardiac, smooth, and skeletal muscle, where it is developmentally regulated. Recently, a detailed study (Molt et al., 2014) established a role for PGM5 (also known as aciculin) in myofibril assembly, remodeling, and maintenance and described its interactions with filamin C (FLNc) and the actin-binding repeat-containing proteins XinA and XinB. PGM5 has also been identified as a binding partner for dystrophin, the protein involved in inherited muscle dystrophy, in skeletal muscle (Belkin & Burridge, 1995).

3. CATALYTIC MECHANISM

3.1 Overview

The sequence conservation of the key catalytic and ligand-binding residues in the PHMs (Fig. 2) suggests that the intramolecular phosphotransfer reaction is highly conserved. To date, with the exception of the glucose 1,6-bisphosphate synthases, it appears that all superfamily members may utilize the multistep reaction first proposed and characterized by Ray and coworkers, based on studies of PGM from rabbit muscle. An overview of the mechanism is illustrated in Fig. 4 with glucose 6-phosphate as substrate. (As the reaction is highly reversible, the 1- and 6-phospho forms of the sugars can be considered as either substrate or product.) The reaction proceeds through the following steps: (1) binding of glucose 6-phosphate to the active, phospho-enzyme; (2) transfer of the enzyme phosphoryl group from the catalytic phosphoserine to the 1-position of the substrate, creating the bis-phosphorylated sugar intermediate and inactive dephospho-enzyme; (3) diffusional reorientation of the intermediate in the active site placing the phosphoryl group at position 6 near the active site serine; and (4) regeneration of the phospho-enzyme by transfer of the phosphoryl group at position 6 of the sugar back to the catalytic serine, and release of the product, glucose 1-phosphate. This scheme comprises the "exchange" model of the reaction, one of two possibilities initially proposed (Ray, Mildvan & Long, 1973) and subsequently confirmed by structural studies of a related enzyme (Regni, Naught, Tipton, & Beamer, 2004).

Studies of rabbit PGM span more than three decades and produced many fundamental insights into enzyme mechanism. PGM was also instrumental in early work on the theory of induced fit (Yankeelov & Koshland, 1965). These seminal contributions to enzymology cannot be adequately reviewed herein; interested readers should refer to the publications of Ray et al. In the following sections, we highlight a few recent developments in understanding the PHM mechanism.

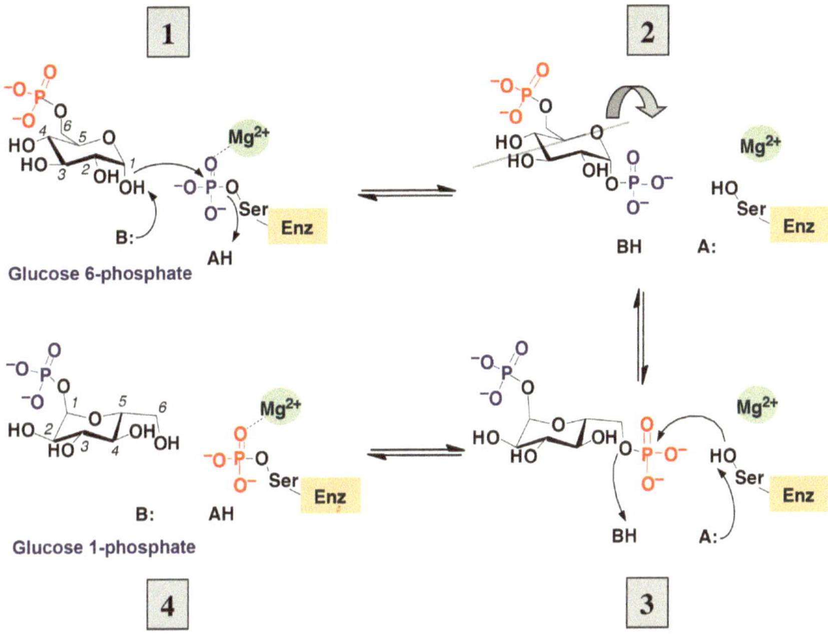

Fig. 4 Detailed reaction mechanism of the PHMs showing the multistep conversion of glucose 6-phosphate to glucose 1-phosphate. The reaction proceeds via transfer of the enzymic phosphoryl group (*blue*) to substrate, to form a bisphosphorylated sugar intermediate. Reorientation of this intermediate (*gray arrow*) places the 6-phospho group (*red*) of the sugar near the conserved phosphoserine residue, allowing for the second phosphoryl transfer to occur back to enzyme. The product is now phosphorylated at the 1-position with the phosphoryl group (*blue*) derived from the enzyme. Roles for residues in acid–base catalysis are also indicated.

3.2 The "Flip"

One of the most mechanistically unique features of the PHM reaction is the required 180 degree reorientation (flip) of the reaction intermediate, which occurs in between the two phosphoryl transfers (between steps 2 and 3 in Fig. 4). As previously noted (Ray & Roscelli, 1964), two kinetic models are consistent with the required chemistry: (1) The intermediate could dissociate from the enzyme each time it is formed and rebind in a randomly determined orientation, via a ping-pong mechanism; or (2) the intermediate could reorient without completely losing its association with the enzyme, which could have advantages in terms of efficiency of turnover. The latter model was elegantly demonstrated for *P. aeruginosa* PMM/PGM using isotope trapping experiments (Naught & Tipton, 2005). This work showed that glucose 1,6-bisphosphate preferentially remains bound to the enzyme throughout the course of the reaction, dissociating only once every 15 catalytic cycles. This result is consistent with prior estimates of the frequency of dissociation of glucose 1,6-bisphosphate from rabbit PGM as occurring once per 20 catalytic cycles (Ray & Roscelli, 1964). Thus, in the two best-studied enzymes in the superfamily, the reaction appears to be a simple example of processivity, as defined by multiple rounds of catalysis without release of substrate. In the case of *P. aeruginosa* PMM/PGM, specific residues in the active site were identified that affect the efficiency of the processive reaction (Regni, Schramm, & Beamer, 2006).

This unique mechanistic pathway poses considerable challenges for enzyme structure and active site evolution. First, it necessitates design of a ligand-binding site that can accommodate both 1- and 6- phosphosugars in *different* orientations (Fig. 1A, far right). Structural studies of *P. aeruginosa* PMM/PGM (Regni et al., 2004) revealed the elegant solution to this problem, which relies on conserved contacts to the phosphate group of bound ligand as well as the specific stereochemistry of glucose-based phosphosugars (Section 4.3). Second, conformational change of the C-terminal domain of the enzyme is needed, to close the active site for tight interactions with bound substrate, and open it for release of product. Third, the mechanism also requires that binding interactions between the enzyme and intermediate temporarily relax, in order to encourage reorientation of the intermediate but not its complete dissociation from the enzyme (Regni et al., 2006). Potential insights into this last challenge have emerged from biophysical studies of *P. aeruginosa* PMM/PGM, which demonstrate an intriguing connection between phosphorylation of the active site serine and protein flexibility.

3.2.1 Phosphorylation-Dependent Protein Flexibility in Enzyme Mechanism

Due to its amenability to rigorous sample and methodological demands, *P. aeruginosa* PMM/PMM has been the subject of extensive structural and biophysical characterizations. During nuclear magnetic resonance (NMR) studies of this enzyme, which are conducted at relatively high temperatures (37°C), it became apparent that multiple residues existed in two states (populations) in solution (Sarma et al., 2012). Additional work implicated >40 residues as being affected by the phosphorylation state of the protein. Subsequently, the phospho- and dephospho-states of the enzyme were characterized by hydrogen–deuterium exchanged (HDX) mass spectrometry (MS) and HDX-NMR (Lee, Villar, Artigues, & Beamer, 2014; Xu, Lee,

Beamer, & Van Doren, 2015). Surprisingly, widespread and significant changes in protein flexibility related to phosphorylation were observed, affecting >100 of the 450 residues in the enzyme, despite the modest chemical difference of a single phosphorylation site. In particular, the dephospho-form of the enzyme, which corresponds to the state of the protein present during the reorientation of the intermediate (steps 2–3 in Fig. 4), was found to be significantly more flexible, based on an overall increase in deuterium exchange (Lee, Villar, et al., 2014). HDX-NMR studies also show a smaller set of residues with decreased exchange, corresponding to a decrease in flexibility (Xu et al., 2015). In addition to these HDX studies, several traditional biochemical assays further support phosphorylation-dependent changes in protein flexibility, with the dephospho-protein being more flexible overall (Xu et al., 2015).

The correlation between dephospho-enzyme and increased structural flexibility was intriguing, since the dephospho-enzyme is present at the point of the reaction when reorientation of the intermediate occurs (between steps 2 and 3 in Fig. 4). Based on this observation, it was proposed that the increased structural flexibility might facilitate release of high affinity interactions with the bound intermediate, thereby encouraging its reorientation (Lee, Villar, et al., 2014). In other members of the superfamily, similar biochemical evidence is mounting for phosphorylation-dependent differences in protein flexibility (Lee, Stiers, et al., 2014; Stiers, Xu, Lee, & Beamer, n.d.), suggesting that this is a conserved feature of the PHMs, and supporting its functional importance in enzyme mechanism.

3.3 Acid–Base Catalysis

The chemistry of phosphoryl transfer reactions has been previously studied in many systems (Knowles, 1980) and may proceed via either an associative or dissociative mechanism. In either case, the hydroxyl group that accepts the phosphoryl group must first be deprotonated, a step that is presumably facilitated by a general base. Transfer of the phosphoryl group may also be aided by protonation of the alkoxide formed by departure of the phosphoryl group through participation of a general acid. Fig. 4 shows the putative roles of the general acid–base catalysts in the conversion of glucose 6-phosphate to glucose 1-phosphate. Typically, such residues are expected to meet several criteria (Guillén Schlippe & Hedstrom, 2005), including being within hydrogen bonding distance of the position where proton abstraction or donation occurs. In addition, mutagenesis of the general acid–base is expected to significantly impair enzyme function (i.e., >1000-fold reduction in activity) in the absence of accompanying structural changes.

Until recently, candidates for residues acting as a general acid or base in the PHM mechanism were not forthcoming. However, mutagenesis and kinetic studies of *P. aeruginosa* PMM/PGM showed that a histidine (His329) in the active site is critical for enzyme activity (Lee, Mehra-Chaudhary, Furdui, & Beamer, 2013). Moreover, analyses of protein–ligand complexes show that this residue is appropriately positioned to abstract a proton from the O1/O6 hydroxyl of the phosphosugar substrates, and thus may serve as the general base in the reaction. Histidine is strongly sequence-conserved at this position in the sequence of many proteins in the superfamily, except for the PGM subgroup (purple shading in Fig. 2). In the PGMs, however, a lysine is often conserved at a corresponding structural position (but not corresponding in sequence). Mutation of this lysine in PGM from *Salmonella enterica* (serovar Typhimurium) also results in significant catalytic impairment (Lee et al., 2013). Thus, candidate residues for the general base appear to be position-conserved within subgroups but may be only structurally conserved across subgroups (PGM vs PMM/PGM) (Fig. 2).

One candidate for the general acid is an arginine residue found near the N-terminus of the PHMs (blue star in Fig. 2) (Lee et al., 2013). Although not often found in this catalytic role, some precedent exists in other enzyme systems (Keenholtz et al., 2013; Silva, Schulz, Jahn, Jahn, & Ramos, 2010). In addition to its proximity to the site of phosphoryl transfer, this residue appears to be absolutely conserved across the PHM superfamily (Fig. 2), consistent with a critical functional role (Lee et al., 2013).

4. THREE-DIMENSIONAL STRUCTURE

4.1 Structures in the PDB

Prior to the year 2000, only one enzyme in the superfamily, rabbit PGM, had been characterized by X-ray crystallography (Dai, Liu, Ray, & Konno, 1992; Lin et al., 1986; Liu, Ray, & Baranidharan, 1997). Fortunately, over the past decade, the number of structurally characterized proteins in the PHM superfamily increased significantly. Currently, approximately 20 crystal structures of independent proteins are known, including several from each of the four major enzyme subgroups (Table 3). In total, these comprise >40 entries in the PDB, representing various versions of the respective proteins (e.g., mutants, ligand complexes, etc.). Fifteen crystal structures are available for *P. aeruginosa* PMM/PGM,

Table 3 Summary of Protein Data Bank Depositions for Members of the PHM Superfamily

Organism	Lineage	UniProt ID	PDB ID	No. Residues	Resolution (Å)[a]	P-Ser	Divalent Metal Ion	Ligand	Oligomer?
PGM									
S. typhimurium	Bacterial	Q8ZQW9	2FUV, 3NA5, 3OLP	546	1.70	N	Mg		Dimer
T. thermophilus	Bacterial	Q5SLE0	2Z0F	524	2.52	N	No (pH 5.4)		Dimer
L. minor	Bacterial	Q4QCF1	4QG5	589	3.50		Mg		
H. sapiens	Eukaryote	P36871	5EPC, 5F9C, 5HSH	562	1.85	N	Mg		
O. cuniculus	Eukaryote	P00949	3PMG, 1VKL, 1JDY, 1LXT, 1C4G, 1C47	561	2.40	Y/N	Mg, Ni, Cd, Co	G1P6V, G16P	
P. tetraurelia	Eukaryote	P47244[b]	1KFI, 1KFQ	572	2.40	N	Zn, Ca		
PNGM									
B. anthracis	Bacterial	Q81VN7	3PDK	448	2.70	N	No (pH 4.5)		Dimer
F. tularensis	Bacterial	Q5NII8	3I3W	443	2.30	Y	Zn		Dimer
PMM/PGM									
B. thailandensis	Bacterial	Q2SYG4	3UW2	485	1.95	Y	Zn		
P. aeruginosa	Bacterial	P26276	1K35, 1K2Y, 1P5D, 1P5G, 1PCJ, 1PCM, 3BKQ, 3C04, 2H4L, 2H5A, 2FKF, 2FKM, 3RSM, 4IL8, 4MRQ	463	1.60	Y/N	Zn, Mn	G1P, G6P, M1P, M6P, R1P, X1P, G16P	
X. citri	Bacterial	Q8PGN7	5BMN, 5KLO, 5BMP	468	1.27	Y	Mg	G16P	

Continued

Table 3 Summary of Protein Data Bank Depositions for Members of the PHM Superfamily—cont'd

Organism	Lineage	UniProt ID	PDB ID	No. Residues	Resolution (Å)	P-Ser	Divalent Metal Ion	Ligand	Oligomer?
PAGM									
C. albicans	Eukaryote	Q9P4V2	2DKA, 2DKC, 2DKD	544	1.93	N	No (pH 4.6), Zn	GlcNAc-6P, GlcNAc-1P	
A. fumigatus	Eukaryote	B0XPI4	4BJU	549	2.35	Y	Mg		
M. musculus	Eukaryote	Q9CYR6	1WJW	112	NMR	—	—		
Ambiguous subgroup									
B. melitensis	Bacterial	D0B6G5	4HJH	481	2.10	Y	Mg	G6P	
T. thermophilus	Bacterial	Q5SKJ3	1TUO	464	1.70		No (pH 4.4)		
P. horikoshii	Archaea	O58651	1WQA	455	2.00	N	Mg		Tetramer
S. tokodaii	Archaea	Q976E4	2F7L	455	2.80	N	No (pH 5.4)		Tetramer

[a]In cases of multiple structures, only the highest resolution listed.
[b]Reported to lack enzyme activity.
Ligand abbreviations: G16P, glucose 1,6-bisphosphate; G1P6V, glucose 1-phosphate-6-vanadate; X1P, xylose 1-phosphate; see also Table 1 legend. Ligands derived from crystallization buffer (e.g., sulfate ions) not included. Oligomers are described in Luebbering et al. (2012) and references therein.

which is also the only full-length protein in the family to be characterized by NMR (Sarma et al., 2012; Xu et al., 2015).

A number of recent structures were targets of structural genomics efforts and represent sequence-diverse members of the superfamily, such as pdb: 1TUO and 4HJH. These proteins do not segregate cleanly into one of the four major subgroups, according to the criteria laid out in Section 2.1, and so are listed separately in Table 3. Some of the unique features seen in these proteins include PGM-like sequences for their sugar-binding loop (-GEESF-), but in the context of a shorter polypeptide chain (450 instead of the typical 550 residues). Several also have an arginine residue at the position of the putative general base (Section 3.3). It seems likely that these proteins represent distinct divisions within the known subgroups, or perhaps even novel subgroups of the superfamily. Clarification of these issues awaits further study.

The growing database of 3D structures provides a broad perspective on the structural similarities and differences among these enzymes. Overall, the structures show that the domain architecture and key active site regions of the enzymes have remained remarkably similar throughout evolution (Sections 4.2–4.3). In addition, a unifying theme of structural flexibility of, or within, domain 4 is also apparent in the superfamily (Section 4.4), although the nature of the conformational change varies. Most strikingly, multiple distinct oligomeric forms have recently been identified for several proteins in the superfamily, some of which appear to be subgroup specific (Section 4.4).

In the following sections, we describe the structural architecture, key regions of the active site, the known oligomeric assemblies of the PHMs, and the conserved conformational mobility of domain 4.

4.2 Overall Architecture

Despite considerable overall sequence diversity, the structural scaffold of enzymes in the PHM superfamily is remarkably conserved. To date, all of the known structures share a 4-domain architecture, arranged in an overall heart-shape (Figs. 1B and 5A). Each structural domain is of approximately similar size—100–150 residues—and contains both α-helices and β-sheets. The active site is found within a large, positively-charged cleft found at the confluence of these four domains. Residues from each of the four structural domains contribute to the formation of the cleft and also have roles in catalysis and/or ligand binding. Depending on the enzyme subgroup, additional

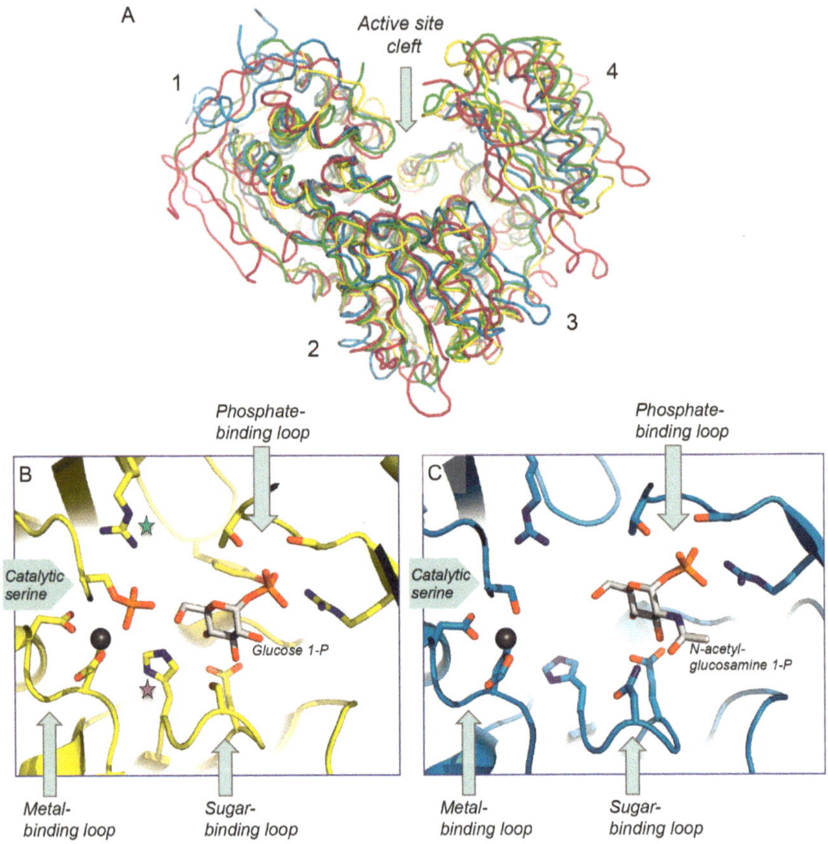

Fig. 5 Overall structure and active site of enzymes in the major PHM subgroups. (A) A superposition of four crystal structures, one from each major subgroup. Shown are human PGM1 (pdb: 5EPC) in *pink*, *C. albicans* PAGM (pdb: 2DKA) in *blue*, *Bacillus anthracis* PNGM (pdb: 3PDK) in *green*, and *P. aeruginosa* PMM/PGM (pdb: 1K35) in *yellow*. Domains 1–4 are numbered. Close-up views of the active site of (B) *P. aeruginosa* PMM/PGM (*yellow*) in complex with glucose 1-phosphate (pdb: 1P5D) and (C) *C. albicans* PAGM (*blue*) in complex with GlcNAc-1P (pdb: 2DKD). Side-by-side comparison shows the highly similar arrangements of the two active sites. Divalent metal ion is shown as *gray sphere*; the catalytic serine is shown in its two possible states: phosphorylated in (B) and unphosphorylated in (C). *Cyan* and *purple stars* mark putative general acid and general base, respectively. Key areas of the active site are labeled. See Fig. 2 for sequence alignments of these two proteins.

secondary structural elements may be present, particularly in the longer PGM and PAGM subgroups. These additional residues generally form secondary structural elements that decorate the periphery of the protein (e.g., pink regions in Fig. 5A).

In the structure of rabbit PGM, it was noted that domains 1–3 (D1–3) share a common topological core (Dai et al., 1992), presumably arising from a series of ancient gene duplications. This common topology for D1–3 occurs in all known structures of the superfamily, regardless of subgroup. The common core consists of a mixed α/β fold, with a central β-sheet flanked on either side by an α-helix. The same loop (between β-strands 3 and 4 of the common core) in D1–3 contributes important residues involved in catalysis or ligand binding (Regni, Tipton, & Beamer, 2002).

As noted previously, proteins in the PAGM subgroup are distinguished by a conserved, circular permutation within domain 1. This effects the sequence order of conserved secondary structural elements (swapped regions in Fig. 2), including the phosphoserine loop (Nishitani et al., 2006), but does not change their three-dimensional location. The circular permutation of domain 1 appears to be both unique to and characteristic of the PAGM subgroup. However, it has no obvious connection to its substrate preference for *N*-acetylglucosamine phosphosugars, as all of the key catalytic and ligand-binding loops are found in similar positions (Nishitani et al., 2006).

The C-terminal domain (D4) of enzymes in the PHM superfamily is structurally unrelated to the fold of D1–3. It has α + β topology and contains a TATA-binding protein-like fold, consisting of one β-sheet and two α-helices (Regni et al., 2002). Typically, the β-sheet in D4 is composed of four or more strands, although for proteins in the PNGM subgroup only three strands are found (Mehra-Chaudhary et al., 2011). In several members of the superfamily, it has been shown that D4 is an independently folded unit, maintaining a well-ordered structure in the absence of the other three domains (Schramm et al., 2010; Wei et al., 2014) (see also pdb: 1WJW). This appears to be consistent with the conformational variability of this domain relative to the others that has been noted in multiple proteins in the superfamily (Luebbering et al., 2012).

4.3 Key Regions of Active Site

The active site of the PHMs can be segregated into four highly conserved regions that contribute to catalysis and substrate binding (Figs. 2, 5B and C). These regions are: (i) the phosphoserine residue that participates in phosphoryl transfer, (ii) the metal-binding loop, (iii) the sugar-binding loop, and (iv) the phosphate-binding site that interacts with the phosphate group of the substrate. It is important to note the difference between the phosphoryl transfer site (i), which involves the phosphorylated serine of the *enzyme*, and the phosphate-binding site (iv), a group of residues that interacts with the phosphate moiety of the *substrate*. Each of these key regions is found on a loop/turn of the polypeptide chain and is located in a different structural domain of the enzyme: region (i) in domain 1, region (ii) in domain 2, etc. (Fig. 1B). Thus, the active site relies on residues from all four structural domains for its integrity.

4.3.1 Phosphoserine Loop

The catalytic reaction of the PHMs proceeds via transfer of a phosphoryl group (PO_3^{2-}) from a conserved serine in domain 1 to the substrate (Fig. 5B and C). Hence this residue and the geometry of the loop in which it resides are critical for efficient catalysis. Indeed, the residues surrounding this serine (TASHNP) are very highly conserved in the α-D-phosphohexomutase superfamily (region I in Fig. 2). In crystal structures of the PHMs, this serine may be phosphorylated or not (Table 3) depending on the protein and conditions used for crystallization. [The phosphorylation level of the serine is known to be affected by ligands and other experimental variables in several proteins in the superfamily (Lee, Furdui, & Beamer, 2017; Lee, Stiers, et al., 2014; Sarma et al., 2012; Xu et al., 2015).]

The catalytic serine is located in a rare, type II β-turn, with backbone angles that fall in the disallowed region of the Ramachandran plot, and are similar to those of a "nucleophilic elbow," a structural motif also found in α/β hydrolases (Ollis et al., 1992; Regni et al., 2002). The loop containing the phosphoserine is anchored by extensive interresidue interactions, which presumably serve to stabilize it in the appropriate conformation for catalysis. Among these interactions are multiple contacts to a highly conserved arginine that participates in a buried salt-bridge with a nearby aspartate (Asp62 and Arg64 in human PGM1) (Beamer, 2015; Liu et al., 1997).

4.3.2 Metal-Binding Loop

All enzymes in the α-D-phosphohexomutase superfamily require a divalent metal ion for maximal activity. In most reports, maximal enzyme activity is achieved with Mg^{2+}, although other metals have been reported to also produce activity, including Mn^{2+} (Akutsu et al., 2005; Namboori & Graham, 2008; Ray, 1969; Ye et al., 1994). The binding site for the metal ion is located proximal to the phosphoserine in the active site (Fig. 5B and C). Three highly conserved aspartate residues (region II in Fig. 2), found in a loop of domain 2, serve as ligands. An oxygen atom from the phosphate group of the phosphoserine also coordinates the metal ion. The positive charge of the metal ion acts as an electron-withdrawing agent, facilitating breakage of the P–O bond necessary for phosphoryl transfer. Hence the phosphoserine and metal-binding loop are intimately connected in the active site and function together during catalysis.

The metal-binding loop is located in a tight turn found deep in the active site cleft. It often includes the sequence motif: -DGDGDR- (Fig. 2). However, new structural representatives have revealed that other residues may take the place of the two glycines (G1 and G2), including alanine and proline for G1, and alanine, phenylalanine, and tyrosine for G2. Regardless of this variation in sequence, the geometry of the loop remains invariant. A number of crystal structures in the PHM superfamily were found to lack a bound metal (Table 3). This appears to result from the low pH of the crystallization conditions, which protonates the aspartates and interferes with metal binding. Somewhat surprisingly, only a small rearrangement of the aspartate side chains are necessary to accommodate loss of metal; the overall geometry of the loop is unperturbed (Mehra-Chaudhary et al., 2011).

4.3.3 Sugar-Binding Loop

Domain 3 contains another key region in the active site of the PHMs, the sugar-binding loop (Fig. 5B and C). Residues in this loop are expected to make direct hydrogen bonds with the sugar moiety of the substrate and have been proposed to play a role in substrate specificity (Shackelford et al., 2004). This loop contains five highly conserved residues that vary somewhat according to subgroup (Table 1; region III in Fig. 2) but typically includes a three-residue submotif of Glu-X-Ser (or Asn in PAGM). For the two proteins in the superfamily with available high-resolution enzyme–ligand complexes, *P. aeruginosa* PMM/PGM and *C. albicans* PAGM, the conserved Glu is seen to make a bidentate interaction with the O3 and O4 hydroxyls of the sugar ring (Fig. 5B and C), while the Ser (or Asn) makes additional contacts to these hydroxyls, depending on whether the 1- or 6-phosphosugar is bound (Section 4.3.5).

4.3.4 Phosphate-Binding Loop

Contacts between the protein and the phosphate group of bound substrate are largely provided by residues in a loop of D4 (region IV in Fig. 2), although one or more residues in domain 1 may also be involved (e.g., see the tyrosine in Fig. 5B). The phosphate-binding site appears designed to anchor bound substrate in the active site cleft via multiple interactions with the phosphate group. Other conserved residues that contact the phosphate include an arginine, serine, and threonine (Fig. 5B and C). These protein–phosphate contacts are nearly identical regardless of whether 1- or 6-phosphosugars are bound (Regni et al., 2004). The phosphate-binding site appears to be a key determinant of ligand recognition, as enzyme–ligand complexes adopting a half-closed conformer (with ligand partially exposed to solvent) retain most of the contacts to phosphate, even while those to the sugar hydroxyls are lost (Regni et al., 2006).

Often, binding of ligand and formation of the phosphate-binding site require motion of, or within, D4. This changes the active site from an open cleft to a deeply buried pocket that encloses substrate and positions it for catalysis. In some members of the superfamily, a rotation of D4 is involved in creating the phosphate-binding site (Regni et al., 2004); in others, movement of the phosphate-binding loop appears to be sufficient (Müller et al., 2002; Nishitani et al., 2006).

4.3.5 Binding of 1- vs 6-Phosphosugars

As noted in Section 3, the exchange mechanism for the PHMs raises questions about how a single catalytic site can accomplish phosphoryl transfer to both the 1- and 6-hydroxyls of the sugar. Enzyme–ligand complexes for both 1-phospho and 6-phospho forms of the sugar substrates, first determined for *P. aeruginosa* PMM/PGM (Regni et al., 2004) and subsequently for *C. albicans* PAGM (Nishitani et al., 2006), showed that this occurs through binding of the sugar ring in two different orientations (Fig. 1A, far right). In this superposition, which shows these two sugars as found in the active site of their respective complexes with *P. aeruginosa* PMM/PGM, it can be seen that the phosphate groups of glucose 1-phosphate and glucose 6-phosphate are in nearly identical positions, resulting in a 180 degree difference in the orientation of the sugar ring.

One key to recognition of the two orientations of the sugar ring is a conserved set of contacts to O3 and O4 hydroxyls (Regni et al., 2004). Because these hydroxyls are both equatorial in their stereochemistry, these atoms exchange places when the sugar ring is rotated by 180 degree. This allows the contacting residues to simply switch from O3 to O4 and vice versa, depending on whether the 1- or 6-phosphosugar is bound. A similar exchange could not occur for sugars where these hydroxyls are in axial positions (e.g., galactose), since they would occupy very different positions after reorientation. It appears that these exchangeable contacts with O3 and O4 are another key to recognition of substrates in the entire PHM superfamily, as the O3 and O4 hydroxyls and their stereochemistry are common to four major hexose substrates of the PHMs (Fig. 1A), and the key residues in the sugar-binding loop (Section 4.3.3) are also conserved.

4.3.6 Subgroup Specific Substrate Specificity

One consequence of the two binding orientations of the 1- and 6-phospho versions of substrate is that the substituent at the 2′-position of the sugar occupies two very different positions depending on which form is bound (two green arrows on glucose 1-phosphate/glucose 6-phosphate superposition in Fig. 1A). As can also be seen on this figure, it is the substituent at the 2′-position of the sugar ring that differs among these four hexose substrates (glucose, mannose, glucosamine, and *N*-acetylglucosamine) (Shackelford et al., 2004). The design of the PHM active site, which allows structural variation at the 2′-position of the sugar, appears to be advantageous for two reasons. First, it accommodates the dual binding mode of the 1- and 6-phosphosugars that is required by enzyme mechanism (Fig. 4). Second, it facilitates use of these four related sugars as substrates, which is key to the functional diversity of the superfamily.

The known crystal structures suggest that the ancestral PHM was designed for glucose-based substrates, or possibly glucose/mannose. With a PGM-like active site as a starting point, only small changes would then be necessary over the course of evolution to relax glucose specificity and create preferences for the related sugars of glucosamine and *N*-acetylglucosamine. This model is consistent with biochemical data showing that most enzymes in the superfamily characterized to date have PGM activity, regardless of subgroup (Section 2.1).

4.4 Quaternary Structure and Conformational Mobility of Domain 4

Until recently, enzymes in the PHM superfamily were only known to function as monomers. However, crystal structures of proteins in the PNGM and PGM subgroups revealed apparent dimers (Fig. 6), which were subsequently confirmed in solution (Mehra-Chaudhary, Mick, Tanner, & Beamer, 2011; Mehra-Chaudhary, Mick, Tanner, Henzl, & Beamer, 2011). These observations led to a search for other potential oligomers in the superfamily, which also identified tetramers for PHMs from *Pyrococcus horikoshii* (pdb: 1WQA) and *Sulfolobus tokodaii* (pdb: 2F7L) (Luebbering et al., 2012).

For the PGM subgroup, a dimeric arrangement was verified for the *Salmonella typhimurium* enzyme (pdb: 3NA5), the interface of which is formed by an N-terminal extension of the protein (interface A in Fig. 6). Sequence

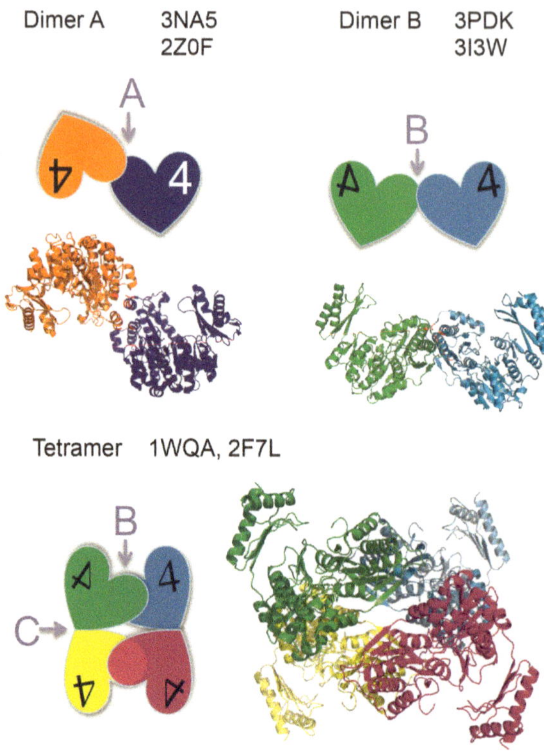

Fig. 6 Structures of experimentally confirmed oligomers in the PHM superfamily. *Schematics* (*heart shapes*) and *ribbon diagrams* are shown for each assembly; relevant PDB IDs are noted. Domain 4 is indicated by "4" on each schematic to indicate relative orientation of the protomers within the assembly. Three different protomer–protomer interfaces are indicated by A, B, and C. Two distinct dimers and one tetramer are known. Note how domain 4 is oriented away from the assembly interfaces, permitting conformational mobility. *Reprinted from Luebbering, E. K., Mick, J., Singh, R. K., Tanner, J. J., Mehra-Chaudhary, R., et al. (2012). Conservation of functionally important global motions in an enzyme superfamily across varying quaternary structures.* Journal of Molecular Biology, 423, *831–846.*

searches confirmed the presence of a similar motif in a subgroup of bacterial species (Mehra-Chaudhary, Mick, Tanner, Henzl et al., 2011). For the two PNGM proteins, both known structures (pdb: 3I3W and 3PDK) share a similar dimeric arrangement, despite only moderate pairwise sequence identity between the proteins (39%). This suggests that many other PNGMs may share this assembly (Mehra-Chaudhary, Mick, Tanner, & Beamer, 2011). The dimers found in the PGM and PNGM subgroups are structurally distinct and appear to have evolved independently (Fig. 6). However, the tetramer comprises two dimers with the PNGM arrangement, suggesting an evolutionary relationship between these two assemblies.

Examination of the confirmed oligomers of the PHMs shows that they are all assembled in a manner that permits conformational freedom of D4 (Fig. 6). This is consistent with the known functional importance of D4 motion, or of loops therein, in several proteins in the superfamily (e.g., for substrate binding/product release and during reorientation of the intermediate). Moreover, normal mode analyses of the various assemblies show that predicted low frequency global motions contribute to the conformational variability of this domain (Luebbering et al., 2012). Hence, it appears that the PHM superfamily has conserved not only residues essential for catalysis but also the conformational flexibility of its C-terminal domain, despite considerable sequence diversity and even across varying oligomeric states. This makes the PHM superfamily an interesting case of evolutionarily conserved protein dynamics.

5. CONCLUSIONS AND FUTURE DIRECTIONS

Recent years have witnessed tremendous gains in our knowledge and appreciation of the PHM superfamily. This includes new connections to inherited human disease, which should grow rapidly due to the availability of sequence information on disease-associated mutations. Another future area of development for these enzymes includes metabolic engineering, which was not addressed herein but is an active area of research (Anasontzis, Kourtoglou, Villas-Boas, Hatzinikolaou, & Christakopoulos, 2016; Lee et al., 2011; Liu et al., 2014; Nic Lochlainn & Caffrey, 2009; Rodríguez-Díaz, Rubio-del-Campo, & Yebra, 2012; Wang et al., 2016; Zabala, Braña, Flórez, Salas, & Méndez, 2013). New biochemical and structural data summarized in this review have shed light on intriguing mechanistic features of the PHMs that were previously unappreciated. These insights should facilitate future investigations of other proteins in the superfamily.

Despite significant progress, many questions remain regarding the PHMs, from the biological to the molecular level. Enzyme representatives from plants remain woefully understudied, despite their critical functional roles. At the cellular level, very little is known about potential protein–protein interactions involving these enzymes, despite their established importance in some systems (Arimura et al., 2009; Belkin & Burridge, 1994, 1995; Gundlach et al., 2015; Molt et al., 2014) and likelihood in others. A role for posttranslation modification in the regulation of enzyme activity, localization, or protein–protein interactions is also a potential new horizon (Dey, Bounelis, Fritz, Bedwell, & Marchase, 1994; Fu et al., 1995; Gururaj, Barnes, Vadlamudi, & Kumar, 2004). Structural representatives for the important phosphopentomutase and glucose 1,6-bisphosphate synthase subgroups are still lacking, as are a number of key enzyme–ligand complexes (e.g., for the PNGMs). Investigations of these poorly studied aspects of the PHMs seem certain to reveal as yet unexpected features/functions of these proteins.

Perhaps the most important underlying issue that remains to be addressed is a systematic characterization of the phosphosugar substrates utilized by the PHMs overall, as well as within and among the various subgroups of the superfamily. This knowledge would help clarify the issue of potential functional overlap between enzymes in different subgroups and may even define additional groupings within the superfamily. Such fundamental information is relevant for the design of inhibitors and also for understanding tissue-specific disease phenotypes. It should also help correct the widespread functional misannotation of these enzymes in databases. This ambitious goal is now feasible using a combination of approaches such as high-throughput substrate profiling and bioinformatic analyses, as has been done with other enzyme families (Burroughs et al., 2006; Huang et al., 2015; Pandya et al., 2013).

ACKNOWLEDGMENT

Studies of PHMs in the Beamer Laboratory have been supported by the National Science Foundation, including Grants MCB-0918389 and MCB-1409898.

REFERENCES

Akutsu, J.-I., Zhang, Z., Tsujimura, M., Sasaki, M., Yhoda, M., & Kawarabayasi, Y. (2005). Characterization of a thermostable enzyme with phosphomannomutase/phosphoglucomutase activities from the hyperthermophilic Archaeon *Pyrococcus horikoshii* OT3. *Journal of Biochemistry*, *138*, 159–166.

Anasontzis, G. E., Kourtoglou, E., Villas-Boas, S. G., Hatzinikolaou, D. G., & Christakopoulos, P. (2016). Metabolic engineering of Fusarium oxysporum to improve its ethanol-producing capability. *Frontiers in Microbiology*, 7, 1–10.

Arimura, T., Inagaki, N., Hayashi, T., Shichi, D., Sato, A., Hinohara, K., et al. (2009). Impaired binding of ZASP/Cypher with phosphoglucomutase 1 is associated with dilated cardiomyopathy. *Cardiovascular Research*, *83*, 80–88.

Bandini, G., Mariño, K., Güther, M. L. S., Wernimont, A. K., Kuettel, S., Qiu, W., et al. (2012). Phosphoglucomutase is absent in *Trypanosoma brucei* and redundantly substituted by phosphomannomutase and phospho-N-acetylglucosamine mutase. *Molecular Microbiology*, *85*, 513–534.

Barreteau, H., Kovac, A., Boniface, A., Sova, M., Gobec, S., & Blanot, D. (2008). Cytoplasmic steps of peptidoglycan biosynthesis. *FEMS Microbiology Reviews*, *32*, 168–207.

Beamer, L. J. (2015). Mutations in hereditary phosphoglucomutase 1 deficiency map to key regions of enzyme structure and function. *Journal of Inherited Metabolic Disease, 38*, 243–256.

Belkin, A. M., & Burridge, K. (1994). Expression and localization of the phosphoglucomutase-related cytoskeletal protein, aciculin, in skeletal muscle. *Journal of Cell Science, 107*, 1993–2003.

Belkin, A. M., & Burridge, K. (1995). Association of aciculin with dystrophin and utrophin. *The Journal of Biological Chemistry, 270*, 6328–6337.

Belkin, A. M., Klimanskaya, I. V., Lukashev, M. E., Lilley, K., Critchley, D. R., & Koteliansky, V. E. (1994). A novel phosphoglucomutase-related protein is concentrated in adherens junctions of muscle and nonmuscle cells. *Journal of Cell Science, 107*, 159–173.

Bizzini, A., Majcherczyk, P., Beggah-Moller, S., Soldo, B., Entenza, J. M., Gaillard, M., et al. (2007). Effects of α-phosphoglucomutase deficiency on cell wall properties and fitness in *Streptococcus gordonii. Microbiology, 153*, 490–498.

Boles, E., Liebetrau, W., Hofmann, M., & Zimmermann, F. K. (1994). A family of hexosephosphate mutases in *Saccharomyces cerevisiae. European Journal of Biochemistry, 220*, 83–96.

Brautaset, T., Petersen, S. B., & Valla, S. (2000). *In vitro* determined kinetic properties of mutant phosphoglucomutases and their effects on sugar catabolism in Escherichia coli. *Metabolic Engineering, 2*, 104–114.

Buchanan, J. T., Stannard, J. a., Lauth, X., Ostland, V. E., Powell, H. C., Westerman, M. E., et al. (2005). *Streptococcus iniae* phosphoglucomutase is a virulence factor and a target for vaccine development. *Infection and Immunity, 73*, 6935–6944.

Burroughs, A. M., Allen, K. N., Dunaway-Mariano, D., & Aravind, L. (2006). Evolutionary genomics of the HAD superfamily: Understanding the structural adaptations and catalytic diversity in a superfamily of phosphoesterases and allied enzymes. *Journal of Molecular Biology, 361*, 1003–1034.

Chiang, S. L., & Mekalanos, J. J. (1999). rfb mutations in *Vibrio cholerae* do not affect surface production of toxin-coregulated pili but still inhibit intestinal colonization. *Infection and Immunity, 67*, 976–980.

Czibener, C., Del Giudice, M. G., Spera, J. M., Fulgenzi, F. R., & Ugalde, J. E. (2016). Delta-pgm, a new live-attenuated vaccine against *Brucella suis. Vaccine, 34*, 1524–1530.

Dai, J.-B., Liu, Y., Ray, W. J., & Konno, M. (1992). The crystal structure of muscle phosphoglucomutase refined at 2.7 Å resolution. *The Journal of Biological Chemistry, 267*, 6322–6337.

Davies, E. J., Tetlow, I. J., Bowsher, C. G., & Emes, M. J. (2003). Molecular and biochemical characterization of cytosolic phosphoglucomutase in wheat endosperm (*Triticum aestivum* L. cv. Axona). *Journal of Experimental Botany, 54*, 1351–1360.

De Reuse, H., Labigne, A., & Mengin-Lecreulx, D. (1997). The *Helicobacter pylori* ureC gene codes for a phosphoglucosamine mutase. *Journal of Bacteriology, 179*, 3488–3493.

Dey, N. B., Bounelis, P., Fritz, T. A., Bedwell, D. M., & Marchase, R. B. (1994). The glycosylation of phosphoglucomutase is modulated by carbon source and heat shock in *Saccharomyces cerevisiae. The Journal of Biological Chemistry, 269*, 27143–27148.

Fang, W., Du, T., Raimi, O. G., Hurtado-Guerrero, R., Urbaniak, M. D., Ibrahim, A. F. M., et al. (2013). Genetic and structural validation of Aspergillus fumigatus UDP-N-acetylglucosamine pyrophosphorylase as an antifungal target. *Molecular Microbiology, 89*, 479–493.

Felek, S., Muszynski, A., Carlson, R. W., Tsang, T. M., Hinnebusch, B. J., & Krukonis, E. S. (2010). Phosphoglucomutase of *Yersinia pestis* is required for autoaggregation and polymyxin B resistance. *Infection and Immunity, 78*, 1163–1175.

Fu, L., Bounelis, P., Dey, N., Browne, B. L., Marchase, R. B., & Bedwell, D. M. (1995). The posttranslational modification of phosphoglucomutase is regulated by galactose induction and glucose repression in *Saccharomyces cerevisiae*. *Journal of Bacteriology*, *177*, 3087–3094.

Galperin, M. Y., Koonin, E. V., & Bairoch, A. (1998). A superfamily of metalloenzymes unifies phosphopentomutase and cofactor-independent phosphoglycerate mutase with alkaline phosphatases and sulfatases. *Protein Science*, *7*, 1829–1835.

Gautam, A., Vyas, R., & Tewari, R. (2011). Peptidoglycan biosynthesis machinery: A rich source of drug targets. *Critical Reviews in Biotechnology*, *31*, 295–336.

Goldberg, J. B., Coyne, M. J., Neely, A. N., & Holder, I. A. (1995). Avirulence of a *Pseudomonas aeruginosa* algC mutant in a burned-mouse model of infection. *Infection and Immunity*, *63*, 4166–4169.

Govan, J. R. W., & Deretic, V. (1996). Microbial pathogenesis in cystic fibrosis: Mucoid *Pseudomonas aeruginosa* and *Burkholderia cepacia*. *Microbiological Reviews*, *60*, 539–574.

Grass, S., Buscher, A. Z., Swords, W. E., Apicella, M. A., Barenkamp, S. J., Ozchlewski, N., et al. (2003). The *Haemophilus influenzae* HMW1 adhesin is glycosylated in a process that requires HMW1C and phosphoglucomutase, an enzyme involved in lipooligosaccharide biosynthesis. *Molecular Microbiology*, *48*, 737–751.

Greig, K. T., Antonchuk, J., Metcalf, D., Morgan, P. O., Krebs, D. L., Zhang, J.-G., et al. (2007). Agm1/Pgm3-mediated sugar nucleotide synthesis is essential for hematopoiesis and development. *Molecular and Cellular Biology*, *27*, 5849–5859.

Gründling, A., & Schneewind, O. (2007). Genes required for glycolipid synthesis and lipoteichoic acid anchoring in *Staphylococcus aureus*. *Journal of Bacteriology*, *189*, 2521–2530.

Guillén Schlippe, Y. V., & Hedstrom, L. (2005). A twisted base? The role of arginine in enzyme-catalyzed proton abstractions. *Archives of Biochemistry and Biophysics*, *433*, 266–278.

Gundlach, J., Mehne, F. M. P., Herzberg, C., Kampf, J., Valerius, O., Kaever, V., et al. (2015). An essential poison: Synthesis and degradation of cyclic Di-AMP in *Bacillus subtilis*. *Journal of Bacteriology*, *197*, 3265–3274.

Gururaj, A., Barnes, C. J., Vadlamudi, R. K., & Kumar, R. (2004). Regulation of phosphoglucomutase 1 phosphorylation and activity by a signaling kinase. *Oncogene*, *23*, 8118–8127.

Hardy, G., Caimano, M., & Yother, J. (2000). Capsule biosynthesis and basic metabolism in *Streptococcus pneumoniae* are linked through the cellular phosphoglucomutase. *Journal of Bacteriology*, *182*, 1854–1863.

Hardy, G., Magee, A., Ventura, C., Caimano, M., & Yother, J. (2001). Essential role for cellular phosphoglucomutase in virulence of type 3 *Streptococcus pneumoniae*. *Infection and Immunity*, *69*, 2309–2317.

Hofmann, M., Boles, E., & Zimmermann, F. K. (1994). Characterization of the essential yeast gene encoding *N*-acetylglucosamine-phosphate mutase. *European Journal of Biochemistry*, *221*, 741–747.

Huang, H., Pandya, C., Liu, C., Al-Obaidi, N. F., Wang, M., Zheng, L., et al. (2015). Panoramic view of a superfamily of phosphatases through substrate profiling. *Proceedings of the National Academy of Sciences of the United States of America*, *112*, E1974–E1983.

Jolly, L., Ferrari, P., Blanot, D., Van Heijenoort, J., Fassy, F., & Mengin-Lecreulx, D. (1999). Reaction mechanism of phosphoglucosamine mutase from *Escherichia coli*. *European Journal of Biochemistry*, *262*, 202–210.

Jolly, L., Wu, S., Van Heijenoort, J., De Lencastre, H., Mengin-Lecreulx, D., & Tomasz, A. (1997). The femR315 gene from *Staphylococcus aureus*, the interruption of which results in reduced methicillin resistance, encodes a phosphoglucosamine mutase. *Journal of Bacteriology*, *179*, 5321–5325.

Kang, J., Xu, L., Yang, S., Yu, W., Liu, S., Xin, Y., et al. (2013). Effect of Phosphoglucosamine mutase on biofilm formation and antimicrobial susceptibilities in *M. smegmatis* glmM Gene knockdown strain. *PloS One8*. http://dx.doi.org/10.1371/journal.pone.0061589.

Kato, N., Mueller, C. R., Wessely, V., Lan, Q., & Christensen, B. M. (2005). *Aedes aegypti* phosphohexomutases and uridine diphosphate-hexose pyrophosphorylases: Comparison of primary sequences, substrate specificities and temporal transcription. *Insect Molecular Biology, 14*, 615–624.

Keenholtz, R. A., Mouw, K. W., Boocock, M. R., Li, N. S., Piccirilli, J. A., & Rice, P. A. (2013). Arginine as a general acid catalyst in serine recombinase-mediated DNA cleavage. *The Journal of Biological Chemistry, 288*, 29206–29214.

Kim, S. H., Ahn, S. H., Lee, J. H., Lee, E. M., Kim, N. H., Park, K. J., et al. (2003). Genetic analysis of phosphomannomutase/phosphoglucomutase from *Vibrio furnissii* and characterization of its role in virulence. *Archives of Microbiology, 180*, 240–250.

Knowles, J. R. (1980). Enzyme-catalyzed phosphoryl transfer. *Annual Review of Biochemistry, 49*, 877–919.

Lee, Y., Furdui, C. M., & Beamer, L. J. (2017). Data on the phosphorylation state of the catalytic serine of enzymes in the α-D-phosphohexomutase superfamily. *Data in Brief, 10*, 398–405.

Lee, K. S., Hong, M. E., Jung, S. C., Ha, S. J., Yu, B. J., Koo, H. M., et al. (2011). Improved galactose fermentation of *Saccharomyces cerevisiae* through inverse metabolic engineering. *Biotechnology and Bioengineering, 108*, 621–631.

Lee, Y., Mehra-Chaudhary, R., Furdui, C., & Beamer, L. J. (2013). Identification of an essential active-site residue in the α-D-phosphohexomutase enzyme superfamily. *FEBS Journal, 280*, 2622–2632.

Lee, Y., Stiers, K. M., Kain, B. N., & Beamer, L. J. (2014b). Compromised catalysis and potential folding defects in *in vitro* studies of missense mutants associated with hereditary phosphoglucomutase 1 deficiency. *The Journal of Biological Chemistry, 289*, 32010–32019.

Lee, Y., Villar, M. T., Artigues, A., & Beamer, L. J. (2014a). Promotion of enzyme flexibility by dephosphorylation and coupling to the catalytic mechanism of a phosphohexomutase. *The Journal of Biological Chemistry, 289*, 4674–4682.

Li, S., Kang, J., Yu, W., Zhou, Y., Zhang, W., Xin, Y., et al. (2012). Identification of *M. tuberculosis* Rv3441c and *M. smegmatis* MSMEG_1556 and essentiality of *M. smegmatis* MSMEG_1556. *PloS One7*. http://dx.doi.org/10.1371/journal.pone.0042769.

Li, C., Rodriguez, M., & Banerjee, D. (2000). Cloning and characterization of complementary DNA encoding human *N*-acetylglucosamine-phosphate mutase protein. *Gene, 242*, 97–103.

Liaw, S. J., Lee, Y. L., & Hsueh, P. R. (2010). Multidrug resistance in clinical isolates of *Stenotrophomonas maltophilia*: Roles of integrons, efflux pumps, phosphoglucomutase (SpgM), and melanin and biofilm formation. *International Journal of Antimicrobial Agents, 35*, 126–130.

Lin, Z. J., Konno, M., Abad-Zapatero, C., Wierenga, R., Murthy, M. R., Ray, W. J., et al. (1986). The structure of rabbit muscle phosphoglucomutase at intermediate resolution. *The Journal of Biological Chemistry, 261*, 264–274.

Liu, X. D., Duan, J., & Guo, L. H. (2009). Role of phosphoglucosamine mutase on virulence properties of *Streptococcus mutans*. *Oral Microbiology and Immunology, 24*, 272–277.

Liu, Y., Ray, W. J., & Baranidharan, S. (1997). Structure of rabbit muscle phosphoglucomutase refined at 2.4 Å resolution. *Acta Crystallographica. Section D, Biological Crystallography, 53*, 392–405.

Liu, Y., Zhu, Y., Li, J., Shin, H. d., Chen, R. R., Du, G., et al. (2014). Modular pathway engineering of *Bacillus subtilis* for improved N-acetylglucosamine production. *Metabolic Engineering, 23*, 42–52.

Lowry, O. H., & Passonneau, J. V. (1969). Phosphoglucomutase kinetics with the phosphates of fructose, glucose, mannose, ribose, and galactose. *Biochemistry, 244*, 910–916.

Lu, M., & Kleckner, N. (1994). Molecular cloning and characterization of the pgm gene encoding phosphoglucomutase of *Escherichia coli*. *Journal of Bacteriology, 176*, 5847–5851.

Luebbering, E. K., Mick, J., Singh, R. K., Tanner, J. J., Mehra-Chaudhary, R., & Beamer, L. J. (2012). Conservation of functionally important global motions in an enzyme superfamily across varying quaternary structures. *Journal of Molecular Biology, 423*, 831–846.

Lundin, K. E., Hamasy, A., Backe, P. H., Moens, L. N., Falk-Sörqvist, E., Elgstøen, K. B., et al. (2015). Susceptibility to infections, without concomitant hyper-IgE, reported in 1976, is caused by hypomorphic mutation in the phosphoglucomutase 3 (PGM3) gene. *Clinical Immunology, 161*, 366–372.

Ma, L., Wang, J., Wang, S., Anderson, E. M., Lam, J. S., Parsek, M. R., et al. (2012). Synthesis of multiple Pseudomonas aeruginosa biofilm matrix exopolysaccharides is post-transcriptionally regulated. *Environmental Microbiology, 14*, 1995–2005.

Maliekal, P., Sokolova, T., Vertommen, D., Veiga-da-Cunha, M., & Van Schaftingen, E. (2007). Molecular identification of mammalian phosphopentomutase and glucose-1,6-bisphosphate synthase, two members of the α-D-phosphohexomutase family. *The Journal of Biological Chemistry, 282*, 31844–31851.

Malinova, I., Kunz, H. H., Alseekh, S., Herbst, K., Fernie, A. R., Gierth, M., et al. (2014). Reduction of the cytosolic phosphoglucomutase in *Arabidopsis* reveals impact on plant growth, seed and root development, and carbohydrate partitioning. *PloS One, 9*, 1–11.

Manjunath, S., Lee, C. H., VanWinkle, P., & Bailey-Serres, J. (1998). Molecular and biochemical characterization of cytosolic phosphoglucomutase in maize. Expression during development and in response to oxygen deprivation. *Plant Physiology, 117*, 997–1006.

McCarthy, T. R., Torrelles, J. B., MacFarlane, A. S., Katawczik, M., Kutzbach, B., DesJardin, L. E., et al. (2005). Overexpression of mycobacterium tuberculosis manB, a phosphomannomutase that increases phosphatidylinositol mannoside biosynthesis in *Mycobacterium smegmatis* and mycobacterial association with human macrophages. *Molecular Microbiology, 58*, 774–790.

McKay, G. A., Woods, D. E., MacDonald, K. L., & Poole, K. (2003). Role of phosphoglucomutase of *Stenotrophomonas maltophilia* in lipopolysaccharide biosynthesis, virulence, and antibiotic resistance. *Infection and Immunity, 71*, 3068–3075.

Mehra-Chaudhary, R., Mick, J., & Beamer, L. J. (2011c). Crystal structure of *Bacillus anthracis* phosphoglucosamine mutase, an enzyme in the peptidoglycan biosynthetic pathway. *Journal of Bacteriology, 193*, 4081–4087.

Mehra-Chaudhary, R., Mick, J., Tanner, J. J., & Beamer, L. J. (2011b). Quaternary structure, conformational variability and global motions of phosphoglucosamine mutase. *FEBS Journal, 278*, 3298–3307.

Mehra-Chaudhary, R., Mick, J., Tanner, J. J., Henzl, M. T., & Beamer, L. J. (2011a). Crystal structure of a bacterial phosphoglucomutase, an enzyme involved in the virulence of multiple human pathogens. *Proteins, 79*, 1215–1229.

Mengin-Lecreulx, D., & Heijenoort, J. V. (1996). Characterization of the essential gene glmM encoding phosphoglucosamine mutase in *Escherichia coli*. *Molecular Biology, 271*, 32–39.

Mio, T., Yamada-Okabe, T., Arisawa, M., & Yamada-Okabe, H. (2000). Functional cloning and mutational analysis of the human cDNA for phosphoacetylglucosamine mutase: Identification of the amino acid residues essential for the catalysis. *Biochimica et Biophysica Acta—Gene Structure and Expression, 1492*, 369–376.

Mitra, S., Cui, J., Robbins, P. W., & Samuelson, J. (2010). A deeply divergent phosphoglucomutase (PGM) of *Giardia lamblia* has both PGM and phosphomannomutase activities. *Glycobiology, 20*, 1233–1240.

Moiseeva, E., Belkin, A., Spurr, N., Koteliansky, V., & Critchley, D. (2004). A novel dystrophin/utrophin-associated protein is an enzymatically inactive member of the phosphoglucomutase superfamily. *European Journal of Biochemistry, 235*, 103–113.
Molt, S., Bührdel, J. B., Yakovlev, S., Schein, P., Orfanos, Z., Kirfel, G., et al. (2014). Aciculin interacts with filamin C and Xin and is essential for myofibril assembly, remodeling and maintenance. *Journal of Cell Science*, 3578–3592.
Monteiro, M. A., Fortuna-Nevin, M., Farley, J., & Pavliak, V. (2003). Phase-variation of the truncated lipo-oligosaccharide of *Neisseria meningitidis* NMB phosphoglucomutase isogenic mutant NMB-R6. *Carbohydrate Research, 338*, 2905–2912.
Müller, S., Diederichs, K., Breed, J., Kissmehl, R., Hauser, K., Plattner, H., et al. (2002). Crystal structure analysis of the exocytosis-sensitive phosphoprotein, pp63/parafusin (phosphoglucomutase), from *Paramecium* reveals significant conformational variability. *Journal of Molecular Biology, 315*, 141–153.
Namboori, S. C., & Graham, D. E. (2008). Acetamido sugar biosynthesis in the euryarchaea. *Journal of Bacteriology, 190*, 2987–2996.
Naught, L., & Tipton, P. (2001). Kinetic mechanism and pH dependence of the kinetic parameters of *Pseudomonas aeruginosa* phosphomannomutase/phosphoglucomutase. *Archives of Biochemistry and Biophysics, 396*, 111–118.
Naught, L. E., & Tipton, P. A. (2005). Formation and reorientation of glucose 1,6-bisphosphate in the PMM/PGM reaction: Transient-state kinetic studies. *Biochemistry, 44*, 6831–6836.
Nic Lochlainn, L., & Caffrey, P. (2009). Phosphomannose isomerase and phosphomannomutase gene disruptions in *Streptomyces nodosus*: Impact on amphotericin biosynthesis and implications for glycosylation engineering. *Metabolic Engineering, 11*, 40–47.
Nishitani, Y., Maruyama, D., Nonaka, T., Kita, A., Fukami, T. A., Mio, T., et al. (2006). Crystal structures of *N*-acetylglucosamine-phosphate mutase, a member of the α-D-phosphohexomutase superfamily, and its substrate and product complexes. *The Journal of Biological Chemistry, 281*, 19740–19747.
Ollis, D. L., Cheah, E., Cygler, M., Dijkstra, B., Frolow, F., Sybille, M., et al. (1992). The α/β hydrolase fold. *Protein Engineering, Design & Selection, 5*, 197–211.
Ortner, S., Binder, M., Scheiner, O., Wiedermann, G., & Duchêne, M. (1997). Molecular and biochemical characterization of phosphoglucomutases from *Entamoeba histolytica* and *Entamoeba* dispar. *Molecular and Biochemical Parasitology, 90*, 121–129.
Paixão, T. A., Roux, C. M., Den Hartigh, A. B., Sankaran-Walters, S., Dandekar, S., Santos, R. L., et al. (2009). Establishment of systemic *Brucella melitensis* infection through the digestive tract requires urease, the type IV secretion system, and lipopolysaccharide O antigen. *Infection and Immunity*, 77, 4197–4208.
Pandya, C., Brown, S., Pieper, U., Sali, A., Dunaway-Mariano, D., Babbitt, P. C., et al. (2013). Consequences of domain insertion on sequence-structure divergence in a superfold. *Proceedings of the National Academy of Sciences of the United States of America, 110*, E3381–E3387.
Paterson, G. K., Cone, D. B., Peters, S. E., & Maskell, D. J. (2009). The enzyme phosphoglucomutase (Pgm) is required by *Salmonella enterica* serovar typhimurium for O-antigen production, resistance to antimicrobial peptides and in vivo fitness. *Microbiology, 155*, 3403–3410.
Pei, J., & Ficht, T. A. (2004). *Brucella abortus* rough mutants are cytopathic for macrophages in culture. *Infection and Immunity, 72*, 440–450.
Periappuram, C., Steinhauer, L., Barton, D. L., Taylor, D. C., Chatson, B., & Zou, J. (2000). The plastidic phosphoglucomutase from *Arabidopsis*. A reversible enzyme reaction with an important role in metabolic control. *Plant Physiology, 122*, 1193–1199.
Rajesh, T., Song, E., Lee, B. R., Park, S. H., Jeon, J. M., Kim, E., et al. (2012). Increased sensitivity to chloramphenicol by inactivation of manB in Streptomyces coelicolor. *Journal of Microbiology and Biotechnology, 22*, 1324–1329.

Rathi, B., Sarangi, A. N., & Trivedi, N. (2009). Genome subtraction for novel target definition in *Salmonella typhi*. *Bioinformation*, *4*, 143–150.

Rauwerdink, A., & Kazlauskas, R. J. (2015). How the same core catalytic machinery catalyzes 17 different reactions: The serine-histidine-aspartate catalytic triad of α/β-hydrolase fold enzymes. *ACS Catalysis*, *5*, 6153–6176.

Ray, W. J. J. (1969). Role of bivalent cations in the phosphoglucomutase system. 1. Characterization of enzyme-metal complexes. *The Journal of Biological Chemistry*, *244*, 3740–3747.

Ray, W. J., Burgner, J. W., & Post, C. B. (1990). Characterization of vanadate-based transition-state-analogue complexes of phosphoglucomutase by spectral and NMR techniques. *Biochemistry*, *29*, 2770–2778.

Ray, W. K., Keith, S. M., Desantis, A. M., Hunt, P., Larson, T. J., Helm, R. F., et al. (2005). A phosphohexomutase from the archaeon Sulfolobus solfataricus is covalently modified by phosphorylation on serine. *Journal of Bacteriology*, *187*, 4270–4275.

Ray, W. J. J., Mildvan, A. S., & Long, J. W. (1973). Arrangement of the phosphate- and metal-binding subsites of phosphoglucomutase. Intersubsite relationships by means of inhibition patterns. *Biochemistry*, (19), 3724–3732.

Ray, W. J. J., & Roscelli, G. A. (1964). A kinetic study of the phosphoglucomutase pathway. *The Journal of Biological Chemistry*, *239*, 1228–1236.

Regni, C., Naught, L., Tipton, P. A., & Beamer, L. J. (2004). Structural basis of diverse substrate recognition by the enzyme PMM/PGM from *P. aeruginosa*. *Structure*, *12*, 55–63.

Regni, C., Schramm, A. M., & Beamer, L. J. (2006). The reaction of phosphohexomutase from *Pseudomonas aeruginosa*: Structural insights into a simple processive enzyme. *The Journal of Biological Chemistry*, *281*, 15564–15571.

Regni, C., Tipton, P. A., & Beamer, L. J. (2002). Crystal structure of PMM/PGM: An enzyme in the biosynthetic pathway of *P. aeruginosa* virulence factors. *Structure*, *10*, 269–279.

Rocchetta, H. L., Burrows, L. L., & Lam, J. S. (1999). Genetics of O-antigen biosynthesis in *Pseudomonas aeruginosa*. *Microbiology and Molecular Biology Reviews*, *63*, 523–553.

Rodríguez-Díaz, J., Rubio-del-Campo, A., & Yebra, M. J. (2012). Regulatory insights into the production of UDP-N-acetylglucosamine by *Lactobacillus casei*. *Bioengineered*, *3*, 339–342.

Sahu, P. K., Iyer, P. S., Barage, S. H., Sonawane, K. D., & Chopade, B. A. (2014). Characterization of the *algC* gene expression pattern in the multidrug resistant *Acinetobacter baumannii* AIIMS 7 and correlation with biofilm development on abiotic surface. *The Scientific World Journal*, *2014*, 593546.

Sandlin, R. C., & Stein, D. C. (1994). Role of phosphoglucomutase in lipooligosaccharide biosynthesis in *Neisseria gonorrhoeae*. *Journal of Bacteriology*, *176*, 2930–2937.

Sarma, A. V. S., Anbanandam, A., Kelm, A., Mehra-Chaudhary, R., Wei, Y., Qin, P., et al. (2012). Solution NMR of a 463-residue phosphohexomutase: Domain 4 mobility, substates, and phosphoryl transfer defect. *Biochemistry*, *51*, 807–819.

Sassi, A., Lazaroski, S., Wu, G., Haslam, S. M., Fliegauf, M., Mellouli, F., et al. (2014). Hypomorphic homozygous mutations in phosphoglucomutase 3 (PGM3) impair immunity and increase serum IgE levels. *The Journal of Allergy and Clinical Immunology*, *133*, 1410–1419. 1419–13.

Schramm, A. M., Karr, D., Mehra-Chaudhary, R., Van Doren, S. R., Furdui, C. M., & Beamer, L. J. (2010). Breaking the covalent connection: Chain connectivity and the catalytic reaction of PMM/PGM. *Protein Science: A Publication of the Protein Society*, *19*, 1235–1242.

Shackelford, G. S., Regni, C. A., & Beamer, L. J. (2004). Evolutionary trace analysis of the α-D-phosphohexomutase superfamily. *Protein Science*, *13*, 2130–2138.

Shimazu, K., Takahashi, Y., Karibe, H., Mitsuhashi, F., & Konishi, K. (2012). Contribution of phosphoglucosamine mutase to determination of bacterial cell morphology in Streptococcus gordonii. *Odontology*, *100*, 28–33.

Shimazu, K., Takahashi, Y., Uchikawa, Y., Shimazu, Y., Yajima, A., Takashima, E., et al. (2008). Identification of the *Streptococcus gordonii* glmM gene encoding phosphoglucosamine mutase and its role in bacterial cell morphology, biofilm formation, and sensitivity to antibiotics. *FEMS Immunology and Medical Microbiology*, *53*, 166–177.

Silva, P. J., Schulz, C., Jahn, D., Jahn, M., & Ramos, M. J. (2010). A tale of two acids: When arginine is a more appropriate acid than H3O+. *Journal of Physical Chemistry B*, *114*, 8994–9001.

Stiers, K. M., Graham, A. C., Kain, B. N., & Beamer, L. J. (2017). Asp263 missense variants perturb the active site of human phosphoglucomutase 1. *FEBS Journal*, *284*, 937–947.

Stiers, K. M., Kain, B. N., Graham, A. C., & Beamer, L. J. (2016). Induced structural disorder as a molecular mechanism for enzyme dysfunction in phosphoglucomutase 1 deficiency. *Journal of Molecular Biology*, *428*, 1493–1505.

Stiers, K. M., Xu, J., Lee, Y., & Beamer, L. J. (n.d.). Phosphorylation dependent effects on the structural flexibility of phosphoglucosamine mutase. To Be Submitted.

Stojkovic, T., Vissing, J., Petit, F., Piraud, M., Orngreen, M. C., Andersen, G., et al. (2009). Muscle glycogenosis due to phosphoglucomutase 1 deficiency. *New England Journal of Medicine*, *361*, 425–427.

Stray-Pedersen, A., Backe, P. H., Sorte, H. S., Mørkrid, L., Chokshi, N. Y., Erichsen, H. C., et al. (2014). PGM3 mutations cause a congenital disorder of glycosylation with severe immunodeficiency and skeletal dysplasia. *American Journal of Human Genetics*, *95*, 96–107.

Swords, W. E., Buscher, B. A., Ver Steeg Li, K., Preston, A., Nichols, W. A., Weiser, J. N., et al. (2000). Non-typeable *Haemophilus influenzae* adhere to and invade human bronchial epithelial cells via an interaction of lipooligosaccharide with the PAF receptor. *Molecular Microbiology*, *37*, 13–27.

Tang, H., DiMango, E., Bryan, R., Gambello, M., Iglewski, B., Goldberg, J., et al. (1996). Contribution of specific *Pseudomonas aeruginosa* virulence factors to pathogenesis of pneumonia in a neonatal mouse model of infection. *Infection and Immunity*, *64*, 37–43.

Tavares, I. M., Leitão, J. H., & Sá-Correia, I. (2003). Chromosomal organization and transcription analysis of genes in the vicinity of *Pseudomonas aeruginosa* glmM gene encoding phosphoglucosamine mutase. *Biochemical and Biophysical Research Communications*, *302*, 363–371.

Tegtmeyer, L. C., Rust, S., van Scherpenzeel, M., Ng, B. G., Losfeld, M.-E., Timal, S., et al. (2014). Multiple phenotypes in phosphoglucomutase 1 deficiency. *New England Journal of Medicine*, *370*, 533–542.

Ugalde, J., Comerci, D., Leguizamón, M., & Ugalde, R. (2003). Evaluation of *Brucella abortus* phosphoglucomutase (pgm) mutant as a new live rough-phenotype vaccine. *Infection and Immunity*, *71*, 6264–6269.

Ugalde, J., Czibener, C., Feldman, M., & Ugalde, R. (2000). Identification and characterization of the *Brucella abortus* phosphoglucomutase gene: Role of lipopolysaccharide in virulence and intracellular multiplication. *Infection and Immunity*, *68*, 5716–5723.

Wang, H., Yang, Y., Lin, L., Zhou, W., Liu, M., Cheng, K., et al. (2016). Engineering *Saccharomyces cerevisiae* with the deletion of endogenous glucosidases for the production of flavonoid glucosides. *Microbial Cell Factories*, *15*, 134.

Wei, Y., Marcink, T. C., Xu, J., Sirianni, A. G., Sarma, A. V. S., Prior, S. H., et al. (2014). Chemical shift assignments of domain 4 from the phosphohexomutase from *Pseudomonas aeruginosa* suggest that freeing perturbs its coevolved domain interface. *Biomolecular NMR Assignments*, *8*, 329–333.

West, N. P., Jungnitz, H., Fitter, J. T., Mcarthur, J. D., Guzman, C. A., & Walker, M. J. (2000). Role of phosphoglucomutase of *Bordetella bronchiseptica* in lipopolysaccharide biosynthesis and virulence. *Infection and Immunity*, *68*, 4673–4680.

Whitehouse, D. B., Tomkins, J., Lovegrove, J. U., Hopkinson, D. A., & McMillian, W. O. (1998). A phylogenetic approach to the identification of phosphoglucomutase genes. *Molecular Biology and Evolution*, *15*, 456–462.

Willems, A. P., van Engelen, B. G. M., & Lefeber, D. J. (2016). Genetic defects in the hexosamine and sialic acid biosynthesis pathway. *Biochimica et Biophysica Acta, 1860*, 1640–1654.

Wright, G. D. (2016). Antibiotic adjuvants: Rescuing antibiotics from resistance. *Trends in Microbiology, 24*, 1–10.

Xu, J., Lee, Y., Beamer, L. J., & Van Doren, S. R. (2015). Phosphorylation in the catalytic cleft stabilizes and attracts domains of a phosphohexomutase. *Biophysical Journal, 108*, 325–337.

Yajima, A., Takahashi, Y., Shimazu, K., Urano-Tashiro, Y., Uchikawa, Y., Karibe, H., et al. (2009). Contribution of phosphoglucosamine mutase to the resistance of *Streptococcus gordonii* DL1 to polymorphonuclear leukocyte killing. *FEMS Microbiology Letters, 297*, 196–202.

Yankeelov, J., & Koshland, D. J. (1965). Evidence for conformational changes induced by substrates of phosphoglucomutase. *The Journal of Biological Chemistry, 240*, 1593–1602.

Ye, R. W., Zielinski, N. A., & Chakrabarty, A. M. (1994). Purification and characterization of Phosphomannomutase/Phosphoglucomutase from Pseudomonas aeruginosa involved in biosynthesis of both alginate and lipopolysaccharide. *Journal of Bacteriology, 176*, 4851–4857.

Zabala, D., Braña, A. F., Flórez, A. B., Salas, J. A., & Méndez, C. (2013). Engineering precursor metabolite pools for increasing production of antitumor mithramycins in *Streptomyces argillaceus*. *Metabolic Engineering, 20*, 187–197.

Zhang, Y., Yu, X., Ichikawa, M., Lyons, J. J., Datta, S., Lamborn, I. T., et al. (2014). Autosomal recessive phosphoglucomutase 3 (PGM3) mutations link glycosylation defects to atopy, immune deficiency, autoimmunity, and neurocognitive impairment. *The Journal of Allergy and Clinical Immunology, 133*, 1400–1409.

Zhou, D., Stephens, D. S., Gibson, B. W., Engstrom, J., McAllister, C., Lee, F., et al. (1994). Lipooligosaccharide biosynthesis in pathogenic. *The Journal of Biological Chemistry, 269*, 11162–11169.

CHAPTER NINE

Mechanistic Insights Into Catalytic RNA–Protein Complexes Involved in Translation of the Genetic Code

Satya B. Routh, Rajan Sankaranarayanan[1]
CSIR–Centre for Cellular and Molecular Biology, Hyderabad, India
[1]Corresponding author: e-mail address: sankar@ccmb.res.in

Contents

1. Introduction 306
2. RNPCs in Translation 307
   2.1 Ribonuclease P 307
   2.2 Spliceosome 313
   2.3 Ribosome 318
   2.4 Aminoacyl-tRNA Synthetases and Their Proofreading Modules 325
3. Conclusion 338
Acknowledgments 339
References 339

Abstract

The contemporary world is an "RNA–protein world" rather than a "protein world" and tracing its evolutionary origins is of great interest and importance. The different RNAs that function in close collaboration with proteins are involved in several key physiological processes, including catalysis. Ribosome—the complex megadalton cellular machinery that translates genetic information encoded in nucleotide sequence to amino acid sequence—epitomizes such an association between RNA and protein. RNAs that can catalyze biochemical reactions are known as ribozymes. They usually employ general acid–base catalytic mechanism, often involving the 2′-OH of RNA that activates and/or stabilizes a nucleophile during the reaction pathway. The protein component of such RNA–protein complexes (RNPCs) mostly serves as a scaffold which provides an environment conducive for the RNA to function, or as a mediator for other interacting partners. In this review, we describe those RNPCs that are involved at different stages of protein biosynthesis and in which RNA performs the catalytic function; the focus of the account is on highlighting mechanistic aspects of these complexes. We also provide a perspective on such associations in the context of proofreading during translation of the genetic code. The latter aspect is not much appreciated and recent works suggest that this is an avenue worth exploring, since an understanding of the subject can provide useful insights into how RNAs collaborate with proteins to ensure fidelity during these essential cellular processes. It may also aid in comprehending evolutionary aspects of such associations.

Advances in Protein Chemistry and Structural Biology, Volume 109
ISSN 1876-1623
http://dx.doi.org/10.1016/bs.apcsb.2017.04.002

1. INTRODUCTION

The origin of life on earth has always intrigued evolutionary biologists, and it is still a matter of much speculation and debate. Of the many theories that have been propounded to explain how life originated on earth, the "RNA world hypothesis" has arguably caught the attention of most and has probably been the most investigated paradigm among all. In 1962, Alexander Rich was the first to posit that ribonucleic acid (RNA) could have been the precursor to life on earth (Rich, 1962). The idea was further propagated in a more formal manner in the second half of 1960s independently by Woese, Crick, and Orgel (Crick, 1968; Orgel, 1968; Woese, 1967). However, it is still a "chicken and egg" paradox as to which came first—nucleic acids or proteins—although it was realized very early that nucleic acids are the hereditary material that code for proteins, but proteins are required to replicate the genetic material for the latter's transference to the next generation. In essence, nucleic acids and proteins were known to shake hands with each other for their own synthesis, sustenance, and maintenance (Orgel, 2000).

The basis for the belief that RNA could have given birth to primordial life forms stemmed from the minimal pieces of evidence that were available then, and also from certain assumptions. The fact that RNA is the genetic blueprint of several viruses such as tobacco mosaic virus was a major indicator for RNA-based origin of life on earth (Pennazio & Roggero, 2000). In addition to this, it was also surmised that nucleotide cofactors (e.g., NAD, NADP, FAD, etc.) or their derivatives (e.g., ATP, CoA, etc.) represent relics from the primitive earth, wherein RNA existed in the absence of proteins (Orgel, 1968; Orgel & Sulston, 1971; White, 1976; Woese, 1967). Nevertheless, it was not until the discovery of ribozymes (catalytic RNAs) in the early 1980s independently by Cech and Altman that the catalytic potential of RNAs was realized and appreciated (Guerrier-Takada, Gardiner, Marsh, Pace, & Altman, 1983; Kruger et al., 1982). This unanticipated finding provided immense strength to the idea that RNA was the foundation on which life came into existence in primeval times. Consequently, the term "RNA world" was coined to signify a "prelife world" comprising various forms of RNA (Gilbert, 1986), and the scientific community delved deeper into this area of research to uncover newer functional regimes of RNA, to discover yet unidentified forms of RNA, and to unearth more hidden proofs about RNAs role in giving "life to earth."

Irrespective of the multiple theories on the origin of life on earth, research over the past two to three decades has firmly established that the present-day world is an "RNA–protein world" in which RNA plays a central role in metabolism (Altman, 2013). The notion that proteins are the only workhorses of the cell has now been rendered untenable, considering the myriad functions that are now known to be performed by RNA apart from just being a genetic message carrier or a delivery molecule. These functions include catalysis of several key biochemical reactions and regulation of a variety of cellular processes including gene expression. Interestingly, many of these functions are carried out by a diverse array of RNA species not in isolation but in close association with proteins. In light of this fact, it has also been theorized that an "RNA–peptide world" rather than an "RNA world" could have predated the earliest life forms on earth (Alva, Söding, & Lupas, 2015; Kunin, 2000; Pascal, 2007; Patel, Percivalle, Ritson, Duffy, & Sutherland, 2015; Söding & Lupas, 2003). According to this model of origin of life, RNA and peptide coexisted and coevolved to give rise to all extant biological systems.

Not only from an evolutionary standpoint but also from the functional angle, it is pertinent to describe the different associations between RNAs and proteins that are meant for the accomplishment of specific physiological goals. In the current account, we describe the mechanistic aspects of those RNA–protein complexes (RNPCs) that have catalytic function ingrained in their RNA component and are involved directly or indirectly in translation of the genetic code. A perspective is also provided implicating the role of such RNPCs in ensuring translational fidelity and to emphasize the need to address this aspect more thoroughly. This holds great significance since it is an area which is still underappreciated, and atomic-level insights are now becoming available on such systems through overall development in the field of structural biology, particularly using cryo-electron microscopy (cryo-EM) (Bai, McMullan, & Scheres, 2015; Fernandez-Leiro & Scheres, 2016; Khoshouei et al., 2016; Punjani, Rubinstein, Fleet, & Brubaker, 2017; Subramaniam, Earl, Falconieri, Milne, & Egelman, 2016).

2. RNPCS IN TRANSLATION

2.1 Ribonuclease P

Historically, ribonuclease P (RNase P) was one of the first two ribozymes to be discovered simultaneously, the other being self-splicing introns (RNase P was discovered in the 1970s, but the fact that its RNA component has catalytic activity was discovered in the early 1980s) (Guerrier-Takada et al., 1983; Kruger et al., 1982; Robertson, Altman, & Smith, 1972). However, the latter is an example wherein RNA is the sole component of the enzymatic system and does not involve or require protein partners. This automatically implies that RNase P was the first RNPC that was found to have its catalytic activity residing in the RNA component rather than the protein component. Given the focus of the current account on catalytic RNPCs, those catalytic forms of RNA (group I and group II introns, and a number of other ribozymes such as *glmS* ribozyme, hairpin ribozyme, hammerhead ribozyme, etc.) that do not associate with proteins stand beyond the scope of the review (Ferré-D'Amaré & Scott, 2010; Jimenez, Polanco, & Lupták, 2015; McNeil, Semper, & Zimmerly, 2016; Nielsen & Johansen, 2009). Furthermore, of the two forms of RNase P, the protein-only RNase P or proteinaceous RNase P (PRORP) (reported to exist in the nucleus and/or mitochondria/chloroplasts of numerous eukaryotic species) that lacks an RNA component naturally lies outside of the review's purview and will thus not be dealt with (Schelcher, Sauter, & Giegé, 2016).

RNase P is a universally conserved RNPC which occurs in all the three domains of life, viz., Bacteria, Archea, and Eukarya. It is also one of the only three known instances of multiple turnover ribozyme, the other two being spliceosome and ribosome (Klemm et al., 2016; Mondragón, 2013). It consists of a single RNA molecule, which is referred to as P RNA, as well as a protein component comprising 1 polypeptide (called RnpA) in bacteria, 4 or 5 in archea, and up to 10 in eukaryotes (Altman, 2011). The biochemical function of this ribozyme is to endonucleolytically cleave off the 5′ leader sequence from a precursor tRNA (pre-tRNA), thereby leading to the latter's maturation (Robertson et al., 1972). It is also involved in the processing of other RNAs, such as 4.5S RNA and transfer-messenger RNA (tmRNA), wherein it cleaves off a single-stranded nucleotide stretch from the 5′ end of a double-stranded region of RNA (Altman, 2011; Esakova & Krasilnikov, 2010; Marvin & Engelke, 2009). The endonucleolytic cleavage reaction carried out by RNase P results in the formation of two products—a single-stranded RNA piece having a 3′-hydroxyl (3′-OH) and a double-stranded RNA harboring a 5′-phosphate (5′-P) (Mondragón, 2013). Moreover, RNase P in human has been demonstrated to be needed for proper transcription of several RNA polymerase III–transcribed small noncoding RNAs, including 5S ribosomal RNA (rRNA), U6 small nuclear RNA (snRNA), and signal recognition particle RNA (Jarrous & Reiner, 2007; Reiner, Ben-Asouli, Krilovetzky, & Jarrous, 2006).

The core of P RNA comprises two independently folded domains, viz., catalytic (C) domain and specificity (S) domain, the former being the larger of the two (Fig. 1A). As the name itself suggests, the C-domain is essential for catalysis, while the S-domain governs substrate specificity (Loria & Pan, 1996; Pan, 1995). Crystal structures of bacterial RNase P have provided useful insights into the relative domain disposition of P RNA, spatial organization of RnpA, and substrate recognition/binding mode. The two domains come together to form a curved bilayered architecture consisting of multiple coaxially stacked helices (Kazantsev et al., 2005; Torres-Larios, Swinger, Krasilnikov, Pan, & Mondragón, 2005). There are five sequentially distant but universally conserved regions (CR) in P RNA which form two universally conserved three-dimensional modules. The C-domain contains CR-I, -IV, and -V which together form one module, whereas CR-II and -III in the S-domain form the other module (Chen & Pace, 1997; Loria & Pan, 1996; Pan, 1995). In a typical bacterial P RNA, there are about 18 helices (designated P1 to P18). Of these, only those that compose the core are also present in the archeal and eukaryotic counterparts; the peripheral parts (e.g., P13, P14, and P18 are absent in most archeal P RNA) are lacking in these P RNAs (Harris, Haas, Williams, Frank, & Brown, 2001). Substrate recognition by the S-domain is effected through its interaction with the TψC and the D loops of pre-tRNA substrate (Fig. 1B). The conserved module in the S-domain forms a T loop which forms an interleaved structure with the T loop of pre-tRNA. It is therefore suggested that the sequence conservation in this module is a manifestation of structural constraint that is crucial for substrate recognition (LaGrandeur, Huttenhofer, Noller, & Pace, 1994; Loria, Niranjanakumari, Fierke, & Pan, 1998; Loria & Pan, 1997; Odell, Huang, Jakacka, & Pan, 1998; Pan, Loria, & Zhong, 1995; Reiter et al., 2010).

RnpA, the protein component of bacterial RNase P, is a small protein which has ~120 residues, constituting ~10% of the bacterial holoenzyme by mass. The crystal structure of RNase P holoenzyme from *Thermotoga maritima* in complex with *Escherichia coli* $tRNA^{Phe}$ has revealed that RnpA could be involved in the recognition and proper positioning of the 5′ leader sequence of the pre-tRNA substrate, although its exact function is still not known (Reiter et al., 2010). RnpA is dispensable in vitro, since bacterial P RNA has been shown to retain catalytic activity, albeit at higher cation concentrations (Guerrier-Takada et al., 1983). However, the protein is necessary for the ribozyme's in vivo activity (Apirion, 1980; Schedl & Primakoff, 1973). Moreover, the catalytic efficiency of P RNA enhances

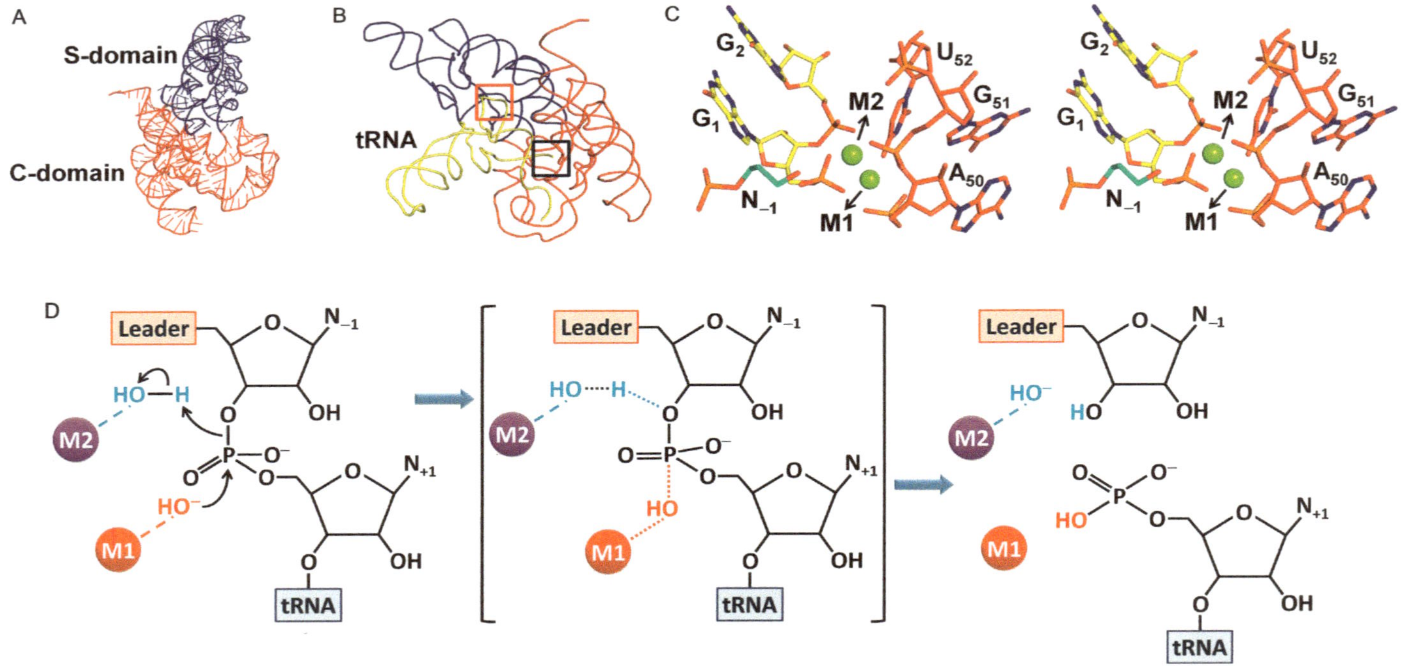

Fig. 1 RNase P and its catalytic mechanism. (A) Structure of P RNA of RNase P from *Thermotoga maritima* (PDB id: 3Q1Q) showing catalytic domain (C-domain, *red*) and specificity domain (S-domain, *blue*). (B) Structure of P RNA in complex with $tRNA^{Phe}$ (*yellow*) (PDB id: 3Q1Q) depicting interaction of S-domain with TψC and D loops of $tRNA^{Phe}$ (*red box*), and catalytic site harboring the 5′ end (*black box*). (C) Stereo image of the catalytic center showing two metal ions M1 and M2 (*green spheres*) and the N_{-1} residue of 5′ leader (*cyan*); P RNA and $tRNA^{Phe}$ residues are rendered in *red* and *yellow*, respectively. (D) Proposed two metal ion mechanism of catalysis.

by at least two orders of magnitude when it is present in complex with RnpA (Buck, Dalby, Poole, Kazantsev, & Pace, 2005; Kurz, Niranjanakumari, & Fierke, 1998). In the case of archeal and eukaryotic RNase P, P RNA has markedly reduced in vitro biochemical activity when it is present alone and, hence, requires the cooperation of complementary proteins to achieve its optimal efficiency. These proteins also stabilize the P RNA structure and are probably required for its proper folding (Esakova & Krasilnikov, 2010; Jarrous & Gopalan, 2010; Marvin & Engelke, 2009; Walker & Engelke, 2006).

As regards catalysis, RNase P is a metalloribozyme which requires divalent cation such as Mg^{2+} for its activity as has been determined through biochemical investigations. The latter have shown that at least two divalent metal ions are needed for both substrate binding and catalysis (Howard, Klemm, & Fierke, 2015). Transient kinetic analyses have revealed that RNase P undergoes a four-step reaction or catalytic cycle. The first step involves binding of the pre-tRNA substrate (S) to the ribozyme (E) forming the ES complex. In the next step, termed "conformational change" step, the ES complex transitions to the activated or catalytically competent state (ES*) in a metal ion-dependent manner. This is followed by the conversion of substrate to products, the latter being still bound to the enzyme (EP). In the final step, the products dissociate from the ribozyme, thereby making the latter available for the next catalytic cycle (Hsieh & Fierke, 2009; Hsieh et al., 2010). The product-release step in the reaction scheme is the rate-limiting step (Beebe & Fierke, 1994). Additional biophysical and biochemical approaches have demonstrated that RNase P uses single ionization producing a metal-hydroxide nucleophile to effect catalysis, as opposed to a general base used by nucleases (Fig. 1D) (Cassano, Anderson, & Harris, 2004a). Furthermore, the transition from ES to ES* is accompanied by minor but crucial repositioning of the substrate in the active site with the activated ES* complex being stabilized by divalent cations. It is believed that the "conformational change" step serves to discriminate between cognate and noncognate substrates, and probably involves unwinding of the two ends of pre-tRNA. Thus, transition to ES* acts as the proofreading step during catalysis (Hsieh et al., 2010; Pomeranz Krummel, Kent, MacMillan, & Altman, 2000; Sun, Campbell, Zahler, & Harris, 2006).

The ligand-bound structures of RNase P have validated some of the biochemical findings, thereby providing useful mechanistic insights (Reiter et al., 2010). In the crystal structures of tRNA-bound holoenzyme, two metal ions are seen situated close to each other in the active site of the enzyme (Fig. 1C). One (M1) is found surrounded by the 5′ end of mature tRNA on the one hand, and the phosphoryl groups of A50 and G51, and O4 of the universally conserved U52 on the other hand (numbering according to RNase P from *T. maritima*). M1 has been proposed to be the primary catalytic ion. The other metal ion (M2) lies in the vicinity of the phosphoryl group of G51 from P RNA, the 3′-OH of the leader, and the 5′ end of the tRNA. Interestingly, there is no involvement of the protein RnpA in coordinating the two metal ions. The crystal structures, combined with a number of other biophysical/biochemical analyses, have suggested that RNase P follows a general two metal ion mechanism for catalysis (Fig. 1D) (Beebe, Kurz, & Fierke, 1996; Christian, Kaye, & Harris, 2000; Christian, Smith, Perera, & Harris, 2006; Hsieh et al., 2010; Kirsebom, 2010; Steitz & Steitz, 1993). Such a mechanism has been proposed for other large ribozymes, nucleases, and polymerases (Mondragón, 2013). Additional metal ion-binding sites have been identified on both P RNA and pre-tRNA through single-atom modifications and functional group substitutions (Chen, Li, & Gegenheimer, 1997; Christian, Kaye, & Harris, 2002; Crary, Kurz, & Fierke, 2002; Persson, Cuzic, & Hartmann, 2003; Pfeiffer et al., 2000; Smith & Pace, 1993; Warnecke, Fürste, Hardt, Erdmann, & Hartmann, 1996; Warnecke, Held, Busch, & Hartmann, 1999). These data have implicated the auxiliary metal ions along with M1 and M2 in orchestrating catalysis. Experiments have further suggested that the 2′-OH of N_{-1} nucleotide of pre-tRNA substrate interacts with a metal-bound water molecule, the latter being proposed to be the catalytic water (Kazantsev & Pace, 1998; Persson et al., 2003). Kinetic isotope effect analysis has indicated that the catalytic water molecule, likely to be activated by M1 and stabilized by M2, mounts a nucleophilic attack on the scissile phosphodiester bond, while another water molecule is coordinated by M2 to protonate the 3′ oxyaninon leaving group, thereby causing hydrolysis or cleavage of the bond. It is also plausible that both M1 and N_{-1} 2′-OH together with M2 are involved in stabilizing the transition state during the reaction pathway. Catalysis by RNase P is thought to proceed in a concerted fashion involving a pentacoordinate transition state (Cassano et al., 2004a; Cassano, Anderson, & Harris, 2004b). Interestingly, even PRORP appears to employ the two metal ion-based catalytic mechanism like its ribozymic counterpart (Schelcher et al., 2016).

Despite such in-depth understanding, a lot remains to be addressed and deciphered about this RNPC. The transient kinetic studies require further complementation from structural studies, probably using a combination of time-resolved crystallography and cryo-EM, to gain deeper mechanistic insights. Moreover, these approaches can provide a better comprehension of the dynamics involved during the catalytic cycle. The kinetics of catalysis by archeal and eukaryotic RNase P, and the mechanistic details thereof remain to be elucidated. Another aspect that needs to be probed is whether RNase P uses the same mechanism for nonpre-tRNA substrate. The roles of RnpA and individual protein subunits of archeal or eukaryotic RNase P also require further investigation. Besides, the evolutionary links between RNase P from different domains of life as well as between the different variants of RNase P need to be established.

2.2 Spliceosome

Spliceosome is a large RNPC which, as the name itself indicates, is involved in splicing (i.e., ligating) the coding regions (called exons) of a precursor messenger RNA (pre-mRNA) with the concomitant excision of intervening noncoding regions (termed introns) (Berget, Moore, & Sharp, 1977; Brody & Abelson, 1985; Chow, Gelinas, Broker, & Roberts, 1977; Grabowski, Seiler, & Sharp, 1985). The consequence of splicing is the formation of mature mRNA from pre-mRNA in the nucleus. This multimegadalton complex is specific to eukaryotic organisms, since split genes—wherein exons are interspersed with introns—are characteristic of the domain Eukarya. Splicing also proceeds via another pathway whereby the introns themselves act as ribozymes and catalyze their own removal, hence the name self-splicing introns (one of the first two ribozymes to be discovered!) (Kruger et al., 1982). However, such introns do not form RNPC and therefore will not be discussed in this review. One of the intriguing aspects of spliceosome (which unfortunately also happens to be impeding for a holistic understanding of the complex) is its diversity across different eukaryotic groups (Hudson, Stark, Fast, Russell, & Rader, 2015). Despite this fact, the catalytic core of spliceosome remains the same throughout.

Splicing is a precise event and any error in the excision of introns or joining of exons can have deleterious consequences (Novoyatleva, Tang, Rafalska, & Stamm, 2006; Ward & Cooper, 2010). The information for such precision resides in short stretches of conserved sequences in the intron itself (Will & Lührmann, 2011). These sequences are present at the 5′ splice site (5′SS), the 3′SS, and a branch site (BS). The consensus sequence at the 5′SS in yeast is GURAGU (in metazoa, it is GUAUGU), whereas the one at the 3′SS in all eukaryotes is YAG. BS has a consensus sequence UACUAAC in yeast or YNCURAC in metazoa, the adenosine being the one that is absolutely essential for splicing. BS typically occurs 18–40 nucleotides upstream of 3′SS. A characteristic stretch of tandem pyrimidines, called polypyrimidine tract, is present between BS and 3′SS in metazoa, but not in lower eukaryotes such as fungi. Apart from these consensus sequences, there are other *cis*-acting elements which can modulate splicing by either enhancing or repressing the same. These are referred to as exonic and intronic splicing enhancers (ESEs and ISEs) or exonic and intronic splicing silencers (ESSs and ISSs). The different consensus sequences serve as cues for the assembly of spliceosome on pre-mRNA (Lee & Rio, 2015; Matlin, Clark, & Smith, 2005; Wang & Burge, 2008).

Spliceosome is an extremely dynamic and gargantuan macromolecular complex in which the associated structural rearrangements begin right from its de novo assembly on pre-mRNA and goes on all the way to catalysis and its disassembly (Fig. 2A) (Matera & Wang, 2014; Papasaikas & Valcárcel, 2016; Sperling, 2017; Will & Lührmann, 2011). A typical spliceosome is comprised of five small nuclear ribonucleoprotein particles (snRNPs) which associate with a diverse array of non-snRNP factors whose number varies from tens in certain organisms to hundreds in others (Hudson et al., 2015; Jurica & Moore, 2003; Will & Lührmann, 2011). The snRNPs involved in the assembly of spliceosome are named U1, U2, U4, U5, and U6. Splicing of two exons in a nuclear pre-mRNA by spliceosome takes place via two successive transesterification reactions (Fig. 2C) (Galej, Nguyen, Newman, & Nagai, 2014; Papasaikas & Valcárcel, 2016; Sperling, 2017). The sequence of biochemical events leading to the joining of two exons and simultaneous excision of the intervening intron is as follows. In the first step, the 2′-OH of BS adenosine (A) mounts a nucleophilic attack on the phosphodiester bond at the 5′SS (i.e., the phosphate linking the 5′-terminal guanosine (G) of 5′SS to its preceding nucleoside) using a nucleophilic substitution type 2 (S_N2) reaction mechanism. This results in the separation of the 5′ exon (i.e., exon upstream of the intron) from the rest of the transcript, accompanied by the formation of a lariat intermediate in which the BS adenosine (A) enters into a 2′–5′ phosphodiester bond with the 5′-terminal guanosine (G) of the intron. In the second transesterification reaction, also proceeding via S_N2 mechanism, the 3′-OH of the severed 5′ exon nucleophilically attacks the phosphodiester bond at the 3′SS (i.e., the phosphate linking the 3′-terminal guanosine (G) of 3′SS to its succeeding nucleoside). Consequently, the 5′ exon is joined/ligated to the 3′ exon through a 5′–3′ phosphodiester linkage, whereas the intervening intron is released/excised as a lariat. Such a reaction scheme is also followed by group II introns (found in bacterial, mitochondrial, and chloroplast genomes) and hence it was proposed that spliceosome-mediated splicing has evolved from group II introns.

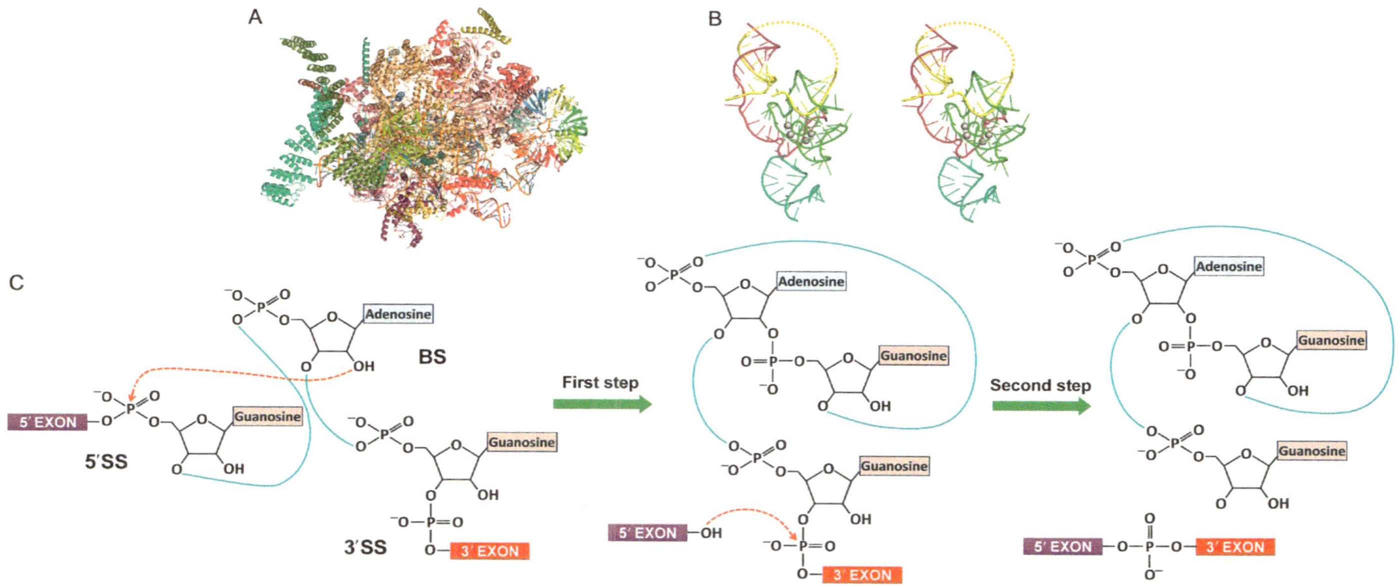

Fig. 2 Spliceosome and splicing mechanism. (A) Structure of C* complex (catalytically competent complex for the second transesterification reaction) of spliceosome from *Saccharomyces cerevisiae* (PDB id: 5MPS). (B) Stereo image of the catalytic site of spliceosome from *Schizosaccharomyces pombe* (PDB id: 3JB9) depicting U2, U5, and U6 snRNAs in *magenta*, *cyan*, and *green*, respectively; lariat is colored *yellow*. The *yellow-dashed line* represents those residues of the lariat that could not be modeled because of lack of electron density. Note the presence of four metal ions (*gray spheres*) in the active site. (C) The two transesterification reactions that occur through S_N2 nucleophilic attack by the 2′- and 3′-OH groups.

In addition to the similarity in the steps during catalysis by spliceosome and group II introns, there are several other compelling proofs which indicate that spliceosome has evolved from group II introns. The two major lines of evidence in this regard are (a) the overall fold or structure of the catalytic centre of spliceosome is strikingly similar to that of group II introns (Fica, Mefford, Piccirilli, & Staley, 2014; Hang, Wan, Yan, & Shi, 2015) and (b) spliceosome-mediated splicing follows the two metal ion catalytic mechanism like group II introns (Fica et al., 2013). These observations have come to the fore through a number of biophysical, biochemical, and structural investigations that different spliceosomal complexes have been subjected to, including the recent cryo-EM structures of U4/U6.U5 tri-snRNP complex from *Saccharomyces cerevisiae* and U2.U6.U5 postspliceosomal complex from *Schizosaccharomyces pombe* (Nguyen et al., 2015; Wan et al., 2016; Yan et al., 2015). The catalytic core of spliceosome comprises helix I of U2/U6 snRNA, intramolecular stem loop of U6 snRNA and U5 snRNA (Fig. 2B). In addition to these elements in the catalytic site of spliceosome, two catalytic Mg^{2+} ions are anchored by various phosphate groups. One of the two Mg^{2+} ions (M1) is held by the phosphates of G48, G66, and U68, whereas the other (M2) is coordinated by the phoshphates of A47, G48, and U68, with all the four nucleotides belonging to U6 snRNA. The two catalytic Mg^{2+} ions participate in activating the 2′-OH and 3′-OH nucleophiles during the two successive nucleophilic attacks as well as in stabilizing the partial negative charges that develop in the transition states. Interestingly, the two ions M1 and M2 switch their roles of activation and stabilization during the two transesterification reactions. The aforesaid elements of the spliceosomal catalytic centre, including the Mg^{2+} ions, are all sheltered in a positively charged pocket of Prp8 protein. In the structure of postspliceosomal complex, the intron lariat is seen bound by base pairing interactions with both U2 and U6 snRNAs. Furthermore, the positions of the two catalytic Mg^{2+} ions in spliceosome correspond to those observed in group II intron (Nguyen et al., 2016; Papasaikas & Valcárcel, 2016; Sperling, 2017).

It was proposed that spliceosome utilizes the same catalytic site for performing the two catalytic steps. As such, a two-state conformation model was put forth (Konarska & Query, 2005; Query & Konarska, 2004; Will & Lührmann, 2011). According to the model, the lariat intermediate that is produced during the first catalytic step moves away from the catalytic site so as to accommodate the 3′SS and make the site available for the second step. Such conformational remodeling of the lariat intermediate in the complex would obviate the need for the free 5′ exon to be repositioned for the second transesterification reaction. Support for the model came from observations that many bonds in spliceosome are broken and reinstated during the dynamics of the "spliceosomal dance" (Hoskins, Gelles, & Moore, 2011; Smith, Query, & Konarska, 2008; Warnasooriya & Rueda, 2014). The interactions with stem II of U2 snRNA present an instance of such remodeling events (Hilliker, Mefford, & Staley, 2007; Perriman & Ares, 2007). The model gained further support when the catalytic steps were shown to be reversible in nature (Tseng & Cheng, 2008). Very recently, three groups working independently have been able to capture the C* complex of spliceosome (the activated conformer for the second transesterification) and solve its structure using cryo-EM (Bertram et al., 2017; Fica et al., 2017; Yan, Wan, Bai, Huang, & Shi, 2017). It is observed there are several structural variations between complex C (the precatalytic complex) and its catalytically competent state complex C*. A comparison of these two complexes has revealed that the branch helix is rotated by 75 degree in C* with respect to that in C. Additionally, the branched lariat intermediate in C* is repositioned, wherein the lariat sits ~20 Å apart from the active site, thereby creating space for the 3′SS. The active site still retains the U6 snRNA catalytic core firmly anchored in the cavity of Prp8 protein. Moreover, the ACAGAGA stretch of U6 snRNA that stabilizes the 5′SS through base pairing is remodeled in C*. In a nutshell, the recent cryo-EM structures of complex C and complex C* have further validated the two-state conformation model, besides providing useful mechanistic insights into the catalytic phase of spliceosome dynamics.

From the wealth of knowledge accumulated over the years, it is quite evident that the huge and variable repertoire of proteins in the spliceosome orchestrates the modulation of the malleable and intricate network of protein–protein, RNA–RNA, and protein–RNA interactions, and therefore the overall spliceosomal dynamics. Chief among them are helicases belonging to DExD/H-box family of RNA unwindases/RNPases, of which the conserved ones are eight in number in *S. cerevisiae* and five additional ones in higher eukaryotes (De, Schmitzová, & Pena, 2016). Besides, the "protein embellishments" on the core spliceosome particle hold the key to the diverse regulation of splicing process which in turn generates the whole gamut of transcripts (alternative forms of mature mRNAs, called isoforms) through alternative splicing in a spatio-temporal manner depending on the signaling cues and metabolic requirement of the cell (Chen & Manley, 2009; Lee & Rio, 2015; Matlin et al., 2005; Nilsen & Graveley, 2010). An extreme case that exemplifies such a scenario is presented by *Dscam* gene in *Drosophila melanogaster* which has 38,016 isoforms that are synthesized through alternative splicing (Schmucker et al., 2000). All these features make spliceosome a protein-dependent metalloribozyme which performs substrate-assisted catalysis. This places spliceosome in a different league from ribosome since in the latter case the constituent proteins are thought to have only peripheral roles.

Alongside endeavors that intend to decipher the underlying mechanisms and dynamics of spliceosome assembly, catalysis and disassembly, efforts are on toward understanding the phenomenon of cotranscriptional splicing (Bentley, 2014). On the basis of several recent studies, it is now believed that most nuclear pre-mRNAs undergo cotranscriptional processing. This is opposed to the earlier notion that transcript maturation is a sequential event, wherein splicing occurs in the end just before the mRNA is exported to the cytoplasm from the nucleus. Evidence for such a phenomenon has been gathered from imaging studies that have identified supraspliceosome, a mega complex comprising multiple spliceosomes associated with a single transcript, analogous to polysome in which multiple ribosomes are present on an mRNA like "beads on a string" (Azubel, Habib, Sperling, & Sperling, 2006; Shefer, Sperling, & Sperling, 2014). Nevertheless, it is certain that a lot remains to be learnt about and from this cellular mega structure.

2.3 Ribosome

Ribosome was the first RNPC that was discovered in the mid 1950s by Palade (1955). However, the fact that it is a ribozyme was realized only in the early part of the last decade, about half-a-century later from the time of its discovery (Ban, Nissen, Hansen, Moore, & Steitz, 2000; Nissen, Hansen, Ban, Moore, & Steitz, 2000), although propositions and indications had been in place for a long time (Barta, Steiner, Brosius, Noller, & Kuechler, 1984; Crick, 1968; Vester & Garrett, 1988). Ribosome is another complex mega intracellular structure like spliceosome and is responsible for protein biosynthesis in all organisms. It has two subunits—a small subunit and a large subunit—which are synthesized and exist independently of each other (Fig. 3A). These two subunits come together and assemble on mRNA to form the complete ribosome during translation (Steitz & Moore, 2003). Bacterial ribosome differs from the eukaryotic and archeal counterparts, the latter two being phylogenetically closer to each other (Dennis, 1997; Greber et al., 2012; Woese & Fox, 1977; Woese, Kandler, & Wheelis, 1990). Moreover, the two Eukarya-specific organelles—mitochondrion (present in all eukaryotic cells) and plastid (found in plants and algae)—possess their

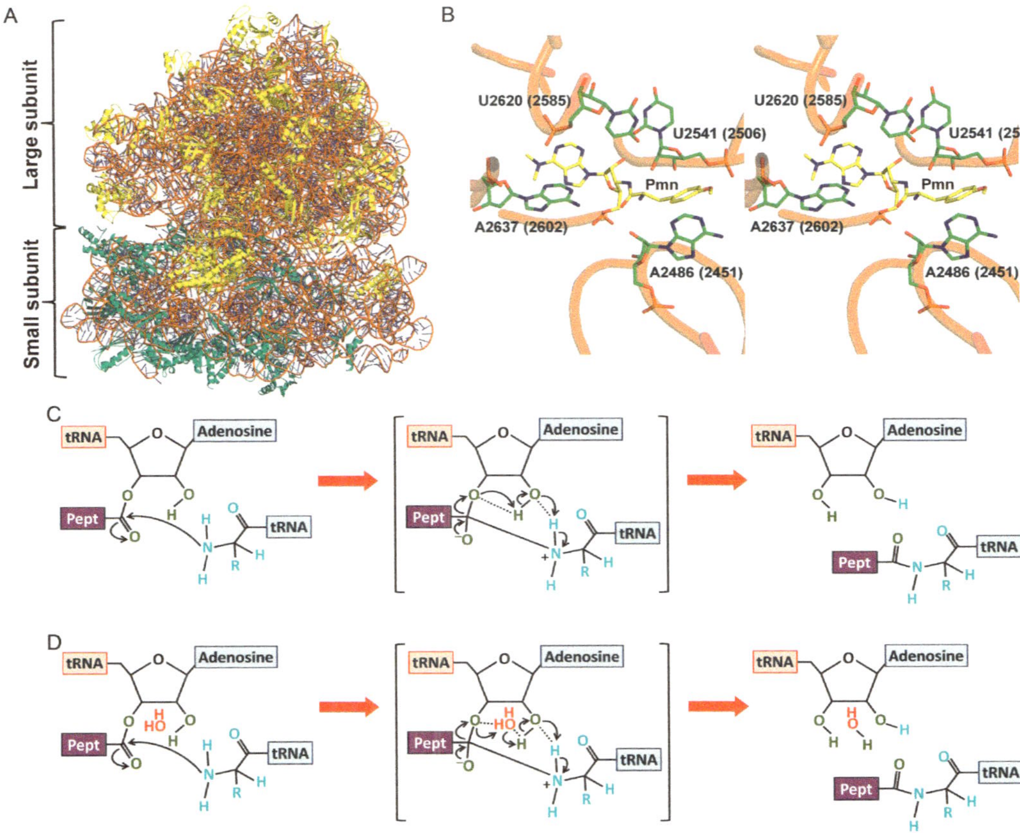

Fig. 3 Ribosome and mechanism of peptide bond formation. (A) Structure of the complete 70S ribosome from *Escherichia coli* (PDB id: 5H5U) showing the small (30S) and the large (50S) subunits. (B) Stereo image of the catalytic center (i.e., PTC) of ribosome from the archaeon *Haloarcula marismortui* (PDB id: 1FG0) in complex with puromycin (Pmn, *yellow*). The *numbers in parentheses* denote those in *E. coli*. (C) Proposed six-membered ring proton shuttle mechanism during catalysis of peptide bond formation at PTC. Note that two protons are relayed simultaneously. (D) Proposed eight-membered ring double-proton shuttle mechanism during catalysis of peptide bond formation at PTC. Note that three protons are transferred in a concerted fashion.

own ribosomes, which share their ancestry with the bacterial ribosome. Nevertheless, the core structure in all these variants remains the same. Various aspects of translation have been thoroughly dissected using a multitude of measures (Ling & Ermolenko, 2016; Moore & Steitz, 2011; Rozov, Demeshkina, Westhof, Yusupov, & Yusupova, 2016; Voorhees & Ramakrishnan, 2013). However, focussing on catalysis, we will describe only the mechanisms involved in peptide bond formation during elongation as well as peptide release during termination.

Biochemical and structural studies performed in the first half of the last decade clarified the long-standing ambiguity about ribosome being a ribozyme or a protein enzyme (Cech, 2000). The heart of ribosome, called peptidyl transferase centre (PTC), that catalyzes peptide bond formation consists only of rRNA elements; no part of the constituent proteins exists in the vicinity of PTC, the nearest protein element being ~18 Å away from the site (Fig. 3B) (Ban et al., 2000; Nissen et al., 2000). During peptide bond formation, the α-amino (α-NH_2) group of aminoacyl-tRNA (aa-tRNA) lodged at ribosomal A site mounts a nucleophilic attack on the carbonyl carbon of peptidyl-tRNA (pt-tRNA) residing at ribosomal P site (Rodnina, 2013). Consequently, the peptidyl moiety of pt-tRNA gets bonded to the aminoacyl moiety of aa-tRNA through a peptide linkage, thereby converting the pt-tRNA to a deacylated tRNA and the aa-tRNA to a pt-tRNA, the peptidyl moiety of which is increased by one residue. The peptidyl transferase reaction at the PTC has been dissected extensively using a multipronged approach comprising kinetic/biochemical analyses, biophysical methods, and structural means. On the basis of the first atomic-resolution structure of the 50S subunit in combination with pK_a analysis of ribosomal nucleotides, it was postulated that the ribosome employs general acid–base catalysis (Muth, Ortoleva-Donnelly, & Strobel, 2000; Nissen et al., 2000). The reaction mechanism proposed that N3 of the universally conserved A2451 (belonging to 23S rRNA; numbering according to *E. coli*) first acts as a base to deprotonate the α-NH_2 group of the A site aa-tRNA and then functions as an acid to protonate the 3′-oxygen leaving group of the P site. Since the pK_a of an efficient acid–base catalyst has to be close to 1 under physiological conditions, it was also posited that the near-neutral pK_a in this case is achieved through a network of charge relay which includes G2447 and A2450. However, later experiments, including mutational studies, rendered this mechanism untenable and suggested that the ribosome is not the source of ionizing groups during the catalytic step, i.e., ribosomal residues are not directly involved in general acid–base catalysis (Bieling, Beringer, Adio, & Rodnina, 2006; Leung, Suslov, Tuttle, Sengupta, & Piccirilli, 2011).

An alternative model has been put forth on the basis of kinetic and structural investigations in conjunction with computational analyses. A substrate-assisted catalytic mechanism has been envisaged, according to which the 2′-OH group of 3′-terminal A76 of P site pt-tRNA is implicated in catalysis (Leung et al., 2011; Rodnina, 2013). A comparison of structures of 50S subunit in complex with various transition state analogs (TSAs) shows that the 2′-OH is involved in hydrogen bonding with the nucleophilic α-NH_2 group, 2′-OH of A2452 and a water molecule in the prereaction state, whereas the interaction between N3 of A2451 and the nucleophilic α-NH_2 group is abolished (Schmeing, Huang, Kitchen, Strobel, & Steitz, 2005). Additional support for the model has come from atomic mutagenesis studies, wherein the 2′-OH of A2451 or pt-tRNA (A76) was supplanted by 2′-H or 2′-F. It has been hypothesized that the 2′-OH of A76 is either involved in the proper positioning of the nucleophile, or it serves as a general acid–base to abstract a proton from the nucleophile and then supply a proton to A76 3′-O (to which the peptide is attached) when the tetrahedral transition state (TS) breaks down (Das, Bhattacharyya, & Burma, 1999; Dorner, Panuschka, Schmid, & Barta, 2003; Erlacher et al., 2006; Schmeing, Huang, Kitchen, et al., 2005; Schmeing, Huang, Strobel, & Steitz, 2005; Weinger, Parnell, Dorner, Green, & Strobel, 2004). However, it is unlikely that the route of proton transfer from 2′-OH to 3′-O would be direct (Carrasco, Hiller, & Strobel, 2011; Huang, Carrasco, Pfund, & Strobel, 2008; Rangelov, Vayssilov, Yomtova, & Petkov, 2006). Intriguingly enough, the 2′-OH of A76 of pt-tRNA has been found to be dispensable for the acyl transfer from the vicinal 3′-O in an in vitro translational system (Koch, Huang, & Sprinzl, 2008). In silico simulations have indicated that such a discrepancy may be due to functional substitution of the 2′-OH by a water molecule, although the basis of this discrepancy is still ambiguous (Åqvist, Lind, Sund, & Wallin, 2012).

Although the above model suggests that the A76 2′-OH may perform general acid–base catalysis, experiments and molecular dynamics simulations have demonstrated that catalysis at PTC proceeds through a reduction in entropy rather than general acid–base catalysis (Sharma, Xiang, Kato, & Warshel, 2005; Sievers, Beringer, Rodnina, & Wolfenden, 2004). The PTC achieves this by cradling the substrate in an optimum orientation as the active site preorganizes itself for catalysis and stabilization of the tetrahedral TS using an induced-fit mechanism (Schmeing, Huang, Strobel, et al., 2005; Zaher, Shaw, Strobel, & Green, 2011). It is now believed that a number of water molecules as well as residues in the PTC not only provide a rigid framework to facilitate the stabilization of the tetrahedral TS, but they also participate in the relay or shuttling of protons during the reaction (Trobro & Åqvist, 2005, 2006; Wallin & Åqvist, 2010). On the basis of these revelations, two mechanisms of proton shuttle have been proposed, even though in either model, the A76 2′-OH of P site-bound pt-tRNA serves to bridge the nucleophilic α-NH_2 group of aa-tRNA and the 3′-O of pt-tRNA during the course of proton relay. The first model, referred to as concerted six-membered ring proton shuttle, posits that the α-NH_2 group of A site aa-tRNA mounts a nucleophilic attack on the carbonyl carbon of P site pt-tRNA, thereby forming the tetrahedral TS. Resolution of the TS ensues through the transfer of a proton from the α-NH_2 group of aa-tRNA to the 2′-OH of A76 of pt-tRNA, and a concomitant protonation of the 3′-O by the 2′-OH. Thus, two protons "move" simultaneously in this model (Fig. 3C) (Trobro & Åqvist, 2005, 2006). The second model is known as eight-membered ring double-proton shuttle. According to this model, the formation of tetrahedral TS is the same as proposed in the first model. However, resolution of the TS involves a proton transfer from the nucleophilic α-NH_2 group of aa-tRNA to the A76 2′-OH of pt-tRNA, and a proton relay from the 2′-OH to the vicinal 3′-O of A76 via a water molecule. Therefore, this model envisages "movement" of three protons instead of two (Fig. 3D) (Wallin & Åqvist, 2010).

Proposal of the double-proton shuttle mechanism has originated from analyses of the structure of 50S subunit with bound TSA (Schmeing, Huang, Kitchen, et al., 2005). In the structure, it is observed that a bridging active site water molecule is properly poised between the 2′-OH and 3′-O of A76 of pt-tRNA. Moreover, the water molecule is also found to have hydrogen bond interactions with the carbonyl oxygen and 2′-OH of C2063 (belonging to 23S rRNA). The structure also shows that another active site water molecule is bound by U2584 and A2602, and it has been implicated in the stabilization of the oxyanion tetrahedral TS. Such a function of this water molecule has been predicted to be true for both models of catalysis (Åqvist et al., 2012). Nevertheless, it has been argued that stabilization of the oxyanion is not essential for peptide bond formation (Carrasco et al., 2011). Kinetic analyses and computational studies have garnered evidence favoring the eight-membered double-proton shuttle mechanism (Kuhlenkoetter, Wintermeyer, & Rodnina, 2011; Wallin & Åqvist, 2010). Moreover, studies using heavy-atom kinetic isotope effects have demonstrated that the rate-limiting step involves both TS formation and proton relay, whereas resolution/breakdown of the TS is a separate and rapid step during ribosomal catalysis (Hiller, Singh, Zhong, & Strobel, 2011). This pathway is significantly different from the uncatalyzed peptide bond formation in solution, wherein the reaction progresses via two intermediates, viz., a zwitterion tetrahedral intermediate (T±) followed by its deprotonation to generate an anionic intermediate (T −); catalysis by ribosome does not seem to involve the formation and/or accumulation of the zwitterionic intermediate (Leung et al., 2011). Thus, ribosome not only disposes the substrate in an orientation which is conducive for catalysis, but also provides an environment which alters the rate-limiting transition state and, hence, the reaction trajectory (Hiller et al., 2011). On the basis of structural interpretations and mutagenesis experiments, several 23S rRNA nucleotides such as C2603, A2451, and A2602 have also been postulated to have an important role in catalysis, although a direct involvement of any of the 23S rRNA residues in catalysis has not been established yet (Leung et al., 2011). What is, however, evident is that the ribosomal elements, especially 23S rRNA, are crucial for proper and optimum positioning of the substrates during catalysis. For instance, the structures have revealed that A- and P-loops of 23S rRNA interact with the 3′-CCA end of aa-tRNA and pt-tRNA, respectively, to orient them precisely on the ribosome for catalysis (Schmeing, Huang, Kitchen, et al., 2005; Schmeing, Huang, Strobel, et al., 2005).

Once ribosome has reached a stop codon, translation (elongation) stops and the polypeptide is released upon hydrolysis of the ester bond that links the C-terminal residue of the polypeptide to the last tRNA (i.e., the final pt-tRNA) in P site (Schmeing & Ramakrishnan, 2009). Decoding of stop codons (UAA, UAG, and UGA) happens differently from other (sense) codons, since tRNA is not involved in this process. Translation release factors (RFs) are instead required for the recognition of stop codons. Bacteria have two class I RFs, viz., RF1 and RF2, but the two appear to have differences in their specificity and mechanism. In contrast, archea and eukaryotes have only one class I RF, called eRF1. Class I RFs are involved not only in stop codon recognition, but also in triggering polypeptide release. On the other hand, the class II RF, named RF3 (eRF3 in archea and eukaryotes), induces the removal of class I RFs from the posttermination ribosomal complex. Several structures of ribosome in complex with RFs have revealed many structural differences in stop codon decoding compared to sense codon decoding by tRNAs (Korostelev, 2011).

As far as catalysis is concerned, there are similarities as well as differences between polypeptide release during translation termination and peptide bond formation during elongation. As in the case of elongation, PTC is also involved in catalysis during termination. A universally conserved GGQ motif in RF1/2 is found inserted in PTC and is situated close to the 3′-CCA end of pt-tRNA in P site (Laurberg et al., 2008). The presence of glycines in the motif gives the latter a unique conformation which is not possible if they are replaced by any other residue; hence, mutation of glycines causes significant reduction in activity (Frolova et al., 1999; Shaw & Green, 2007). However, glutamine is not directly involved in catalysis, since substitution of a residue (except proline, which changes the backbone conformation) for the glutamine of GGQ motif has only marginal effect on catalysis (Mora et al., 2003; Shaw & Green, 2007). By contrast, the main chain amide of glutamine is located near the A76 3′-O of pt-tRNA and probably has a more important role than the side chain of glutamine. The main chain amide is thought to stabilize the oxyanion tetrahedral transition state and deacylated tRNA during the reaction pathway. The structures also explain why substitution of proline for glutamine has drastic effects on activity, as proline at that position would prevent the orientation of main chain amide favorable for catalysis (Jin, Kelley, Loakes, & Ramakrishnan, 2010; Korostelev et al., 2008; Laurberg et al., 2008). Nevertheless, the glutamine residue has been shown to discriminate between water and bulkier nucleophiles by a proposed exclusion of the latter (Jin, Kelley, & Ramakrishnan, 2011; Shaw & Green, 2007). Unlike peptide bond catalysis, in which the α-NH_2 group of aa-tRNA acts as a nucleophile, pt-tRNA hydrolysis involves a nucleophilic water molecule. In the crystal structure, a water molecule is seen coordinated by the A76 2′-OH of pt-tRNA, A2451 of 23S rRNA, and main chain amide and side chain of glutamine of GGQ motif (Jin et al., 2010). Interestingly, the 2′-OH has been demonstrated to be important for catalysis of polypeptide release (Shaw, Trobro, He, Åqvist, & Green, 2012). In light of these findings, a proton shuttle mechanism has been proposed for catalysis of translation termination as well. However, unlike in peptide bond formation, a concerted proton relay does not seem to work in this case; instead, shuttling of only one proton appears to be involved in the hydrolytic cleavage of polypeptide from tRNA (Kuhlenkoetter et al., 2011; Trobro & Åqvist, 2009).

Given the conservation of the core structure of ribosome from different domains of life as well as different organelles, it is quite possible that the basic mechanism and fundamental principles for all types of ribosomes are similar, if not identical. Nevertheless, the differences that do exist among these are likely to confer specific variations to the common theme of action, at least in terms of their regulation, since different ribosomes function in conditions which are rather varying from one another (most archea dwell in extremes of environmental conditions, whereas organelles have their own specific microenvironment). The advent of new techniques and remarkable advancements in the existing ones have revolutionized the field of ribosomal research. They have helped and are continuing to help address such intricate aspects of this glorified cellular mega machinery that were otherwise recalcitrant to probing. One among these is ribosome-mediated quality control of protein biosynthesis, wherein polypeptide tagging (such as ubiquitination) plays a crucial role (Simms, Thomas, & Zaher, 2017; Thommen, Holtkamp, & Rodnina, 2016). Unanticipated discoveries are still springing up new surprises for the investigators. For instance, a specialized elongation factor, named EF-P, has been recently discovered which alleviates ribosomal stalling during translation of polyproline codons (Doerfel et al., 2013; Ude et al., 2013; Woolstenhulme et al., 2013). We have come a long way in deciphering numerous aspects of ribosome. However, there are still many gray areas and "grounds of controversy" which need to be addressed and resolved.

2.4 Aminoacyl-tRNA Synthetases and Their Proofreading Modules

Aminoacyl-tRNA synthetases (aaRSs) and their proofreading domains are, in a way, nonconventional or noncanonical members of the league of catalytic RNPCs, since they are essentially proteins and collaborate with RNAs (tRNAs or aa-tRNAs, which also happen to be their substrates!) only transiently. Unlike other RNPCs discussed so far, aaRSs or their editing modules do not usually form stable/permanent complexes with RNAs. Moreover, not all synthetases seem to conform to this category as they do not perform RNA-mediated catalysis. There are two structurally distinct and nonhomologous classes of aaRSs which are responsible for the very first step of translation, i.e., coupling of a genetically encoded amino acid to its cognate tRNA (Eriani, Delarue, Poch, Gangloff, & Moras, 1990; Ibba & Söll, 2000). Thus, aaRSs play a pivotal role in defining the genetic code in all domains of life. Not only do the two classes differ from each other in core structural features, but also in several other aspects, including mechanistic traits as well as dynamics. However, there are certain exceptions, rarities and overlaps found in and between both classes. Apart from the class-defining features of aaRSs, specific phylogenetic characteristics have led to their further division into subclasses, viz., IA to IE for class I and IIA to IIC for class II. Till date, 11 aaRSs families are known to belong to class I, whereas 13 to class II (Perona & Gruic-Sovulj, 2014; Yadavalli & Ibba, 2012).

All aaRSs necessarily possess a synthetic/aminoacylation domain whose active site pocket tethers the amino acid to its cognate tRNA through an ester bond in a two-step reaction, a process which is termed aminoacylation or charging (Fig. 4A) (Ling, Reynolds & Ibba, 2009). The first step, known as activation, involves the linking of AMP moiety of ATP to the amino acid to form aminoacyl-AMP (aa-AMP, also called aminoacyl-adenylate), thus releasing pyrophosphate (PPi) in the process. During this step, one of the α-carboxyl oxygens of the amino acid nucleophilically attacks the α-phosphorus of Mg-ATP (ATP stabilized by Mg^{2+}), resulting in the formation of a mixed anhydride bond between AMP and amino acid (Fersht & Kaethner, 1976). Such a mechanism has also been supported by structural studies (Brick, Bhat, & Blow, 1989; Desogus, Todone, Brick, & Onesti, 2000; Perona, Rould, & Steitz, 1993). However, the characteristics of transition state still remain to be elucidated. Computational analyses have suggested that class I and class II aaRSs differ in their mode of usage of α-carboxyl oxygen during the reaction (Banik & Nandi, 2012). Whereas class I aaRSs use the carboxyl oxygen in *syn* conformation, aaRSs belonging to class II utilize the *anti* conformation. Such a difference in the conformation of the attacking carboxylate group has been attributed to the difference in the binding mode of ATP. Structures have demonstrated that binding of ATP in class I aaRSs occurs in extended conformation, while in the case of the other class of aaRSs, ATP assumes a bent conformation (Perona & Gruic-Sovulj, 2014). Nevertheless, steady-state kinetics has revealed that there is no specific order of binding of ATP and amino acid for both classes of aaRS. Interestingly, the presence of cognate tRNA is absolutely essential for certain aaRSs—arginyl-tRNA synthetase (ArgRS), glutaminyl-tRNA synthetase (GlnRS), glutamyl-tRNA synthetase (GluRS), and class I lysyl-tRNA synthetase (LysRS-I)—to catalyze the activation step of aminoacylation reaction (Ibba et al., 1999; Kern & Lapointe, 1979; Mehler & Mitra, 1967; Ravel, Wang, Heinemeyer, & Shive, 1965). In these synthetases, although ATP and amino acid can bind to the enzyme, crystal structures in complex with various ligands have shown how tRNA binding induces subtle structural rearrangements in the aminoacylation pocket to make the enzyme catalytically competent (Desogus et al., 2000; Perona et al., 1993; Sekine et al., 2003, 2006; Sherlin & Perona, 2003). Such a tRNA-mediated modulation of active site architecture in these aaRSs has also been demonstrated to be crucial for the rejection of noncognate amino acids from the pocket (Bullock, Rodríguez-Hernández, Corigliano, & Perona, 2008; Rodríguez-Hernández, Bhaskaran, Hadd, & Perona, 2010).

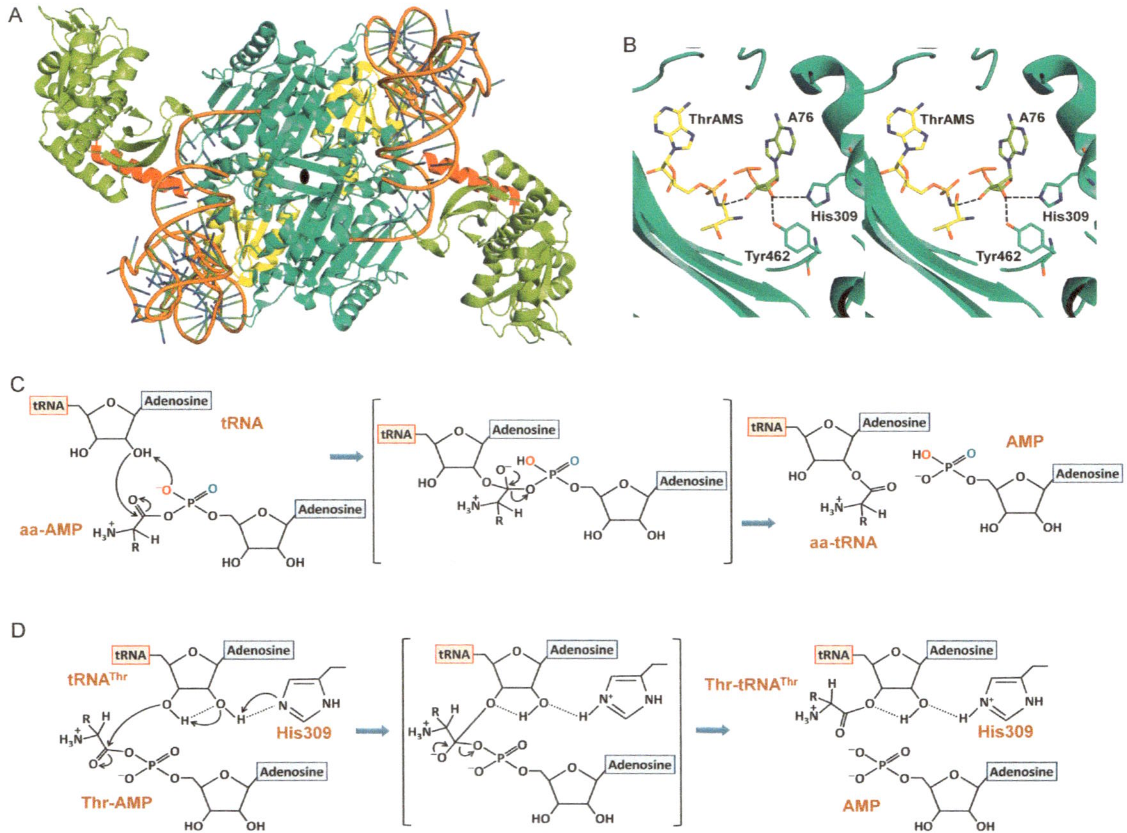

Fig. 4 See legend on next page.

The second step of aminoacylation reaction, referred to as transfer, includes esterification of the aminoacyl moiety of aa-AMP with the 2′- or 3′-OH group of A76 of tRNA; the consequence of this step is generation of aa-tRNA and AMP as final products. During this step, one of the two vicinal hydroxyl groups, depending on the class, mounts a nucleophilic attack on the carbonyl carbon of amino acid bound to AMP. This step has been proposed to be preceded by the abstraction of a proton from the 2′- or 3′-OH of A76, usually by the phosphoryl group of aa-AMP. Evidence collected from structural, biochemical, and simulation studies performed on a number of aaRSs belonging to both classes has implicated nonbridging oxygens of the phosphoryl group of aa-AMP, especially the *pro-S* (S_p) rather than *pro-R* (R_p) oxygen, in the second (i.e., transfer) step of aminoacylation reaction (Fig. 4C) (Cavarelli et al., 1994; Eiler, Dock-Bregeon, Moulinier, Thierry, & Moras, 1999; Guth, Connolly, Bovee, & Francklyn, 2005; Liu & Gauld, 2008; Perona et al., 1993). Intriguingly, threonyl-tRNA synthetase (ThrRS) from *E. coli* does not employ the aforesaid mechanism. Structural and biochemical evidence suggests that it requires the vicinal 2′-OH of the *cis*-diol of A76 to carry out the second step of aminoacylation (i.e., joining of threonine to the 3′-OH of A76 of $tRNA^{Thr}$) (Fig. 4B). A proton shuttle mechanism involving the 2′-OH has been therefore proposed for *E. coli* ThrRS, similar to that for ribosomal peptidyl transfer catalysis (Fig. 4D) (Minajigi & Francklyn, 2008). However, its authenticity has been challenged by computational analyses which indicate that the enzyme utilizes the α-NH_2 group of Thr-AMP to abstract a proton (Huang, Bushnell, Francklyn, & Gauld, 2011). Nevertheless, the mechanism needs further

---

Fig. 4 AaRS and mechanism of aminoacyl transfer. (A) The homodimeric structure of ThrRS from *Escherichia coli* in complex with $tRNA^{Thr}$ (PDB id: 1QF6). N-terminal editing domain, aminoacylation domain, and C-terminal anticodon-binding domain are rendered in *green*, *cyan*, and *yellow*, respectively; a linker helix that connects editing and aminoacylation domains is depicted in *red*. The *black oval* in the center represents the dimeric axis perpendicular to the plane of the image. (B) Stereo image of aminoacylation site of *E. coli* ThrRS in complex with $tRNA^{Thr}$ (PDB id: 1QF6) showing the terminal A76 of tRNA (*green*) and two active site residues, viz., His309 and Tyr462 (*cyan*). The other ligand (ThrAMS, which is the analog of Thr-AMP), colored *yellow*, has been modeled after superimposing two different ligand-bound ThrRS structures (1EVL on 1QF6). Note that the 2′-OH of A76 interacts with His309 and Tyr462, whereas the 3′-OH of A76 is oriented toward the carbonyl carbon of threonyl moiety of ThrAMS. (C) Proposed general mechanism of aminoacyl transfer involving S_p oxygen during aminoacylation by aaRSs. (D) Proposed mechanism of threonyl transfer involving a concerted proton shuttle during threonylation by *E. coli* ThrRS.

probing with regard to the identification and characterization of the transition state. Just like the case of 2′-OH, the role of nonbridging phosphoryl oxygens needs to be investigated in other systems to ascertain the universality or exclusivity of the two mechanistic paradigms. In either scenario, be it 2′-OH or nonbridging oxygen, the enzymes employ a substrate-assisted catalytic mechanism, the generic nature of which is yet to be established for aminoacylation by aaRSs.

Every aaRS utilizes three substrates—amino acid, ATP, and tRNA—during catalysis of aa-tRNA synthesis. While ATP is common to all, amino acid and tRNA are specific to the aaRS. Compared to amino acid, tRNA is an enormous molecule and is thus able to provide a large surface area for recognition by aaRS. Besides, tRNAs also possess one or more idiosyncratic features and contact points distributed over the tRNA body for specific identification by their respective aaRSs (Giegé, Sissler, & Florentz, 1998). For example, alanyl-tRNA synthetase (AlaRS) specifically recognizes G3•U70 wobble base pair on the acceptor stem of $tRNA^{Ala}$ (Hou & Schimmel, 1988; McClain, Chen, Foss, & Schneider, 1988; McClain & Foss, 1988). Due to these characteristics, aaRSs usually do not cross-react with noncognate tRNAs. On the other hand, the small size of amino acids makes their selection by synthetases quite a challenge. The situation is further aggravated by the presence of many proteinogenic as well as nonproteinogenic amino acids in the cellular milieu which bear marked structural and/or chemical resemblance to the cognate amino acid (Bullwinkle, Lazazzera, & Ibba, 2014). For instance, valine and isoleucine differ by just a methyl group. This led Pauling in 1958, even before the discovery of aaRSs, to surmise that for processes involving selection of an amino acid from a set of remarkably similar ones, discrimination based on purely chemical considerations would not be sufficient; this would naturally entail editing/proofreading activities to overcome the errors and account for the the latter's observed frequency during protein biosynthesis (Loftfield & Vanderjagt, 1972; Pauling, 1958).

Approximately, half of the known aaRSs face amino acid discrimination problem and have thus evolved various proofreading strategies to rectify the mistakes (Perona & Gruic-Sovulj, 2014). As aminoacylation by aaRSs is a two-step phenomenon, errors can be cleared at either of the two steps. If error correction happens after the activation step but prior to the transfer step, it is referred to as pretransfer editing. If, however, the same happens after aa-tRNA formation, it is called posttransfer editing. Thus, pretransfer editing clears misactivated noncognate aa-AMP, whereas posttransfer editing eliminates mischarged noncognate aa-tRNA. A "double sieve model" for such aaRSs was, therefore, invoked by Fersht in 1977, wherein it was envisaged that these synthetases consist of two "sieves" (Fersht, 1977a, 1977b, 1998). The first "coarse sieve" serves to exclude amino acids that are larger than the cognate through steric exclusion, while allowing the cognate and structurally/chemically similar noncognate amino acids to "pass through." The second "fine sieve" allows only the noncognate amino acids but not the cognate one to "pass through," thereby eliminating them in the process. The 10 synthetases that are known to employ editing mechanisms have been enumerated in Table 1 (Jakubowski, 2012; Perona & Gruic-Sovulj, 2014; Yadavalli & Ibba, 2012). Interestingly, all of them are capable of catalyzing the activation step in the absence of tRNA.

Synthetases can perform pretransfer editing in one of the following ways. In the case of seryl-tRNA synthetase (SerRS) and mitochondrial ThrRS, the noncognate adenylate species (Thr-AMP and Ser-AMP, respectively) is selectively released in the solution (i.e., kinetic proofreading) and is then hydrolyzed spontaneously (nonenzymatically) (Gruic-Sovulj, Rokov-Plavec, & Weygand-Durasevic, 2007; Ling, Peterson, Simonovic, Söll, & Simonovic, 2012; Rokov-Plavec et al., 2013). In all the other cases, pre-transfer editing is effected in the aminocylation site of aaRS, and its rate is significantly higher than spontaneous hydrolysis of noncognate adenylate. Almost all of these synthetases (e.g., LeuRS, MetRS, LysRS-II) do not need the participation of cognate tRNA for pretransfer editing, the exception being isoleucyl-tRNA synthetase (IleRS) (Cvetesic, Perona, & Gruic-Sovulj, 2012; Dulic, Cvetesic, Perona, & Gruic-Sovulj, 2010). In the case of IleRS, the unexpectedly slow transfer step during aminoacylation is believed to allow a water molecule to compete with $tRNA^{Ile}$ to attack the anhydride bond of Val-AMP. It has also been shown that IleRS does not require the 2′- or 3′-OH of A76 of $tRNA^{Ile}$ to carry out pretrnasfer editing of Val-AMP (Cvetesic, Bilus, & Gruic-Sovulj, 2015). However, mechanistic underpinnings for the anomalous behavior of IleRS are completely lacking.

Posttransfer editing happens in seven aaRSs belonging to both classes. It has been shown that posttransfer editing occurs in a site distinct from the aminoacylation site and it is located in a separate domain about 35–40 Å away from the aminoacylation site (Fig. 4A) (Cusack, Yaremchuk, & Tukalo, 2000; Dock-Bregeon et al., 2000; Fukai et al., 2000; Nureki et al., 1998; Silvian, Wang, & Steitz, 1999). In the "double sieve model" of Fersht, the first "coarse sieve" corresponds to the aminoacylation site

Table 1 AaRSs Which Make Mistakes and Their Editing Pathways, Editing Substrates, and Editing Sites/Domains

AaRS Class	AaRS	Editing Pathway	Pretransfer Editing Substrate (aa-AMP)	Posttransfer Editing Substrate (aa-tRNA)	*cis*-Editing Domain	*trans*-Editing Factor
Class I	IleRS	Pre- & posttransfer	Hcy, Val	Val	CP1 domain	—
	LeuRS	Pre- & posttransfer	Hcy, Ile, γ-hIle, δ-hIle, γ-hLeu, δ-hLeu, Met, Nva	Hcy, Ile, γ-hLeu, Met, Nva	CP1 domain	—
	MetRS	Pretransfer	Hcy	—	—	—
	ValRS	Posttransfer	—	α-Abu, Hcy, Thr	CP1 domain	—
Class II	AlaRS	Posttransfer	—	Gly, Ser	AlaX-like	AlaXp (AlaX-S, AlaX-M, AlaX-L)
	LysRS-II	Pretransfer	Hcy, Hse, Orn	—	—	—
	PheRS	Posttransfer	—	Ile, Tyr	B3/B4	—
	ProRS	Pre- & posttransfer	Ala	Ala, Cys, 4-hPro	INS, N- and/or C-terminal extension[a]	YbaK, ProX[b]
	SerRS	Pretransfer	Cys, SerHX, Thr	—	—	—
	ThrRS	Pre- & posttransfer	Ser	Ser	N2, NTD[c]	ThrRS-ed[d]

[a]INS denotes insertion domain which is present only in ProRS from bacteria and edits only Ala-tRNA^{Pro}; eukaryotes have N- and/or C-terminal extension instead. The N-terminal extension found in lower eukaryotes need not necessarily be active in posttransfer editing.
[b]YbaK and ProX deacylate Cys-tRNA^{Pro} and Ala-tRNA^{Pro}, respectively. YbaK and ProX are homologs of INS.
[c]N2 and NTD are N-terminal editing domains of ThrRS from bacteria/eukaryotes and archea, respectively. N2 and NTD are nonhomologous. N2 is a structural homolog of AlaX-like domain, whereas NTD belongs to DTD-like fold.
[d]ThrRS-ed is found in several crenarchaeal species and is homologous to NTD.
The amino acids that are underlined are nonstandard (nonproteinogenic) amino acids. Abbreviations: *4-hPro*, 4-hydroxyproline; *Hcy*, homocysteine; *Hse*, homoserine; *Nva*, norvaline; *Orn*, ornithine; *SerHX*, serine hydroxamate; *α-Abu*, α-aminobutyrate; *γ-hIle*, γ-hydroxyisoleucine; *γ-hLeu*, γ-hydroxyleucine; *δ-hIle*, δ-hydroxyisoleucine; *δ-hLeu*, δ-hydroxyleucine.

of aaRS, whereas the second "fine sieve" represents the editing site. MetRS, LysRS-II, and SerRS are three known cases of synthetases that lack a domain for posttransfer editing and thus rely exclusively on pretransfer editing for error correction. The editing domains of class I aaRSs belong to a common homologous fold termed connective peptide 1 (CP1) domain (Chen, Guo, Li, Wang, & Wang, 2000; Lin, Hale, & Schimmel, 1996; Schmidt & Schimmel, 1994, 1995). The editing domains of class II synthetases are diverse in structure and not all of them share a common ancestry. These proofreading domains which are covalently coupled to the aminoacylation domains of aaRS are called *cis*-editing domains. Apart from the latter, several freestanding proofreading modules also exist which assist the *cis*-editing function of synthetases. These standalone modules are referred to as *trans*-editing factors and have been projected to extend the "double sieve model" by constituting the additional "third sieve" (An & Musier-Forsyth, 2004). Interestingly, only class II aaRSs are known to have *trans*-editing domains. An interesting facet of pre- and posttransfer editing is that the two pathways can exist simultaneously in a single synthetase (e.g., LeuRS, IleRS), but either one is functional or one dominates the other if both are functional. However, if one is disrupted, the other pathway can be activated to compensate for the loss of the first (Martinis & Boniecki, 2010). Such a phenomenon was elegantly shown in the case of *E. coli* LeuRS, wherein deletion of the CP1 domain induced the otherwise dormant tRNA-dependent pretransfer editing of Ile-AMP (Boniecki, Vu, Betha, & Martinis, 2008).

Two models have been proposed for posttransfer editing pathway, viz., translocation model and dissociation–reassociation model (Yadavalli & Ibba, 2012). According to the translocation model, upon charging of tRNA with amino acid in the aminoacylation site, the 3′-CCA tail of tRNA harboring the amino acid flips to move to the editing site of aaRS. This flip is accompanied by conformational changes and relative movements of different domains of aaRS. While the cognate amino acid is sterically excluded from the editing site, thereby releasing the cognate aa-tRNA in solution, the noncognate amino acid is edited out in the editing site, thus decoupling the noncognate amino acid from the tRNA. Evidence for the model has come from structural snapshots which have captured tRNAs with the 3′-CCA arm oriented toward both aminoacylation site (i.e., in aminoacylation mode) and editing site (i.e., editing mode) (Fukunaga & Yokoyama, 2005a, 2005b; Palencia et al., 2012; Tukalo, Yaremchuk, Fukunaga, Yokoyama, & Cusack, 2005). This pathway is thought to be operational in class I aaRSs, since product release in these synthetases is known to be rate-limiting, thereby providing ample residence time for such structural rearrangements to take place (Zhang, Perona, Ryu, Francklyn, & Hou, 2006). However, the dynamics of the movement have so far remained elusive. The dissociation–reassociation model posits that aa-tRNA is released in the solution immediately after its formation in the aminoacylation site. The proofreading domain of aaRSs then recaptures the noncognate aa-tRNA species and eliminates them. This model is supported by resampling experiments, wherein it has been demonstrated that the editing domains of aaRSs can compete with elongation factor thermo unstable (EF-Tu, which delivers aa-tRNAs to ribosome) for seizing noncognate aa-tRNAs (Ling, So, et al., 2009). This mode of posttransfer editing is thought to occur in class II synthetases as product release in this case is not rate-limiting (Zhang et al., 2006). Moreover, the rate of reassociation of class II aaRSs with noncognate aa-tRNA substrates is also very high compared to the binding rate of aa-tRNAs with EF-Tu (Ling, So, et al., 2009). The presence of several *trans*-editing factors for class II synthetases also provides additional backing for the dissociation–reassociation pathway of posttransfer editing (Ahel, Korencic, Ibba, & Söll, 2003; An & Musier-Forsyth, 2004; Korencic et al., 2004; Wong, Beuning, Silvers, & Musier-Forsyth, 2003).

As regards the mechanism of catalysis (hydrolysis) of noncognate aa-tRNA during posttransfer editing, the 2′- or 3′-OH of A76 of tRNA has been implicated for several synthetases. The very first indication for the role of vicinal hydroxyls in aaRS-catalyzed deacylation of aa-tRNA came from Cramers group (von der Haar & Cramer, 1975, 1976). The seeds for the idea were born from a previous observation by the same group that the disruption of *cis*-diols has a stabilizing effect on the nonenzymatic hydrolysis of aa-tRNA, i.e., the lack of vicinal hydroxyl decreases the rate of spontaneous cleavage of the scissile ester bond (von der Haar, Schlimme, Gómez-Guillen, & Cramer, 1971). Much later, it was shown that the 3′-OH of A76 of $tRNA^{Ile}$ is essential for IleRS to clear the mischarged Val-$tRNA^{Ile}$ species; but such a requirement of 3′-OH was not observed for ValRS to edit Thr-$tRNA^{Val}$, although the amount of ValRS needed to deacylate 3′-H- and 3′-F-substituted substrates was 10- and 100-fold, respectively, compared to the wild-type substrate (Cvetesic et al., 2015; Nordin & Schimmel, 2002). It was suggested that the 3′-OH is involved in transacylation during posttransfer editing by IleRS. However, the plausibility of 3′-OH being involved in activating a catalytic water molecule during posttransfer editing could not be ruled out. A structural basis was provided for the latter phenomenon, since a water molecule was seen to be positioned by the

3′-OH near the scissile bond in the posttransfer ligand-bound complex of LeuRS (Lincecum et al., 2003). Structural studies on the editing domain of archeal ThrRS provided further evidence in favor of this mechanism, wherein the vicinal hydroxyl was proposed to have a catalytic role in proofreading (Hussain, Kamarthapu, Kruparani, Deshmukh, & Sankaranarayanan, 2010; Hussain et al., 2006). Subsequently, such a substrate-assisted posttransfer editing was demonstrated for PheRS through atomic substitutions of the 2′-OH of A76 of corresponding tRNAs (Ling, Roy, & Ibba, 2007). On the basis of computational analyses, it has been posited that even LeuRS may employ this mechanism of posttransfer editing (Hagiwara, Field, Nureki, & Tateno, 2010). Nevertheless, the universality and the intricacies of this mechanism (e.g., nature of transition state) in posttransfer editing by aaRSs are yet to be established.

2.4.1 DTD-Like Fold

D-Aminoacyl-tRNA deacylase (DTD) was the first *trans*-editing factor to be discovered when it was shown that the enzyme acts on D-Tyr-$tRNA^{Tyr}$ and D-Phe-$tRNA^{Phe}$ (Calendar & Berg, 1967). It was later demonstrated that the enzyme deacylates multiple D-aa-tRNAs (Soutourina, Plateau, & Blanquet, 2000; Soutourina, Plateau, Delort, Peirotes, & Blanquet, 1999). DTD was thus demonstrated to rectify "chiral errors" committed by some of the aaRSs (Calendar & Berg, 1966, 1967; Soutourina et al., 2000; Takayama et al., 2005). In conjunction with other factors like EF-Tu and ribosome which have a preference for L-aa-tRNAs (Englander et al., 2015; Pingoud & Urbanke, 1980; Yamane, Miller, & Hopfield, 1981), DTD aids in excluding D-amino acids from the translational apparatus and has therefore been implicated in enforcing and perpetuating homochirality in the protein world. Such an exclusionary process is important, since a heterochiral polypeptide (one that is composed of both D- and L-amino acids) is likely to undergo spontaneous misfolding because of geometric constraints and steric incompatibilities between residues of two chiralities.

DTD is present in almost all bacterial and eukaryotic organisms but is absent in archea (Wydau, van der Rest, Aubard, Plateau, & Blanquet, 2009). Surprisingly, a DTD-like fold is found appended to the aminoacylation domain of archeal ThrRS as a distinct N-terminal domain (NTD) (Dwivedi, Kruparani, & Sankaranarayanan, 2005; Rigden, 2004). However, *E. coli* DTD bears only ~14% sequence identity with NTD from *Pyrococcus abyssi* (PabNTD). Biochemical studies have revealed that NTD hydrolyzes the misacylated L-Ser-$tRNA^{Thr}$ species; hence, NTD constitutes the *cis*-editing domain of archeal ThrRS (Beebe, Merriman, Ribas De Pouplana, & Schimmel, 2004; Korencic et al., 2004). Interestingly, NTD does not share sequence similarity or structural homology with the *cis*-editing domain of bacterial/eukaryotic ThrRS. Collation of these findings has led to the proposal of a model for perpetuation of homochirality in proteins. According to this model, a DTD-like D-amino acid-editing module would have initially assisted the primordial aaRSs—which were probably weak in chiral discrimination of substrates—to prevent D-amino acid infiltration in proteins. During the course of evolution, it would have retained the function of decoupling D-amino acids from tRNAs in the bacterial and eukaryal domains, whereas it would have covalently linked to ThrRS and acquired the specialized function of eliminating L-Ser-tRNA^{Thr} in the archeal domain (Dwivedi et al., 2005). Notably, archea harbor a distinct protein, unrelated or nonhomologous to DTD in sequence and structure, which performs the same biochemical function as done by DTD; this archeal enzyme has been therefore named DTD2 (Ferri-Fioni et al., 2006). Intriguingly, DTD2 is also found in several plant species (Wydau, Ferri-Fioni, Blanquet, & Plateau, 2007).

Structural and biophysical studies on NTD with substrate-mimicking analogs—L-seryl- and L-threonyl-3′-aminoadenosine (L-Ser3AA and L-Thr3AA)—have shown that the editing site can not only accommodate and bind the noncognate L-serine, but also the cognate L-threonine (Hussain et al., 2010, 2006). This represents an aberration or deviation from the general theme of posttransfer editing by aaRSs, which follow the "double sieve model" and employ steric exclusion for the rejection of cognate amino acid from the editing site, i.e., the second "fine sieve" (Fersht, 1977a, 1977b). On the basis of structural analyses of PabNTD, it was initially proposed that the movement of certain residues in the active site upon binding of cognate threonine constricts the space for a catalytic water molecule, thereby expelling the latter from the pocket and preventing cognate deacylation. Thus, the presence of catalytic water molecule, or the lack of it, was thought to be the underlying principle of substrate discrimination by NTD (Hussain et al., 2010). However, this mechanism has been found to be incomplete upon testing it rigorously through extensive mutational, biochemical, biophysical, and structural approaches using NTDs from multiple organisms. It has been demonstrated that none of the side chains in the active site of NTD is involved in governing substrate specificity. Moreover, it has been observed that the presence of a catalytic water molecule brings about subtle restructuring of the active site such that it does not allow the cognate threonine to be oriented or positioned properly for hydrolysis of the labile ester bond. Therefore, NTD operates in a side chain-independent mode through active site remodeling to ensure substrate specificity (Ahmad et al., 2015).

In the case of DTD, biochemical studies have revealed that the enzyme is absolutely enantioselective, since it does not act on even L-alanine, the smallest L-chiral amino acid, but efficiently deacylates multiple D-aa-tRNAs. Thus, DTD's activity is referred to as "chiral proofreading" (Ahmad et al., 2013; Calendar & Berg, 1967; Routh et al., 2016; Soutourina et al., 2000, 1999). Crystal structures of DTD in complex with the analog D-tyrosyl-3′-aminoadenosine (D-Tyr3AA) have shown that there is no recognition of the side chain chemistry of the D-amino acid substrate by DTD as the side chain of D-tyrosyl moiety of the ligand is seen protruding from the active site pocket. This provides a logical explanation for DTD's promiscuous activity on multiple substrates with D-chirality. An invariant cross-subunit Gly-*cis*Pro motif was proposed to select the D-chirality of the susbtrate using the parallel orientation of the two carbonyl groups of the motif (Ahmad et al., 2013). However, recent biochemical and structural evidence clearly suggests that the active site of DTD is porous to the achiral amino acid glycine, as DTD can effectively deacylate Gly-$tRNA^{Gly}$. Hence, L-chiral rejection and not D-chiral selection is the only mechanistic design principle of DTD's chiral proofreading site (Routh et al., 2016). Notably, the Gly-*cis*Pro motif forms the structural basis for such an exclusionary mechanism of the enzyme. DTD's unwarranted activity on the achiral substrate engenders glycine "misediting paradox," which is, nevertheless, resolved through protection of Gly-$tRNA^{Gly}$ by EF-Tu. Interestingly, just as in the case of NTD, none of the active site side chains is involved in defining DTD's chiral specificity.

Crystal structures of NTD, especially those in complex with L-Ser3AA, have shown that a water molecule (catalytic water) is located between the 2′-OH of ribose and carbonyl carbon of the aminoacyl moiety of the ligand (Fig. 5) (Ahmad et al., 2015; Hussain et al., 2010, 2006). On the basis of these crystal structures, it was proposed that the 2′-OH of A76 of tRNA could be involved in activating a water molecule to carry out catalysis. In none of the ligand-bound crystal structures of DTD is this water molecule seen (Ahmad et al., 2013; Bhatt, Yogavel, Wydau, Berwal, & Sharma, 2010; Routh et al., 2016). However, since both NTD and DTD belong to the same fold, it was envisioned that DTD would also employ a similar catalytic mechanism as NTD (Hussain et al., 2006). Moreover, there is adequate space in DTD's chiral proofreading pocket for accommodation of a water molecule

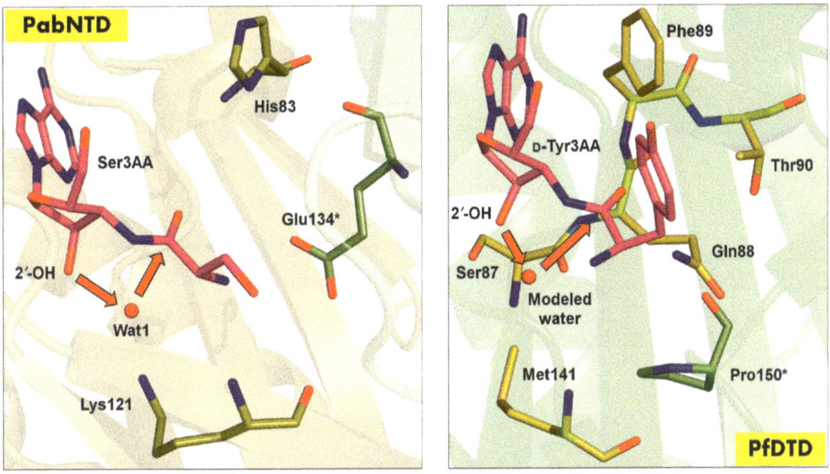

Fig. 5 Catalytic mechanism of DTD-like fold. Crystal structure of PabNTD (PDB id: 3PD2) showing a water molecule (Wat1) positioned between the 2′-OH and the carbonyl carbon of the ligand Ser3AA. Crystal structure of DTD from *Plasmodium falciparum* (PfDTD; PDB id: 4NBI) showing sufficient space for a water molecule to be modeled (i.e., accommodated) between the 2′-OH and the carbonyl carbon of the ligand D-Tyr3AA, although electron density for the water molecule is not observed. * indicates residue from the dimeric counterpart.

(Fig. 5). Although initial structural and biochemical investigations implicated an invariant threonine in the chiral proofreading site in DTD's proposed catalytic mechanism (Bhatt et al., 2010; Ferri-Fioni et al., 2001; Kemp et al., 2007; Lim et al., 2003), recent studies have ruled out this possibility. Active site mutations followed by biochemical analyses in the case of NTD as well as DTD have demonstrated that active site residues are dispensable for catalysis (Ahmad et al., 2015, 2013; Hussain et al., 2010). Nevertheless, single-atom substitution supplanting the 2′-OH with 2′-H or 2′-F completely abolished the activity of both NTD and DTD (Ahmad et al., 2015; Routh et al., 2016). Moreover, most of the interactions with the ligand are mediated by main chain atoms, as seen in the crystal structures of NTD and DTD complexed with their respective ligands (Ahmad et al., 2015, 2013; Hussain et al., 2010, 2006; Routh et al., 2016). All these clearly indicate that DTD-like fold (encompassing DTD and NTD) is primordial in nature as it functions at the RNA–protein interface and does not employ protein side chains for either substrate specificity or catalysis. The fold rather acts as a scaffold for properly orienting the aa-tRNA substrate such that the 2′-OH of A76 of tRNA performs (substrate-assisted) catalysis. The involvement of 2′-OH in catalysis

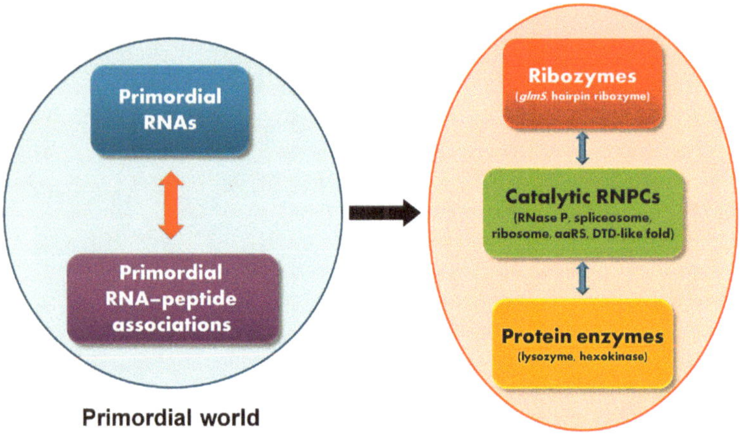

Fig. 6 Model for evolution of contemporary world from primordial world. In the primordial world, ancient RNAs could have existed independently or in collaboration with peptides. During the course of evolution, the primordial world consisting of these entities gave rise to the contemporary world, which comprises modern ribozymes (e.g., *glmS*, hairpin ribozyme), catalytic RNPCs (e.g., RNase P, spliceosome, ribosome, aaRS, DTD-like fold), and protein enzymes (e.g., lysozyme, hexokinase). The catalytic RNPCs occupy an intermediate position between the contemporary RNA-only ribozymes and protein enzymes. Thus, today's world is an "RNA–protein world" rather than a "protein world."

notwithstanding, the details of the catalytic mechanism of the DTD-like fold still remain to be elucidated. Furthermore, these characteristics of the DTD-like fold place it at an intermediate stage (alongside other catalytic RNPCs such as ribosome, spliceosome, and RNase P) between RNA-only ribozymes and protein enzymes in the contemporary world (Fig. 6).

As more research endeavors are going into understanding the mechanistic intricacies of aaRSs and their proofreading modules, the role of *cis*-diols, especially in the context of editing, is becoming more apparent. However, as it is evident from the earlier description, there are many variations in different aspects of aaRS. It would be therefore important to decipher the generic and/or specific roles of vicinal hydroxyls in various systems. Moreover, the details of mechanisms, such as the nature of transition state, are lacking in most of the cases.

3. CONCLUSION

The discovery of catalytic potential of RNA in the early 1980s marked the dawn of a new era in RNA biology (Guerrier-Takada et al., 1983;

Kruger et al., 1982). With the advent of modern tools and techniques, and advancements in the existing ones in the last couple of decades, the field of RNA research has progressed by leaps and bounds. We are now able to explore newer realms and regimes of RNA catalysis and coalitions between catalytic RNAs and proteins, discovering new horizons in the process which are many a time unexpected and unanticipated. They have opened the portals to an entirely new microworld and provided us with a handle to peep into the microcosms of active sites and RNA–protein interfaces. An important aspect and a recurrent theme that has come to the fore about RNA catalysis is the crucial role of vicinal hydroxyl groups (i.e., *cis*-diols) of the ribose moiety of catalytic RNAs, especially the 2′-OH group (Weinger & Strobel, 2006). It appears to be involved in some way or the other, directly or indirectly, in almost all catalytic mechanisms involving RNA. Notably, RNA differs from DNA in having a 2′-OH rather than 2′-H (besides, of course, substitution of uracil (U) for thymine (T)). It is remarkable how such a small and subtle change can lead to such an incredible spectrum of chemical as well as structural differences and diversities between the two kinds of nucleic acids. Our understanding of the mechanistic aspects of various RNA–protein associations and RNA-based catalysis has increased immensely following elucidation of structures of numerous complexes, and all the endeavors being made in this direction are only strengthening the idea that the contemporary world is an "RNA–protein world." Interestingly, ribozymic RNase P and PRORP not only provide an extreme case of convergent evolution, but also possibly a unique "living" instance of transition toward a "protein world" from an "RNA world."

ACKNOWLEDGMENTS

S.B.R. and R.S. acknowledge funding from 12th Five Year Plan Project BSC0113 of CSIR, India. R.S. also acknowledges funding from JC Bose Fellowship of SERB, India, and Centre of Excellence Project of Department of Biotechnology, India.

REFERENCES

Ahel, I., Korencic, D., Ibba, M., & Söll, D. (2003). Trans-editing of mischarged tRNAs. *Proceedings of the National Academy of Sciences of the United States of America*, *100*, 15422–15427.

Ahmad, S., Muthukumar, S., Kuncha, S. K., Routh, S. B., Yerabham, A. S., Hussain, T., et al. (2015). Specificity and catalysis hardwired at the RNA–protein interface in a translational proofreading enzyme. *Nature Communications*, *6*, 7552.

Ahmad, S., Routh, S. B., Kamarthapu, V., Chalissery, J., Muthukumar, S., Hussain, T., et al. (2013). Mechanism of chiral proofreading during translation of the genetic code. *eLife*, *2*, e01519.

Altman, S. (2011). Ribonuclease P. *Philosophical Transactions of the Royal Society of London. Series B, Biological Sciences*, *366*, 2936–2941.

Altman, S. (2013). The RNA–protein world. *RNA*, *19*, 589–590.

Alva, V., Söding, J., & Lupas, A. N. (2015). A vocabulary of ancient peptides at the origin of folded proteins. *eLife*, *4*, e09410.

An, S., & Musier-Forsyth, K. (2004). Trans-editing of Cys-tRNA^{Pro} by *Haemophilus influenzae* YbaK protein. *The Journal of Biological Chemistry*, *279*, 42359–42362.

Apirion, D. (1980). Genetic mapping and some characterization of the rnpA49 mutation of *Escherichia coli* that affects the RNA-processing enzyme ribonuclease P. *Genetics*, *94*, 291–299.

Åqvist, J., Lind, C., Sund, J., & Wallin, G. (2012). Bridging the gap between ribosome structure and biochemistry by mechanistic computations. *Current Opinion in Structural Biology*, *22*, 815–823.

Azubel, M., Habib, N., Sperling, J., & Sperling, R. (2006). Native spliceosomes assemble with pre-mRNA to form supraspliceosomes. *Journal of Molecular Biology*, *356*, 955–966.

Bai, X. C., McMullan, G., & Scheres, S. H. (2015). How cryo-EM is revolutionizing structural biology. *Trends in Biochemical Sciences*, *40*, 49–57.

Ban, N., Nissen, P., Hansen, J., Moore, P. B., & Steitz, T. A. (2000). The complete atomic structure of the large ribosomal subunit at 2.4 Å resolution. *Science*, *289*, 905–920.

Banik, S. D., & Nandi, N. (2012). Mechanism of the activation step of the aminoacylation reaction: A significant difference between class I and class II synthetases. *Journal of Biomolecular Structure & Dynamics*, *30*, 701–715.

Barta, A., Steiner, G., Brosius, J., Noller, H. F., & Kuechler, E. (1984). Identification of a site on 23S ribosomal RNA located at the peptidyl transferase center. *Proceedings of the National Academy of Sciences of the United States of America*, *81*, 3607–3611.

Beebe, J. A., & Fierke, C. A. (1994). A kinetic mechanism for cleavage of precursor tRNA^{Asp} catalyzed by the RNA component of *Bacillus subtilis* ribonuclease P. *Biochemistry*, *33*, 10294–10304.

Beebe, J. A., Kurz, J. C., & Fierke, C. A. (1996). Magnesium ions are required by *Bacillus subtilis* ribonuclease P RNA for both binding and cleaving precursor tRNA^{Asp}. *Biochemistry*, *35*, 10493–10505.

Beebe, K., Merriman, E., Ribas De Pouplana, L., & Schimmel, P. (2004). A domain for editing by an archaebacterial tRNA synthetase. *Proceedings of the National Academy of Sciences of the United States of America*, *101*, 5958–5963.

Bentley, D. L. (2014). Coupling mRNA processing with transcription in time and space. *Nature Reviews. Genetics*, *15*, 163–175.

Berget, S. M., Moore, C., & Sharp, P. A. (1977). Spliced segments at the 5′ terminus of adenovirus 2 late mRNA. *Proceedings of the National Academy of Sciences of the United States of America*, *74*, 3171–3175.

Bertram, K., Agafonov, D. E., Liu, W. T., Dybkov, O., Will, C. L., Hartmuth, K., et al. (2017). Cryo-EM structure of a human spliceosome activated for step 2 of splicing. *Nature*, *542*, 318–323.

Bhatt, T. K., Yogavel, M., Wydau, S., Berwal, R., & Sharma, A. (2010). Ligand-bound structures provide atomic snapshots for the catalytic mechanism of D-amino acid deacylase. *The Journal of Biological Chemistry*, *285*, 5917–5930.

Bieling, P., Beringer, M., Adio, S., & Rodnina, M. V. (2006). Peptide bond formation does not involve acid–base catalysis by ribosomal residues. *Nature Structural & Molecular Biology*, *13*, 423–428.

Boniecki, M. T., Vu, M. T., Betha, A. K., & Martinis, S. A. (2008). CP1-dependent partitioning of pretransfer and posttransfer editing in leucyl-tRNA synthetase. *Proceedings of the National Academy of Sciences of the United States of America*, *105*, 19223–19228.

Brick, P., Bhat, T. N., & Blow, D. M. (1989). Structure of tyrosyl-tRNA synthetase refined at 2.3 Å resolution. Interaction of the enzyme with the tyrosyl adenylate intermediate. *Journal of Molecular Biology*, *208*, 83–98.

Brody, E., & Abelson, J. (1985). The "spliceosome": Yeast pre-messenger RNA associates with a 40S complex in a splicing-dependent reaction. *Science*, *228*, 963–967.

Buck, A. H., Dalby, A. B., Poole, A. W., Kazantsev, A. V., & Pace, N. R. (2005). Protein activation of a ribozyme: The role of bacterial RNase P protein. *The EMBO Journal*, *24*, 3360–3368.

Bullock, T. L., Rodríguez-Hernández, A., Corigliano, E. M., & Perona, J. J. (2008). A rationally engineered misacylating aminoacyl-tRNA synthetase. *Proceedings of the National Academy of Sciences of the United States of America*, *105*, 7428–7433.

Bullwinkle, T., Lazazzera, B., & Ibba, M. (2014). Quality control and infiltration of translation by amino acids outside of the genetic code. *Annual Review of Genetics*, *48*, 149–166.

Calendar, R., & Berg, P. (1966). The catalytic properties of tyrosyl ribonucleic acid synthetases from *Escherichia coli* and *Bacillus subtilis*. *Biochemistry*, *5*, 1690–1695.

Calendar, R., & Berg, P. (1967). D-Tyrosyl RNA: Formation, hydrolysis and utilization for protein synthesis. *Journal of Molecular Biology*, *26*, 39–54.

Carrasco, N., Hiller, D. A., & Strobel, S. A. (2011). Minimal transition state charge stabilization of the oxyanion during peptide bond formation by the ribosome. *Biochemistry*, *50*, 10491–10498.

Cassano, A. G., Anderson, V. E., & Harris, M. E. (2004a). Analysis of solvent nucleophile isotope effects: Evidence for concerted mechanisms and nucleophilic activation by metal coordination in nonenzymatic and ribozyme-catalyzed phosphodiester hydrolysis. *Biochemistry*, *43*, 10547–10559.

Cassano, A. G., Anderson, V. E., & Harris, M. E. (2004b). Understanding the transition states of phosphodiester bond cleavage: Insights from heavy atom isotope effects. *Biopolymers*, *73*, 110–129.

Cavarelli, J., Eriani, G., Rees, B., Ruff, M., Boeglin, M., Mitschler, A., et al. (1994). The active site of yeast aspartyl-tRNA synthetase: Structural and functional aspects of the aminoacylation reaction. *The EMBO Journal*, *13*, 327–337.

Cech, T. R. (2000). Structural biology. The ribosome is a ribozyme. *Science*, *289*, 878–879.

Chen, J. F., Guo, N. N., Li, T., Wang, E. D., & Wang, Y. L. (2000). CP1 domain in *Escherichia coli* leucyl-tRNA synthetase is crucial for its editing function. *Biochemistry*, *39*, 6726–6731.

Chen, Y., Li, X., & Gegenheimer, P. (1997). Ribonuclease P catalysis requires Mg^{2+} coordinated to the *pro-R*$_P$ oxygen of the scissile bond. *Biochemistry*, *36*, 2425–2438.

Chen, M., & Manley, J. L. (2009). Mechanisms of alternative splicing regulation: Insights from molecular and genomics approaches. *Nature Reviews. Molecular Cell Biology*, *10*, 741–754.

Chen, J. L., & Pace, N. R. (1997). Identification of the universally conserved core of ribonuclease P RNA. *RNA*, *3*, 557–560.

Chow, L. T., Gelinas, R. E., Broker, T. R., & Roberts, R. J. (1977). An amazing sequence arrangement at the 5′ ends of adenovirus 2 messenger RNA. *Cell*, *12*, 1–8.

Christian, E. L., Kaye, N. M., & Harris, M. E. (2000). Helix P4 is a divalent metal ion binding site in the conserved core of the ribonuclease P ribozyme. *RNA*, *6*, 511–519.

Christian, E. L., Kaye, N. M., & Harris, M. E. (2002). Evidence for a polynuclear metal ion binding site in the catalytic domain of ribonuclease P RNA. *The EMBO Journal*, *21*, 2253–2262.

Christian, E. L., Smith, K. M., Perera, N., & Harris, M. E. (2006). The P4 metal binding site in RNase P RNA affects active site metal affinity through substrate positioning. *RNA*, *12*, 1463–1467.

Crary, S. M., Kurz, J. C., & Fierke, C. A. (2002). Specific phosphorothioate substitutions probe the active site of *Bacillus subtilis* ribonuclease P. *RNA*, *8*, 933–947.

Crick, F. H. (1968). The origin of the genetic code. *Journal of Molecular Biology*, *38*, 367–379.

Cusack, S., Yaremchuk, A., & Tukalo, M. (2000). The 2 Å crystal structure of leucyl-tRNA synthetase and its complex with a leucyl-adenylate analogue. *The EMBO Journal*, *19*, 2351–2361.

Cvetesic, N., Bilus, M., & Gruic-Sovulj, I. (2015). The tRNA A76 hydroxyl groups control partitioning of the tRNA-dependent pre- and post-transfer editing pathways in class I tRNA synthetase. *The Journal of Biological Chemistry*, *290*, 13981–13991.

Cvetesic, N., Perona, J. J., & Gruic-Sovulj, I. (2012). Kinetic partitioning between synthetic and editing pathways in class I aminoacyl-tRNA synthetases occurs at both pre-transfer and post-transfer hydrolytic steps. *The Journal of Biological Chemistry*, *287*, 25381–25394.

Das, G. K., Bhattacharyya, D., & Burma, D. P. (1999). A possible mechanism of peptide bond formation on ribosome without mediation of peptidyl transferase. *Journal of Theoretical Biology*, *200*, 193–205.

De, I., Schmitzová, J., & Pena, V. (2016). The organization and contribution of helicases to RNA splicing. *Wiley Interdisciplinary Reviews. RNA*, 7, 259–274.

Dennis, P. P. (1997). Ancient ciphers: Translation in archaea. *Cell*, *89*, 1007–1010.

Desogus, G., Todone, F., Brick, P., & Onesti, S. (2000). Active site of lysyl-tRNA synthetase: Structural studies of the adenylation reaction. *Biochemistry*, *39*, 8418–8425.

Dock-Bregeon, A., Sankaranarayanan, R., Romby, P., Caillet, J., Springer, M., Rees, B., et al. (2000). Transfer RNA-mediated editing in threonyl-tRNA synthetase. The class II solution to the double discrimination problem. *Cell*, *103*, 877–884.

Doerfel, L. K., Wohlgemuth, I., Kothe, C., Peske, F., Urlaub, H., & Rodnina, M. V. (2013). EF-P is essential for rapid synthesis of proteins containing consecutive proline residues. *Science*, *339*, 85–88.

Dorner, S., Panuschka, C., Schmid, W., & Barta, A. (2003). Mononucleotide derivatives as ribosomal P-site substrates reveal an important contribution of the 2′-OH to activity. *Nucleic Acids Research*, *31*, 6536–6542.

Dulic, M., Cvetesic, N., Perona, J. J., & Gruic-Sovulj, I. (2010). Partitioning of tRNA-dependent editing between pre- and post-transfer pathways in class I aminoacyl-tRNA synthetases. *The Journal of Biological Chemistry*, *285*, 23799–23809.

Dwivedi, S., Kruparani, S. P., & Sankaranarayanan, R. (2005). A D-amino acid editing module coupled to the translational apparatus in archaea. *Nature Structural & Molecular Biology*, *12*, 556–557.

Eiler, S., Dock-Bregeon, A., Moulinier, L., Thierry, J. C., & Moras, D. (1999). Synthesis of aspartyl-$tRNA^{Asp}$ in *Escherichia coli*—A snapshot of the second step. *The EMBO Journal*, *18*, 6532–6541.

Englander, M. T., Avins, J. L., Fleisher, R. C., Liu, B., Effraim, P. R., Wang, J., et al. (2015). The ribosome can discriminate the chirality of amino acids within its peptidyl-transferase center. *Proceedings of the National Academy of Sciences of the United States of America*, *112*, 6038–6043.

Eriani, G., Delarue, M., Poch, O., Gangloff, J., & Moras, D. (1990). Partition of tRNA synthetases into two classes based on mutually exclusive sets of sequence motifs. *Nature*, *347*, 203–206.

Erlacher, M. D., Lang, K., Wotzel, B., Rieder, R., Micura, R., & Polacek, N. (2006). Efficient ribosomal peptidyl transfer critically relies on the presence of the ribose 2′-OH at A2451 of 23S rRNA. *Journal of the American Chemical Society*, *128*, 4453–4459.

Esakova, O., & Krasilnikov, A. S. (2010). Of proteins and RNA: The RNase P/MRP family. *RNA*, *16*, 1725–1747.

Fernandez-Leiro, R., & Scheres, S. H. (2016). Unravelling biological macromolecules with cryo-electron microscopy. *Nature*, *537*, 339–346.

Ferré-D'Amaré, A. R., & Scott, W. G. (2010). Small self-cleaving ribozymes. *Cold Spring Harbor Perspectives in Biology, 2,* a003574.
Ferri-Fioni, M. L., Fromant, M., Bouin, A. P., Aubard, C., Lazennec, C., Plateau, P., et al. (2006). Identification in archaea of a novel D-Tyr-tRNA^{Tyr} deacylase. *The Journal of Biological Chemistry, 281,* 27575–27585.
Ferri-Fioni, M. L., Schmitt, E., Soutourina, J., Plateau, P., Mechulam, Y., & Blanquet, S. (2001). Structure of crystalline D-Tyr-tRNATyr deacylase. A representative of a new class of tRNA-dependent hydrolases. *The Journal of Biological Chemistry, 276,* 47285–47290.
Fersht, A. R. (1977a). Editing mechanisms in protein synthesis. Rejection of valine by the isoleucyl-tRNA synthetase. *Biochemistry, 16,* 1025–1030.
Fersht, A. R. (1977b). *Enzyme structure and mechanism* (2nd ed.). New York (USA): W. H. Freeman and Company.
Fersht, A. R. (1998). Sieves in sequence. *Science, 280,* 541.
Fersht, A. R., & Kaethner, M. M. (1976). Mechanism of aminoacylation of tRNA. Proof of the aminoacyl adenylate pathway for the isoleucyl- and tyrosyl-tRNA synthetases from *Escherichia coli* K12. *Biochemistry, 15,* 818–823.
Fica, S. M., Mefford, M. A., Piccirilli, J. A., & Staley, J. P. (2014). Evidence for a group II intron-like catalytic triplex in the spliceosome. *Nature Structural & Molecular Biology, 21,* 464–471.
Fica, S. M., Oubridge, C., Galej, W. P., Wilkinson, M. E., Bai, X. C., Newman, A. J., et al. (2017). Structure of a spliceosome remodelled for exon ligation. *Nature, 542,* 377–380.
Fica, S. M., Tuttle, N., Novak, T., Li, N. S., Lu, J., Koodathingal, P., et al. (2013). RNA catalyses nuclear pre-mRNA splicing. *Nature, 503,* 229–234.
Frolova, L. Y., Tsivkovskii, R. Y., Sivolobova, G. F., Oparina, N. Y., Serpinsky, O. I., Blinov, V. M., et al. (1999). Mutations in the highly conserved GGQ motif of class 1 polypeptide release factors abolish ability of human eRF1 to trigger peptidyl-tRNA hydrolysis. *RNA, 5,* 1014–1020.
Fukai, S., Nureki, O., Sekine, S., Shimada, A., Tao, J., Vassylyev, D. G., et al. (2000). Structural basis for double-sieve discrimination of L-valine from L-isoleucine and L-threonine by the complex of tRNA^{Val} and valyl-tRNA synthetase. *Cell, 103,* 793–803.
Fukunaga, R., & Yokoyama, S. (2005a). Aminoacylation complex structures of leucyl-tRNA synthetase and tRNA^{Leu} reveal two modes of discriminator-base recognition. *Nature Structural & Molecular Biology, 12,* 915–922.
Fukunaga, R., & Yokoyama, S. (2005b). Crystal structure of leucyl-tRNA synthetase from the archaeon *Pyrococcus horikoshii* reveals a novel editing domain orientation. *Journal of Molecular Biology, 346,* 57–71.
Galej, W. P., Nguyen, T. H., Newman, A. J., & Nagai, K. (2014). Structural studies of the spliceosome: Zooming into the heart of the machine. *Current Opinion in Structural Biology, 25,* 57–66.
Giegé, R., Sissler, M., & Florentz, C. (1998). Universal rules and idiosyncratic features in tRNA identity. *Nucleic Acids Research, 26,* 5017–5035.
Gilbert, W. (1986). The RNA world. *Nature, 319,* 618.
Grabowski, P. J., Seiler, S. R., & Sharp, P. A. (1985). A multicomponent complex is involved in the splicing of messenger RNA precursors. *Cell, 42,* 345–353.
Greber, B. J., Boehringer, D., Godinic-Mikulcic, V., Crnkovic, A., Ibba, M., Weygand-Durasevic, I., et al. (2012). Cryo-EM structure of the archaeal 50S ribosomal subunit in complex with initiation factor 6 and implications for ribosome evolution. *Journal of Molecular Biology, 418,* 145–160.
Gruic-Sovulj, I., Rokov-Plavec, J., & Weygand-Durasevic, I. (2007). Hydrolysis of non-cognate aminoacyl-adenylates by a class II aminoacyl-tRNA synthetase lacking an editing domain. *FEBS Letters, 581,* 5110–5114.

Guerrier-Takada, C., Gardiner, K., Marsh, T., Pace, N., & Altman, S. (1983). The RNA moiety of ribonuclease P is the catalytic subunit of the enzyme. *Cell*, *35*, 849–857.

Guth, E., Connolly, S. H., Bovee, M., & Francklyn, C. S. (2005). A substrate-assisted concerted mechanism for aminoacylation by a class II aminoacyl-tRNA synthetase. *Biochemistry*, *44*, 3785–3794.

Hagiwara, Y., Field, M. J., Nureki, O., & Tateno, M. (2010). Editing mechanism of aminoacyl-tRNA synthetases operates by a hybrid ribozyme/protein catalyst. *Journal of the American Chemical Society*, *132*, 2751–2758.

Hang, J., Wan, R., Yan, C., & Shi, Y. (2015). Structural basis of pre-mRNA splicing. *Science*, *349*, 1191–1198.

Harris, J. K., Haas, E. S., Williams, D., Frank, D. N., & Brown, J. W. (2001). New insight into RNase P RNA structure from comparative analysis of the archaeal RNA. *RNA (New York, N.Y.)*, 7, 220–232.

Hiller, D. A., Singh, V., Zhong, M., & Strobel, S. A. (2011). A two-step chemical mechanism for ribosome-catalysed peptide bond formation. *Nature*, *476*, 236–239.

Hilliker, A. K., Mefford, M. A., & Staley, J. P. (2007). U2 toggles iteratively between the stem IIa and stem IIc conformations to promote pre-mRNA splicing. *Genes & Development*, *21*, 821–834.

Hoskins, A. A., Gelles, J., & Moore, M. J. (2011). New insights into the spliceosome by single molecule fluorescence microscopy. *Current Opinion in Chemical Biology*, *15*, 864–870.

Hou, Y. M., & Schimmel, P. (1988). A simple structural feature is a major determinant of the identity of a transfer RNA. *Nature*, *333*, 140–145.

Howard, M. H., Klemm, B. P., & Fierke, C. A. (2015). Mechanistic studies reveal similar catalytic strategies for phosphodiester bond hydrolysis by protein-only and RNA-dependent ribonuclease P. *The Journal of Biological Chemistry*, *290*, 13454–13464.

Hsieh, J., & Fierke, C. A. (2009). Conformational change in the *Bacillus subtilis* RNase P holoenzyme–pre-tRNA complex enhances substrate affinity and limits cleavage rate. *RNA*, *15*, 1565–1577.

Hsieh, J., Rueda, D., Koutmou, K. S., Koutmos, M., Walter, N. G., & Fierke, C. A. (2010). A divalent cation stabilizes the active conformation of the *B. subtilis* RNase P pre-tRNA complex: A role for an inner-sphere metal ion in RNase P. *Journal of Molecular Biology*, *400*, 28–51.

Huang, W., Bushnell, E. A., Francklyn, C. S., & Gauld, J. W. (2011). The α-amino group of the threonine substrate as the general base during tRNA aminoacylation: A new version of substrate-assisted catalysis predicted by hybrid DFT. *The Journal of Physical Chemistry. A*, *115*, 13050–13060.

Huang, K. S., Carrasco, N., Pfund, E., & Strobel, S. A. (2008). Transition state chirality and role of the vicinal hydroxyl in the ribosomal peptidyl transferase reaction. *Biochemistry*, *47*, 8822–8827.

Hudson, A. J., Stark, M. R., Fast, N. M., Russell, A. G., & Rader, S. D. (2015). Splicing diversity revealed by reduced spliceosomes in *C. merolae* and other organisms. *RNA Biology*, *12*, 1–8.

Hussain, T., Kamarthapu, V., Kruparani, S. P., Deshmukh, M. V., & Sankaranarayanan, R. (2010). Mechanistic insights into cognate substrate discrimination during proofreading in translation. *Proceedings of the National Academy of Sciences of the United States of America*, *107*, 22117–22121.

Hussain, T., Kruparani, S. P., Pal, B., Dock-Bregeon, A. C., Dwivedi, S., Shekar, M. R., et al. (2006). Post-transfer editing mechanism of a D-aminoacyl-tRNA deacylase-like domain in threonyl-tRNA synthetase from archaea. *The EMBO Journal*, *25*, 4152–4162.

Ibba, M., Losey, H. C., Kawarabayasi, Y., Kikuchi, H., Bunjun, S., & Söll, D. (1999). Substrate recognition by class I lysyl-tRNA synthetases: A molecular basis for gene displacement. *Proceedings of the National Academy of Sciences of the United States of America*, *96*, 418–423.

Ibba, M., & Söll, D. (2000). Aminoacyl-tRNA synthesis. *Annual Review of Biochemistry, 69*, 617–650.

Jakubowski, H. (2012). Quality control in tRNA charging. *Wiley Interdisciplinary Reviews. RNA, 3*, 295–310.

Jarrous, N., & Gopalan, V. (2010). Archaeal/eukaryal RNase P: Subunits, functions and RNA diversification. *Nucleic Acids Research, 38*, 7885–7894.

Jarrous, N., & Reiner, R. (2007). Human RNase P: A tRNA-processing enzyme and transcription factor. *Nucleic Acids Research, 35*, 3519–3524.

Jimenez, R. M., Polanco, J. A., & Lupták, A. (2015). Chemistry and biology of self-cleaving ribozymes. *Trends in Biochemical Sciences, 40*, 648–661.

Jin, H., Kelley, A. C., Loakes, D., & Ramakrishnan, V. (2010). Structure of the 70S ribosome bound to release factor 2 and a substrate analog provides insights into catalysis of peptide release. *Proceedings of the National Academy of Sciences of the United States of America, 107*, 8593–8598.

Jin, H., Kelley, A. C., & Ramakrishnan, V. (2011). Crystal structure of the hybrid state of ribosome in complex with the guanosine triphosphatase release factor 3. *Proceedings of the National Academy of Sciences of the United States of America, 108*, 15798–15803.

Jurica, M. S., & Moore, M. J. (2003). Pre-mRNA splicing: Awash in a sea of proteins. *Molecular Cell, 12*, 5–14.

Kazantsev, A. V., Krivenko, A. A., Harrington, D. J., Holbrook, S. R., Adams, P. D., & Pace, N. R. (2005). Crystal structure of a bacterial ribonuclease P RNA. *Proceedings of the National Academy of Sciences of the United States of America, 102*, 13392–13397.

Kazantsev, A. V., & Pace, N. R. (1998). Identification by modification-interference of purine N-7 and ribose 2′-OH groups critical for catalysis by bacterial ribonuclease P. *RNA, 4*, 937–947.

Kemp, M., Bae, B., Yu, J. P., Ghosh, M., Leffak, M., & Nair, S. K. (2007). Structure and function of the *c-myc* DNA-unwinding element-binding protein DUE-B. *The Journal of Biological Chemistry, 282*, 10441–10448.

Kern, D., & Lapointe, J. (1979). Glutamyl transfer ribonucleic acid synthetase of *Escherichia coli*. Study of the interactions with its substrates. *Biochemistry, 18*, 5809–5818.

Khoshouei, M., Radjainia, M., Phillips, A. J., Gerrard, J. A., Mitra, A. K., Plitzko, J. M., et al. (2016). Volta phase plate cryo-EM of the small protein complex Prx3. *Nature Communications, 7*, 10534.

Kirsebom, L. A. (2010). Roles of metal ions in RNase P catalysis. In F. Liu & S. Altman (Eds.), *RNase P* (pp. 113–134). New York (USA): Springer.

Klemm, B. P., Wu, N., Chen, Y., Liu, X., Kaitany, K. J., Howard, M. J., et al. (2016). The diversity of ribonuclease P: Protein and RNA catalysts with analogous biological functions. *Biomolecules, 6*.

Koch, M., Huang, Y., & Sprinzl, M. (2008). Peptide-bond synthesis on the ribosome: No free vicinal hydroxyl group required on the terminal ribose residue of peptidyl-tRNA. *Angewandte Chemie (International Ed. in English), 47*, 7242–7245.

Konarska, M. M., & Query, C. C. (2005). Insights into the mechanisms of splicing: More lessons from the ribosome. *Genes & Development, 19*, 2255–2260.

Korencic, D., Ahel, I., Schelert, J., Sacher, M., Ruan, B., Stathopoulos, C., et al. (2004). A freestanding proofreading domain is required for protein synthesis quality control in archaea. *Proceedings of the National Academy of Sciences of the United States of America, 101*, 10260–10265.

Korostelev, A. A. (2011). Structural aspects of translation termination on the ribosome. *RNA, 17*, 1409–1421.

Korostelev, A., Asahara, H., Lancaster, L., Laurberg, M., Hirschi, A., Zhu, J., et al. (2008). Crystal structure of a translation termination complex formed with release factor RF2. *Proceedings of the National Academy of Sciences of the United States of America, 105*, 19684–19689.

Kruger, K., Grabowski, P. J., Zaug, A. J., Sands, J., Gottschling, D. E., & Cech, T. R. (1982). Self-splicing RNA: Autoexcision and autocyclization of the ribosomal RNA intervening sequence of *Tetrahymena*. *Cell, 31*, 147–157.

Kuhlenkoetter, S., Wintermeyer, W., & Rodnina, M. V. (2011). Different substrate-dependent transition states in the active site of the ribosome. *Nature, 476*, 351–354.

Kunin, V. (2000). A system of two polymerases—A model for the origin of life. *Origins of Life and Evolution of the Biosphere: The Journal of the International Society for the Study of the Origin of Life, 30*, 459–466.

Kurz, J. C., Niranjanakumari, S., & Fierke, C. A. (1998). Protein component of *Bacillus subtilis* RNase P specifically enhances the affinity for precursor-tRNA^{Asp}. *Biochemistry, 37*, 2393–2400.

LaGrandeur, T. E., Huttenhofer, A., Noller, H. F., & Pace, N. R. (1994). Phylogenetic comparative chemical footprint analysis of the interaction between ribonuclease P RNA and tRNA. *The EMBO Journal, 13*, 3945–3952.

Laurberg, M., Asahara, H., Korostelev, A., Zhu, J., Trakhanov, S., & Noller, H. F. (2008). Structural basis for translation termination on the 70S ribosome. *Nature, 454*, 852–857.

Lee, Y., & Rio, D. C. (2015). Mechanisms and regulation of alternative pre-mRNA splicing. *Annual Review of Biochemistry, 84*, 291–323.

Leung, E. K., Suslov, N., Tuttle, N., Sengupta, R., & Piccirilli, J. A. (2011). The mechanism of peptidyl transfer catalysis by the ribosome. *Annual Review of Biochemistry, 80*, 527–555.

Lim, K., Tempczyk, A., Bonander, N., Toedt, J., Howard, A., Eisenstein, E., et al. (2003). A catalytic mechanism for D-Tyr-tRNA^{Tyr} deacylase based on the crystal structure of *Hemophilus influenzae* HI0670. *The Journal of Biological Chemistry, 278*, 13496–13502.

Lin, L., Hale, S. P., & Schimmel, P. (1996). Aminoacylation error correction. *Nature, 384*, 33–34.

Lincecum, T. L., Tukalo, M., Yaremchuk, A., Mursinna, R. S., Williams, A. M., Sproat, B. S., et al. (2003). Structural and mechanistic basis of pre- and posttransfer editing by leucyl-tRNA synthetase. *Molecular Cell, 11*, 951–963.

Ling, C., & Ermolenko, D. N. (2016). Structural insights into ribosome translocation. *Wiley Interdisciplinary Reviews RNA*, 7, 620–636.

Ling, J., Peterson, K. M., Simonovic, I., Söll, D., & Simonovic, M. (2012). The mechanism of pre-transfer editing in yeast mitochondrial threonyl-tRNA synthetase. *The Journal of Biological Chemistry, 287*, 28518–28525.

Ling, J., Reynolds, N., & Ibba, M. (2009a). Aminoacyl-tRNA synthesis and translational quality control. *Annual Review of Microbiology, 63*, 61–78.

Ling, J., Roy, H., & Ibba, M. (2007). Mechanism of tRNA-dependent editing in translational quality control. *Proceedings of the National Academy of Sciences of the United States of America, 104*, 72–77.

Ling, J., So, B. R., Yadavalli, S. S., Roy, H., Shoji, S., Fredrick, K., et al. (2009b). Resampling and editing of mischarged tRNA prior to translation elongation. *Molecular Cell, 33*, 654–660.

Liu, H., & Gauld, J. W. (2008). Substrate-assisted catalysis in the aminoacyl transfer mechanism of histidyl-tRNA synthetase: A density functional theory study. *The Journal of Physical Chemistry. B, 112*, 16874–16882.

Loftfield, R. B., & Vanderjagt, D. (1972). The frequency of errors in protein biosynthesis. *Biochemical Journal, 128*, 1353–1356.

Loria, A., Niranjanakumari, S., Fierke, C. A., & Pan, T. (1998). Recognition of a pre-tRNA substrate by the *Bacillus subtilis* RNase P holoenzyme. *Biochemistry, 37*, 15466–15473.

Loria, A., & Pan, T. (1996). Domain structure of the ribozyme from eubacterial ribonuclease P. *RNA, 2*, 551–563.

Loria, A., & Pan, T. (1997). Recognition of the T stem-loop of a pre-tRNA substrate by the ribozyme from *Bacillus subtilis* ribonuclease P. *Biochemistry, 36*, 6317–6325.

Martinis, S. A., & Boniecki, M. T. (2010). The balance between pre- and post-transfer editing in tRNA synthetases. *FEBS Letters*, *584*, 455–459.
Marvin, M. C., & Engelke, D. R. (2009). RNase P: Increased versatility through protein complexity? *RNA Biology*, *6*, 40–42.
Matera, A. G., & Wang, Z. (2014). A day in the life of the spliceosome. *Nature Reviews. Molecular Cell Biology*, *15*, 108–121.
Matlin, A. J., Clark, F., & Smith, C. W. (2005). Understanding alternative splicing: Towards a cellular code. *Nature Reviews. Molecular Cell Biology*, *6*, 386–398.
McClain, W., Chen, Y., Foss, K., & Schneider, J. (1988). Association of transfer RNA acceptor identity with a helical irregularity. *Science*, *242*, 1681–1684.
McClain, W., & Foss, K. (1988). Changing the identity of a tRNA by introducing a G-U wobble pair near the 3′ acceptor end. *Science*, *240*, 793–796.
McNeil, B. A., Semper, C., & Zimmerly, S. (2016). Group II introns: Versatile ribozymes and retroelements. *Wiley Interdisciplinary Reviews. RNA*, 7, 341–355.
Mehler, A. H., & Mitra, S. K. (1967). The activation of arginyl transfer ribonucleic acid synthetase by transfer ribonucleic acid. *The Journal of Biological Chemistry*, *242*, 5495–5499.
Minajigi, A., & Francklyn, C. S. (2008). RNA-assisted catalysis in a protein enzyme: The 2′-hydroxyl of $tRNA^{Thr}$ A76 promotes aminoacylation by threonyl-tRNA synthetase. *Proceedings of the National Academy of Sciences of the United States of America*, *105*, 17748–17753.
Mondragón, A. (2013). Structural studies of RNase P. *Annual Review of Biophysics*, *42*, 537–557.
Moore, P. B., & Steitz, T. A. (2011). The roles of RNA in the synthesis of protein. *Cold Spring Harbor Perspectives in Biology*, *3*, a003780.
Mora, L., Heurgue-Hamard, V., Champ, S., Ehrenberg, M., Kisselev, L. L., & Buckingham, R. H. (2003). The essential role of the invariant GGQ motif in the function and stability in vivo of bacterial release factors RF1 and RF2. *Molecular Microbiology*, *47*, 267–275.
Muth, G. W., Ortoleva-Donnelly, L., & Strobel, S. A. (2000). A single adenosine with a neutral pK_a in the ribosomal peptidyl transferase center. *Science*, *289*, 947–950.
Nguyen, T. H., Galej, W. P., Bai, X. C., Savva, C. G., Newman, A. J., Scheres, S. H., et al. (2015). The architecture of the spliceosomal U4/U6.U5 tri-snRNP. *Nature*, *523*, 47–52.
Nguyen, T. H., Galej, W. P., Fica, S. M., Lin, P. C., Newman, A. J., & Nagai, K. (2016). CryoEM structures of two spliceosomal complexes: Starter and dessert at the spliceosome feast. *Current Opinion in Structural Biology*, *36*, 48–57.
Nielsen, H., & Johansen, S. D. (2009). Group I introns: Moving in new directions. *RNA Biology*, *6*, 375–383.
Nilsen, T. W., & Graveley, B. R. (2010). Expansion of the eukaryotic proteome by alternative splicing. *Nature*, *463*, 457–463.
Nissen, P., Hansen, J., Ban, N., Moore, P. B., & Steitz, T. A. (2000). The structural basis of ribosome activity in peptide bond synthesis. *Science*, *289*, 920–930.
Nordin, B. E., & Schimmel, P. (2002). Plasticity of recognition of the 3′-end of mischarged tRNA by class I aminoacyl-tRNA synthetases. *The Journal of Biological Chemistry*, *277*, 20510–20517.
Novoyatleva, T., Tang, Y., Rafalska, I., & Stamm, S. (2006). Pre-mRNA missplicing as a cause of human disease. *Progress in Molecular and Subcellular Biology*, *44*, 27–46.
Nureki, O., Vassylyev, D. G., Tateno, M., Shimada, A., Nakama, T., Fukai, S., et al. (1998). Enzyme structure with two catalytic sites for double-sieve selection of substrate. *Science*, *280*, 578–582.
Odell, L., Huang, V., Jakacka, M., & Pan, T. (1998). Interaction of structural modules in substrate binding by the ribozyme from *Bacillus subtilis* RNase P. *Nucleic Acids Research*, *26*, 3717–3723.

Orgel, L. E. (1968). Evolution of the genetic apparatus. *Journal of Molecular Biology, 38*, 381–393.
Orgel, L. E. (2000). Origin of life. A simpler nucleic acid. *Science, 290*, 1306–1307.
Orgel, L., & Sulston, J. (1971). Polynucleotide replication and the origin of life. In A. Kimball & J. Oro (Eds.), *Prebiotic and biochemical evolution* (pp. 89–94). Amsterdam: North-Holland Publishing Company.
Palade, G. E. (1955). A small particulate component of the cytoplasm. *The Journal of Biophysical and Biochemical Cytology, 1*, 59–68.
Palencia, A., Crépin, T., Vu, M. T., Lincecum, T. L., Jr., Martinis, S. A., & Cusack, S. (2012). Structural dynamics of the aminoacylation and proofreading functional cycle of bacterial leucyl-tRNA synthetase. *Nature Structural & Molecular Biology, 19*, 677–684.
Pan, T. (1995). Higher order folding and domain analysis of the ribozyme from *Bacillus subtilis* ribonuclease P. *Biochemistry, 34*, 902–909.
Pan, T., Loria, A., & Zhong, K. (1995). Probing of tertiary interactions in RNA: 2′-hydroxyl-base contacts between the RNase P RNA and pre-tRNA. *Proceedings of the National Academy of Sciences of the United States of America, 92*, 12510–12514.
Papasaikas, P., & Valcárcel, J. (2016). The spliceosome: The ultimate RNA chaperone and sculptor. *Trends in Biochemical Sciences, 41*, 33–45.
Pascal, R. (2007). A scenario starting from the first chemical building blocks. In J. Reisse (Ed.), *From suns to life: A chronological approach to the history of life on earth* (pp. 163–166). New York: Springer Science & Business Media.
Patel, B. H., Percivalle, C., Ritson, D. J., Duffy, C. D., & Sutherland, J. D. (2015). Common origins of RNA, protein and lipid precursors in a cyanosulfidic protometabolism. *Nature Chemistry, 7*, 301–307.
Pauling, L. (1958). The probability of errors in the process of synthesis of protein molecules [Festschrift für Prof. Dr. Arthur Stoll Siebzigsten Geburtstag, 8 Jan. 1957]. In *Arbeiten aus dem Gebiet der Naturstoffe* (pp. 597–602). Basel (Switzerland): Birkhäuser Verlag.
Pennazio, S., & Roggero, P. (2000). Tobacco mosaic virus RNA as genetic determinant: Genesis of a discovery. *Rivista di Biologia, 93*, 431–455.
Perona, J. J., & Gruic-Sovulj, I. (2014). Synthetic and editing mechanisms of aminoacyl-tRNA synthetases. *Topics in Current Chemistry, 344*, 1–41.
Perona, J. J., Rould, M. A., & Steitz, T. A. (1993). Structural basis for transfer RNA aminoacylation by *Escherichia coli* glutaminyl-tRNA synthetase. *Biochemistry, 32*, 8758–8771.
Perriman, R. J., & Ares, M., Jr. (2007). Rearrangement of competing U2 RNA helices within the spliceosome promotes multiple steps in splicing. *Genes & Development, 21*, 811–820.
Persson, T., Cuzic, S., & Hartmann, R. K. (2003). Catalysis by RNase P RNA: Unique features and unprecedented active site plasticity. *The Journal of Biological Chemistry, 278*, 43394–43401.
Pfeiffer, T., Tekos, A., Warnecke, J. M., Drainas, D., Engelke, D. R., Séraphin, B., et al. (2000). Effects of phosphorothioate modifications on precursor tRNA processing by eukaryotic RNase P enzymes. *Journal of Molecular Biology, 298*, 559–565.
Pingoud, A., & Urbanke, C. (1980). Aminoacyl transfer ribonucleic acid binding site of the bacterial elongation factor Tu. *Biochemistry, 19*, 2108–2112.
Pomeranz Krummel, D. A., Kent, O., MacMillan, A. M., & Altman, S. (2000). Evidence for helical unwinding of an RNA substrate by the RNA enzyme RNase P: Use of an inter-strand disulfide crosslink in substrate. *Journal of Molecular Biology, 295*, 1113–1118.
Punjani, A., Rubinstein, J. L., Fleet, D. J., & Brubaker, M. A. (2017). cryoSPARC: Algorithms for rapid unsupervised cryo-EM structure determination. *Nature Methods, 14*, 290–296.

Query, C. C., & Konarska, M. M. (2004). Suppression of multiple substrate mutations by spliceosomal *prp8* alleles suggests functional correlations with ribosomal ambiguity mutants. *Molecular Cell, 14*, 343–354.

Rangelov, M. A., Vayssilov, G. N., Yomtova, V. M., & Petkov, D. D. (2006). The syn-oriented 2-OH provides a favorable proton transfer geometry in 1,2-diol monoester aminolysis: Implications for the ribosome mechanism. *Journal of the American Chemical Society, 128*, 4964–4965.

Ravel, J. M., Wang, S. F., Heinemeyer, C., & Shive, W. (1965). Glutamyl and glutaminyl ribonucleic acid synthetases of *Escherichia coli* W. Separation, properties, and stimulation of adenosine triphosphate-pyrophosphate exchange by acceptor ribonucleic acid. *The Journal of Biological Chemistry, 240*, 432–438.

Reiner, R., Ben-Asouli, Y., Krilovetzky, I., & Jarrous, N. (2006). A role for the catalytic ribonucleoprotein RNase P in RNA polymerase III transcription. *Genes & Development, 20*, 1621–1635.

Reiter, N. J., Osterman, A., Torres-Larios, A., Swinger, K. K., Pan, T., & Mondragón, A. (2010). Structure of a bacterial ribonuclease P holoenzyme in complex with tRNA. *Nature, 468*, 784–789.

Rich, A. (1962). On the problems of evolution and biochemical information transfer. In M. Kasha & B. Pullman (Eds.), *Horizons in biochemistry* (pp. 103–126). New York: Academic Press.

Rigden, D. J. (2004). Archaea recruited D-Tyr-tRNA^{Tyr} deacylase for editing in Thr-tRNA synthetase. *RNA, 10*, 1845–1851.

Robertson, H. D., Altman, S., & Smith, J. D. (1972). Purification and properties of a specific *Escherichia coli* ribonuclease which cleaves a tyrosine transfer ribonucleic acid precursor. *The Journal of Biological Chemistry, 247*, 5243–5251.

Rodnina, M. V. (2013). The ribosome as a versatile catalyst: Reactions at the peptidyl transferase center. *Current Opinion in Structural Biology, 23*, 595–602.

Rodríguez-Hernández, A., Bhaskaran, H., Hadd, A., & Perona, J. J. (2010). Synthesis of Glu-tRNA^{Gln} by engineered and natural aminoacyl-tRNA synthetases. *Biochemistry, 49*, 6727–6736.

Rokov-Plavec, J., Lesjak, S., Gruic-Sovulj, I., Mocibob, M., Dulic, M., & Weygand-Durasevic, I. (2013). Substrate recognition and fidelity of maize seryl-tRNA synthetases. *Archives of Biochemistry and Biophysics, 529*, 122–130.

Routh, S. B., Pawar, K. I., Ahmad, S., Singh, S., Suma, K., Kumar, M., et al. (2016). Elongation factor Tu prevents misediting of Gly-tRNA(Gly) caused by the design behind the chiral proofreading site of D-aminoacyl-tRNA deacylase. *PLoS Biology, 14*, e1002465.

Rozov, A., Demeshkina, N., Westhof, E., Yusupov, M., & Yusupova, G. (2016). New structural insights into translational miscoding. *Trends in Biochemical Sciences, 41*, 798–814.

Schedl, P., & Primakoff, P. (1973). Mutants of *Escherichia coli* thermosensitive for the synthesis of transfer RNA. *Proceedings of the National Academy of Sciences of the United States of America, 70*, 2091–2095.

Schelcher, C., Sauter, C., & Giegé, P. (2016). Mechanistic and structural studies of protein-only RNase P compared to ribonucleoproteins reveal the two faces of the same enzymatic activity. *Biomolecules, 6*.

Schmeing, T. M., Huang, K. S., Kitchen, D. E., Strobel, S. A., & Steitz, T. A. (2005a). Structural insights into the roles of water and the 2′ hydroxyl of the P site tRNA in the peptidyl transferase reaction. *Molecular Cell, 20*, 437–448.

Schmeing, T. M., Huang, K. S., Strobel, S. A., & Steitz, T. A. (2005b). An induced-fit mechanism to promote peptide bond formation and exclude hydrolysis of peptidyl-tRNA. *Nature, 438*, 520–524.

Schmeing, T. M., & Ramakrishnan, V. (2009). What recent ribosome structures have revealed about the mechanism of translation. *Nature*, *461*, 1234–1242.

Schmidt, E., & Schimmel, P. (1994). Mutational isolation of a sieve for editing in a transfer RNA synthetase. *Science*, *264*, 265–267.

Schmidt, E., & Schimmel, P. (1995). Residues in a class I tRNA synthetase which determine selectivity of amino acid recognition in the context of tRNA. *Biochemistry*, *34*, 11204–11210.

Schmucker, D., Clemens, J. C., Shu, H., Worby, C. A., Xiao, J., Muda, M., et al. (2000). *Drosophila* Dscam is an axon guidance receptor exhibiting extraordinary molecular diversity. *Cell*, *101*, 671–684.

Sekine, S., Nureki, O., Dubois, D. Y., Bernier, S., Chenevert, R., Lapointe, J., et al. (2003). ATP binding by glutamyl-tRNA synthetase is switched to the productive mode by tRNA binding. *The EMBO Journal*, *22*, 676–688.

Sekine, S., Shichiri, M., Bernier, S., Chênevert, R., Lapointe, J., & Yokoyama, S. (2006). Structural bases of transfer RNA-dependent amino acid recognition and activation by glutamyl-tRNA synthetase. *Structure*, *14*, 1791–1799.

Sharma, P. K., Xiang, Y., Kato, M., & Warshel, A. (2005). What are the roles of substrate-assisted catalysis and proximity effects in peptide bond formation by the ribosome? *Biochemistry*, *44*, 11307–11314.

Shaw, J. J., & Green, R. (2007). Two distinct components of release factor function uncovered by nucleophile partitioning analysis. *Molecular Cell*, *28*, 458–467.

Shaw, J. J., Trobro, S., He, S. L., Åqvist, J., & Green, R. (2012). A role for the 2′ OH of peptidyl-tRNA substrate in peptide release on the ribosome revealed through RF-mediated rescue. *Chemistry & Biology*, *19*, 983–993.

Shefer, K., Sperling, J., & Sperling, R. (2014). The supraspliceosome—A multi-task-machine for regulated pre-mRNA processing in the cell nucleus. *Computational and Structural Biotechnology Journal*, *11*, 113–122.

Sherlin, L. D., & Perona, J. J. (2003). tRNA-dependent active site assembly in a class I aminoacyl-tRNA synthetase. *Structure*, *11*, 591–603.

Sievers, A., Beringer, M., Rodnina, M. V., & Wolfenden, R. (2004). The ribosome as an entropy trap. *Proceedings of the National Academy of Sciences of the United States of America*, *101*, 7897–7901.

Silvian, L. F., Wang, J., & Steitz, T. A. (1999). Insights into editing from an Ile-tRNA synthetase structure with $tRNA^{Ile}$ and mupirocin. *Science*, *285*, 1074–1077.

Simms, C. L., Thomas, E. N., & Zaher, H. S. (2017). Ribosome-based quality control of mRNA and nascent peptides. *Wiley Interdisciplinary Reviews. RNA*, *8*.

Smith, D., & Pace, N. R. (1993). Multiple magnesium ions in the ribonuclease P reaction mechanism. *Biochemistry*, *32*, 5273–5281.

Smith, D. J., Query, C. C., & Konarska, M. M. (2008). "Nought may endure but mutability": Spliceosome dynamics and the regulation of splicing. *Molecular Cell*, *30*, 657–666.

Söding, J., & Lupas, A. N. (2003). More than the sum of their parts: On the evolution of proteins from peptides. *BioEssays*, *25*, 837–846.

Soutourina, J., Plateau, P., & Blanquet, S. (2000). Metabolism of D-aminoacyl-tRNAs in *Escherichia coli* and *Saccharomyces cerevisiae* cells. *The Journal of Biological Chemistry*, *275*, 32535–32542.

Soutourina, J., Plateau, P., Delort, F., Peirotes, A., & Blanquet, S. (1999). Functional characterization of the D-Tyr-$tRNA^{Tyr}$ deacylase from *Escherichia coli*. *The Journal of Biological Chemistry*, *274*, 9109–19114.

Sperling, R. (2017). The nuts and bolts of the endogenous spliceosome. *Wiley Interdisciplinary Reviews. RNA*, *8*.

Steitz, T. A., & Moore, P. B. (2003). RNA, the first macromolecular catalyst: The ribosome is a ribozyme. *Trends in Biochemical Sciences*, *28*, 411–418.

Steitz, T. A., & Steitz, J. A. (1993). A general two-metal-ion mechanism for catalytic RNA. *Proceedings of the National Academy of Sciences of the United States of America*, *90*, 6498–6502.
Subramaniam, S., Earl, L. A., Falconieri, V., Milne, J. L., & Egelman, E. H. (2016). Resolution advances in cryo-EM enable application to drug discovery. *Current Opinion in Structural Biology*, *41*, 194–202.
Sun, L., Campbell, F. E., Zahler, N. H., & Harris, M. E. (2006). Evidence that substrate-specific effects of C5 protein lead to uniformity in binding and catalysis by RNase P. *The EMBO Journal*, *25*, 3998–4007.
Takayama, T., Ogawa, T., Hidaka, M., Shimizu, Y., Ueda, T., & Masaki, H. (2005). Esterification of *Escherichia coli* tRNAs with D-histidine and D-lysine by aminoacyl-tRNA synthetases. *Bioscience, Biotechnology, and Biochemistry*, *69*, 1040–1041.
Thommen, M., Holtkamp, W., & Rodnina, M. V. (2016). Co-translational protein folding: Progress and methods. *Current Opinion in Structural Biology*, *42*, 83–89.
Torres-Larios, A., Swinger, K. K., Krasilnikov, A. S., Pan, T., & Mondragón, A. (2005). Crystal structure of the RNA component of bacterial ribonuclease P. *Nature*, *437*, 584–587.
Trobro, S., & Åqvist, J. (2005). Mechanism of peptide bond synthesis on the ribosome. *Proceedings of the National Academy of Sciences of the United States of America*, *102*, 12395–12400.
Trobro, S., & Åqvist, J. (2006). Analysis of predictions for the catalytic mechanism of ribosomal peptidyl transfer. *Biochemistry*, *45*, 7049–7056.
Trobro, S., & Åqvist, J. (2009). Mechanism of the translation termination reaction on the ribosome. *Biochemistry*, *48*, 11296–11303.
Tseng, C. K., & Cheng, S. C. (2008). Both catalytic steps of nuclear pre-mRNA splicing are reversible. *Science*, *320*, 1782–1784.
Tukalo, M., Yaremchuk, A., Fukunaga, R., Yokoyama, S., & Cusack, S. (2005). The crystal structure of leucyl-tRNA synthetase complexed with $tRNA^{Leu}$ in the post-transfer editing conformation. *Nature Structural & Molecular Biology*, *12*, 923–930.
Ude, S., Lassak, J., Starosta, A. L., Kraxenberger, T., Wilson, D. N., & Jung, K. (2013). Translation elongation factor EF-P alleviates ribosome stalling at polyproline stretches. *Science*, *339*, 82–85.
Vester, B., & Garrett, R. A. (1988). The importance of highly conserved nucleotides in the binding region of chloramphenicol at the peptidyl transfer centre of *Escherichia coli* 23S ribosomal RNA. *The EMBO Journal*, *7*, 3577–3587.
von der Haar, F., & Cramer, F. (1975). Isoleucyl-tRNA synthetase from baker's yeast: The 3′-hydroxyl group of the 3′-terminal ribose is essential for preventing misacylation of $tRNA^{Ile}$-C-C-A with misactivated valine. *FEBS Letters*, *56*, 215–217.
von der Haar, F., & Cramer, F. (1976). Hydrolytic action of aminoacyl-tRNA synthetases from baker's yeast: "Chemical proofreading" preventing acylation of $tRNA^{Ile}$ with misactivated valine. *Biochemistry*, *15*, 4131–4138.
von der Haar, F., Schlimme, E., Gómez-Guillen, M., & Cramer, F. (1971). Substrate properties of yeast $tRNA^{Phe}$ oxidized and reduced at the 3′-terminal ribose. *European Journal of Biochemistry*, *24*, 296–302.
Voorhees, R. M., & Ramakrishnan, V. (2013). Structural basis of the translational elongation cycle. *Annual Review of Biochemistry*, *82*, 203–236.
Walker, S. C., & Engelke, D. R. (2006). Ribonuclease P: The evolution of an ancient RNA enzyme. *Critical Reviews in Biochemistry and Molecular Biology*, *41*, 77–102.
Wallin, G., & Åqvist, J. (2010). The transition state for peptide bond formation reveals the ribosome as a water trap. *Proceedings of the National Academy of Sciences of the United States of America*, *107*, 1888–1893.
Wan, R., Yan, C., Bai, R., Wang, L., Huang, M., Wong, C. C., et al. (2016). The 3.8 Å structure of the U4/U6.U5 tri-snRNP: Insights into spliceosome assembly and catalysis. *Science*, *351*, 466–475.

Wang, Z., & Burge, C. B. (2008). Splicing regulation: From a parts list of regulatory elements to an integrated splicing code. *RNA*, *14*, 802–813.

Ward, A. J., & Cooper, T. A. (2010). The pathobiology of splicing. *Journal of Pathology*, *220*, 152–163.

Warnasooriya, C., & Rueda, D. (2014). Single-molecule fluorescence-based studies on the dynamics, assembly and catalytic mechanism of the spliceosome. *Biochemical Society Transactions*, *42*, 1211–1218.

Warnecke, J. M., Fürste, J. P., Hardt, W. D., Erdmann, V. A., & Hartmann, R. K. (1996). Ribonuclease P (RNase P) RNA is converted to a Cd2+-ribozyme by a single Rp-phosphorothioate modification in the precursor tRNA at the RNase P cleavage site. *Proceedings of the National Academy of Sciences of the United States of America*, *93*, 8924–8928.

Warnecke, J. M., Held, R., Busch, S., & Hartmann, R. K. (1999). Role of metal ions in the hydrolysis reaction catalyzed by RNase P RNA from *Bacillus subtilis*. *Journal of Molecular Biology*, *290*, 433–445.

Weinger, J. S., Parnell, K. M., Dorner, S., Green, R., & Strobel, S. A. (2004). Substrate-assisted catalysis of peptide bond formation by the ribosome. *Nature Structural & Molecular Biology*, *11*, 1101–1106.

Weinger, J. S., & Strobel, S. A. (2006). Participation of the tRNA A76 hydroxyl groups throughout translation. *Biochemistry*, *45*, 5939–5948.

White, H. B. (1976). Coenzymes as fossils of an earlier metabolic state. *Journal of Molecular Evolution*, 7, 101–104.

Will, C. L., & Lührmann, R. (2011). Spliceosome structure and function. *Cold Spring Harbor Perspectives in Biology*, *3*, a003707.

Woese, C. R. (1967). *The genetic code: The molecular basis for genetic expression*. New York: Harper & Row.

Woese, C. R., & Fox, G. E. (1977). Phylogenetic structure of the prokaryotic domain: The primary kingdoms. *Proceedings of the National Academy of Sciences of the United States of America*, *74*, 5088–5090.

Woese, C. R., Kandler, O., & Wheelis, M. L. (1990). Towards a natural system of organisms: Proposal for the domains archaea, bacteria, and Eucarya. *Proceedings of the National Academy of Sciences of the United States of America*, *87*, 4576–4579.

Wong, F. C., Beuning, P. J., Silvers, C., & Musier-Forsyth, K. (2003). An isolated class II aminoacyl-tRNA synthetase insertion domain is functional in amino acid editing. *The Journal of Biological Chemistry*, *278*, 52857–52864.

Woolstenhulme, C. J., Parajuli, S., Healey, D. W., Valverde, D. P., Petersen, E. N., Starosta, A. L., et al. (2013). Nascent peptides that block protein synthesis in bacteria. *Proceedings of the National Academy of Sciences of the United States of America*, *110*, E878–E887.

Wydau, S., Ferri-Fioni, M. L., Blanquet, S., & Plateau, P. (2007). GEK1, a gene product of *Arabidopsis thaliana* involved in ethanol tolerance, is a D-aminoacyl-tRNA deacylase. *Nucleic Acids Research*, *35*, 930–938.

Wydau, S., van der Rest, G., Aubard, C., Plateau, P., & Blanquet, S. (2009). Widespread distribution of cell defense against D-aminoacyl-tRNAs. *The Journal of Biological Chemistry*, *284*, 14096–14104.

Yadavalli, S. S., & Ibba, M. (2012). Quality control in aminoacyl-tRNA synthesis its role in translational fidelity. *Advances in Protein Chemistry and Structural Biology*, *86*, 1–43.

Yamane, T., Miller, D. L., & Hopfield, J. J. (1981). Discrimination between D- and L-tyrosyl transfer ribonucleic acids in peptide chain elongation. *Biochemistry*, *20*, 7059–7064.

Yan, C., Hang, J., Wan, R., Huang, M., Wong, C. C., & Shi, Y. (2015). Structure of a yeast spliceosome at 3.6-angstrom resolution. *Science*, *349*, 1182–1191.

Yan, C., Wan, R., Bai, R., Huang, G., & Shi, Y. (2017). Structure of a yeast step II catalytically activated spliceosome. *Science*, *355*, 149–155.

Zaher, H. S., Shaw, J. J., Strobel, S. A., & Green, R. (2011). The 2′-OH group of the peptidyl-tRNA stabilizes an active conformation of the ribosomal PTC. *The EMBO Journal*, *30*, 2445–2453.

Zhang, C. M., Perona, J. J., Ryu, K., Francklyn, C., & Hou, Y. M. (2006). Distinct kinetic mechanisms of the two classes of aminoacyl-tRNA synthetases. *Journal of Molecular Biology*, *361*, 300–311.

Kolibri 360 GmbH
Ferdinand-Jühlke-Straße 7
99095 Erfurt, Deutschland
produktsicherheit@kolibri360.de